Hybrid Intelligent Modeling Technology and Optimization Strategy for Industrial Energy Consumption Processes

Hybrid Intelligent Modeling Technology and Optimization Strategy for Industrial Energy Consumption Processes

Guest Editors

Sheng Du
Xiongbo Wan
Li Jin
Zixin Huang

MDPI • Basel • Beijing • Wuhan • Barcelona • Belgrade • Novi Sad • Cluj • Manchester

Guest Editors

Sheng Du
School of Automation
China University of Geosciences
Wuhan
China

Xiongbo Wan
School of Automation
China University of Geosciences
Wuhan
China

Li Jin
School of Automation
China University of Geosciences
Wuhan
China

Zixin Huang
School of Electrical and Information Engineering
Wuhan Institute of Technology
Wuhan
China

Editorial Office
MDPI AG
Grosspeteranlage 5
4052 Basel, Switzerland

This is a reprint of the Special Issue, published open access by the journal *Energies* (ISSN 1996-1073), freely accessible at: https://www.mdpi.com/journal/energies/special_issues/L1Z3P42UOA.

For citation purposes, cite each article independently as indicated on the article page online and as indicated below:

Lastname, A.A.; Lastname, B.B. Article Title. *Journal Name* **Year**, *Volume Number*, Page Range.

ISBN 978-3-7258-3883-7 (Hbk)
ISBN 978-3-7258-3884-4 (PDF)
https://doi.org/10.3390/books978-3-7258-3884-4

Contents

About the Editors

Sheng Du

Sheng Du received a B.S. degree in Measurement and Control Technology and Instrument and a Ph.D. degree in Control Science and Engineering from China University of Geosciences, Wuhan, China, in 2016 and 2021, respectively. He was a joint Ph.D. student with the Department of Electrical and Computer Engineering, University of Alberta, Edmonton, Canada, from 2019 to 2021. He was a Postdoctoral Fellow in Control Science and Engineering, China University of Geosciences, Wuhan, China, from 2021 to 2023. He is currently a Professor with the School of Automation, China University of Geosciences, Wuhan, China. His research interests include process control, intelligent control, intelligent optimization, computational intelligence, and artificial Dr. Du is a Senior Member of the Chinese Association of Automation (CAA) and a winner of the Outstanding Doctoral Thesis Nomination Award of the CAA. He is an Associate Editor of Measurement and Control.

Xiongbo Wan

Xiongbo Wan received a B.S. degree in information and computing science from Huazhong Agricultural University, Wuhan, China, in 2006, an M.S. degree in probability theory and mathematical statistics and a Ph.D. degree in control theory and engineering from the Huazhong University of Science and Technology, Wuhan, in 2008 and 2011, respectively. From 2012 to 2013, he was a Visiting Scholar with Akita Prefectural University, Akita, Japan. From 2011 to 2014, he was a Lecturer and then an Associate Professor with the College of Engineering, Huazhong Agricultural University. From 2016 to 2017, he was a Visiting Scholar with the Department of Computer Science, Brunel University London, Uxbridge, U.K. Since 2015, he has been with the China University of Geosciences, Wuhan, where he is currently a Professor with the School of Automation. His research interests include fault diagnosis, state estimation, and model predictive control of complex industrial systems.

Li Jin

Li Jin received a B.S. degree in Automation and a Ph.D. degree in Control Science and Engineering from China University of Geosciences, Wuhan, China, in 2016 and 2021, respectively. She was a joint Ph.D. student with the Department of Electrical Engineering and Electronics, University of Liverpool, Liverpool, U.K., from 2018 to 2020. She was a Postdoctoral Fellow in Control Science and Engineering, China University of Geosciences, Wuhan, China, from 2022 to 2024. She is currently a Professor in the School of Automation, China University of Geosciences, Wuhan, China. Her current research interests include power system stability analysis and control, time-delay system, robust theory and application.

Zixin Huang

Zixin Huang received his B.S. and M.S. degrees in engineering from Wuhan Textile University, Wuhan, China, in 2013 and 2016, respectively, received a Ph.D. degree in the School of Automation, China University of Geosciences, Wuhan, China, and in 2020, received the postdoctoral fellow in the Nankai University, Tianjin, China, in 2023. He is currently a postdoctoral fellow with the Shanghai Jiao Tong University, Shanghai, China. He is a senior visiting scholar at university of Science and Technology of China, Anhui, China. He is currently an associate professor with the School of Electrical and Information Engineering, Wuhan Institute of Technology. His current research interests include nonlinear system control and intelligent control. Dr. Huang is a senior member of the Chinese Association of Automation.

energies

MDPI

Editorial

Recent Progress in Hybrid Intelligent Modeling Technology and Optimization Strategy for Industrial Energy Consumption Processes

Sheng Du [1,2,3], Li Jin [1,2,3,*], Zixin Huang [4] and Xiongbo Wan [1,2,3]

1 School of Automation, China University of Geosciences, Wuhan 430074, China; dusheng@cug.edu.cn (S.D.); xbwan23@cug.edu.cn (X.W.)
2 Hubei Key Laboratory of Advanced Control and Intelligent Automation for Complex Systems, Wuhan 430074, China
3 Engineering Research Center of Intelligent Technology for Geo-Exploration, Ministry of Education, Wuhan 430074, China
4 School of Electrical and Information Engineering, Wuhan Institute of Technology, Wuhan 430205, China; huangzx@wit.edu.cn
* Correspondence: jinli@cug.edu.cn

Abstract: This editorial discusses recent progress in hybrid intelligent modeling technology and optimization strategy for industrial energy consumption processes. With the increasing emphasis on sustainable practices, efficient management of industrial energy consumption has become a critical concern. This editorial aims to explore innovative approaches that use artificial intelligence to model and optimize energy use in industrial processes. The integration of advanced technologies such as machine learning, artificial intelligence, and data analytics play a pivotal role in achieving energy efficiency, reducing environmental impacts and ensuring the sustainability of industrial operations. These studies collectively contribute to the body of knowledge on hybrid intelligent modeling technology and optimization strategy, offering practical solutions and theoretical frameworks to address energy conservation and consumption reduction.

Keywords: hybrid intelligent modeling; industrial processes; optimization strategy; artificial intelligence; energy conservation and consumption reduction

Received: 17 March 2025
Accepted: 9 April 2025
Published: 10 April 2025

Citation: Du, S.; Jin, L.; Huang, Z.; Wan, X. Recent Progress in Hybrid Intelligent Modeling Technology and Optimization Strategy for Industrial Energy Consumption Processes. *Energies* **2025**, *18*, 1939. https://doi.org/10.3390/en18081939

1. Introduction

Efficient management of industrial energy consumption has become increasingly vital with the global shift towards sustainable practices. Recent advancements in hybrid intelligent modeling technologies and optimization strategies offer promising avenues for addressing energy conservation and reducing industrial energy consumption. Hybrid intelligent modeling, which combines machine learning algorithms and data-driven techniques, has emerged as an innovative solution to address the complexities associated with variable renewable energy sources and dynamic industrial demands [1–4].

Recent developments in artificial intelligence and machine learning, such as CatBoost, XGBoost [3], adaptive neuro-fuzzy inference systems, and genetic programming techniques [4], have demonstrated significant potential in accurately modeling and predicting energy consumption patterns. By integrating these intelligent models with evolutionary optimization algorithms like Non-dominated Sorting Genetic Algorithms, researchers have successfully tackled multi-objective optimization challenges across various energy-intensive industries, including ironmaking, steel production, and vacuum-pressure swing-adsorption processes [5–7].

Further advancements have been made through integrating data-driven analytics and knowledge-based predictions, complemented by real-time optimization frameworks, to achieve considerable improvements in energy efficiency and operational cost-effectiveness [8,9]. Additionally, intelligent monitoring and anomaly detection techniques, particularly those employing information granules and comprehensive soft-sensing methods, have significantly enhanced operational reliability and the accuracy of predictive maintenance in industrial systems [10–12]. The adoption of optimized fuzzy logic controllers has also provided robust adaptive control solutions, significantly enhancing the performance of complex energy-intensive processes [13].

This editorial discusses recent progress in hybrid intelligent modeling technologies and optimization strategies for industrial energy consumption processes. It examines the innovative integration of machine learning, artificial intelligence, and data analytics, highlighting their crucial roles in enhancing energy efficiency, minimizing environmental impacts, and fostering sustainable industrial operations. Through a critical analysis of contemporary advancements and identification of existing challenges, this study provides valuable insights and theoretical frameworks that contribute practically and theoretically to the ongoing discourse on energy conservation and consumption reduction.

2. Review of Published Articles

In the era of Industry 4.0 and the ever-growing emphasis on sustainable practices, the efficient management of industrial energy consumption has become a critical concern. This Special Issue aims to explore innovative approaches that leverage data-driven intelligence to model and optimize energy use in industrial processes. The integration of advanced technologies such as machine learning, artificial intelligence, and data analytics will play a pivotal role in achieving energy efficiency, reducing environmental impacts and ensuring the sustainability of industrial operations.

The main objective of this Special Issue is to promote research and innovation in the field of hybrid intelligent modeling and optimization for industrial energy consumption processes, especially in the fields of steel metallurgy, chemical engineering, geological drilling, marine exploration, textile, pharmaceutical, and other large-scale industries. In this Special Issue, the research areas include hybrid intelligent modeling techniques, intelligent optimization strategies, case studies and applications, and interdisciplinary approaches. The analysis of the published contributions in the Special Issue is shown in Table 1.

Table 1. Analysis of the published contributions in the Special Issue.

No.	DOI	Research Area	Focus	Type of Research	Industry	Country
[14]	10.3390/en17112770	Optimization algorithms	Integrated energy systems; economically optimized dispatch; landscape uncertainty; carbon trading; P2G-CCS	Mathematical modeling	Energy systems	China
[15]	10.3390/en17195015.	Control strategies	Sustainable transition; energy saving; reinforcement learning; meta-learning	Mathematical modeling	Transportation	China
[16]	10.3390/en17215424	Decision support strategies	Violation dictionary construction; operation risk prediction; random forests; independent component analysis; mutual information	Mathematical modeling	Power systems	China
[17]	10.3390/en17235855	Prediction algorithms	Proton-exchange membrane fuel cells; degradation prediction; durability test; gated recurrent unit; gray wolf optimizer; accuracy; complexity	Mathematical modeling	Energy technology	China

Table 1. *Cont.*

No.	DOI	Research Area	Focus	Type of Research	Industry	Country
[18]	10.3390/en17246345	Process optimization transformer	Transformer insulating oil recovery; data-driven monitoring; thermal cycle of insulating oil; recovery rate	Mathematical modeling	Power systems	China
[19]	10.3390/en17246356	Control strategies	Supercritical unit; coordination control; model predictive control; energy-saving analysis; rapid load change	Mathematical modeling	Energy systems	China
[20]	10.3390/en18010060	Optimization algorithms	Open-pit mine; transportation costs; truck waiting times; excavator boom-and-dipper operation durations; improved genetic algorithm	Mathematical modeling	Mining	China
[21]	10.3390/en18010099	Prediction algorithms	Machine learning; production forecast; data preprocessing; principal component analysis; AutoGluon	Mathematical modeling	Energy extraction	China
[22]	10.3390/en18020252	Energy management strategies	Hybrid energy storage; energy management; state of charge; DC microgrid	Mathematical modeling	Power systems	China
[23]	10.3390/en18061380	Control strategies	Load frequency control; dynamic event-triggered mechanism; distributed model predictive control; cloud-edge-terminal; virtual power plant	Mathematical modeling	Power systems	China

The articles published in this Special Issue comprehensively explore cutting-edge hybrid intelligent modeling technologies and optimization strategies tailored specifically for complex industrial processes. These studies collectively address critical challenges in areas such as energy management, intelligent scheduling, predictive analytics, advanced control strategies, and operational efficiency. Detailed analyses of these articles are summarized below.

Energy management and optimal scheduling are prominently discussed in the literature [14,20,22]. Study [14] presents a sophisticated economic optimal scheduling method for integrated energy systems, explicitly considering uncertainties in wind and solar power generation. By integrating innovative approaches such as power-to-gas and carbon capture and storage technologies, the proposed strategy effectively enhances economic performance while simultaneously achieving significant carbon emission reductions, aligning with sustainable energy objectives. Study [20] introduces an intelligent optimization scheduling approach, applying an enhanced genetic algorithm to reduce energy consumption significantly in open-pit mining equipment operations. This study not only improves energy efficiency but also highlights the adaptability of genetic algorithms in complex industrial environments. Study [22] offers a comprehensive energy management strategy specifically designed for direct current microgrids, emphasizing the optimization of photovoltaic power-tracking. This research addresses crucial issues of renewable energy integration, improving energy utilization efficiency and contributing to the stability of renewable-powered microgrids.

Advanced control methodologies and decision-support strategies form the central focus of the literature [15,19,23]. Study [15] proposes an adaptive multi-objective traffic-signal control model, carefully designed to balance energy savings with enhanced traffic efficiency. The model effectively adapts to dynamic traffic conditions, demonstrating its practical value in urban transportation networks by simultaneously reducing energy consumption and traffic congestion. Study [19] introduces a hybrid predictive-control model combined with proportional-integral-derivative methods tailored specifically for supercritical power units. This novel hybrid control approach effectively maintains operational stability while maximizing energy savings, demonstrating substantial improvements over conventional control strategies. Study [10] explores a distributed model predictive load frequency control approach for virtual power plants, employing an innovative event-based low-delay communication technique under a cloud-edge-terminal integrated framework. The proposed strategy greatly reduces control delays, enhances system reliability, and is particularly valuable for future grid applications involving extensive distributed energy resources.

Predictive modeling, data-driven fault detection, and process optimization techniques are thoroughly investigated in the literature [16–18,21]. Study [16] develops multimodal operation data-mining methods, significantly enhancing the prediction of operational violation risks within power grids. The study effectively leverages complex data sources, thus improving the safety and reliability of grid operations by enabling early identification and prevention of potential disruptions. Study [17] proposes an advanced dynamic degradation prediction model for proton-exchange membrane fuel cells, employing gated recurrent units optimized by gray wolf optimization. This approach substantially enhances predictive accuracy, effectively prolonging the fuel cell lifespan by enabling timely preventive maintenance. Study [18] introduces a comprehensive data-driven system for the full recovery of transformer-insulating oil, leveraging advanced analytics to ensure the reliability and longevity of transformer operations. The innovative integration of data-driven methods significantly contributes to equipment maintenance and operational efficiency. Study [21] implements machine learning techniques, combined effectively with feature selection methods and automatic hyperparameter optimization, to accurately forecast oil well production performance. The research demonstrates the strong capability of intelligent analytics in achieving robust and reliable prediction outcomes, providing valuable decision-support tools for petroleum exploration and production management.

In summary, these articles collectively highlight the profound impact of advanced data-driven intelligent modeling and optimization techniques on enhancing the efficiency, reliability, sustainability, and adaptability of industrial systems. They provide valuable insights into both theoretical innovations and practical implementations, laying a robust foundation for future advancements in intelligent industrial systems under Industry 4.0. Finally, the topics covered in these studies include energy management and optimal scheduling, advanced control strategies, and predictive modeling for complex industrial processes across power systems, transportation, and oil and gas. The authors who have contributed to our Special Issue are all from China.

3. Conclusions

This editorial presents the recent progress in hybrid intelligent modeling technology and optimization strategy for industrial energy consumption processes. These intelligent algorithms leverage data for control, decision-making, and parameter optimization to significantly lower energy usage in industrial systems. In addition, these algorithms not only drive unprecedented levels of efficiency and reliability but also play a critical role in quality improvement. In the era of Industry 4.0, such hybrid intelligent modeling technologies and optimization strategies are proving essential for increasing production efficiency, promoting sustainability, and ensuring high product quality by directly targeting energy reduction.

Funding: This research was funded by the Hubei Provincial Natural Science Foundation of China under Grant No. 2025AFB471, the Natural Science Foundation of Wuhan under Grant No. 2024040801020280, the 111 Project under Grant No. B17040, and in part by the Fundamental Research Funds for the Central Universities, China University of Geosciences under Grant No. 2021237.

Conflicts of Interest: The authors declare no conflicts of interest.

References

1. Akbulut, O.; Cavus, M.; Cengiz, M.; Allahham, A.; Giaouris, D.; Forshaw, M. Hybrid intelligent control system for adaptive microgrid optimization: Integration of rule-based control and deep learning techniques. *Energies* **2024**, *17*, 2260. [CrossRef]
2. Du, J.; Cao, H.; Li, Y.; Yang, Z.; Eslamimanesh, A.; Fakhroleslam, M.; Mansouri, S.S.; Shen, W. Development of hybrid surrogate model structures for design and optimization of CO_2 capture processes: Part I. Vacuum pressure swing adsorption in a confined space. *Chem. Eng. Sci.* **2024**, *283*, 119379. [CrossRef]
3. Li, X.; Wang, Z.; Yang, C.; Bozkurt, A. An advanced framework for net electricity consumption prediction: Incorporating novel machine learning models and optimization algorithms. *Energy* **2024**, *296*, 131259. [CrossRef]
4. Bakare, M.S.; Abdulkarim, A.; Shuaibu, A.N.; Muhamad, M.M. A hybrid long-term industrial electrical load forecasting model using optimized ANFIS with gene expression programming. *Energy Rep.* **2024**, *11*, 5831–5844. [CrossRef]
5. Zhuang, X.; Wang, W.; Su, Y.; Yan, B.; Li, Y.; Li, L.; Hao, Y. Multi-objective optimization of reservoir development strategy with hybrid artificial intelligence method. *Expert Syst. Appl.* **2024**, *241*, 122707. [CrossRef]
6. Jiabao, W.; Jianliang, Z.; Yaozu, W.; Zhengjian, L.; Qingke, S.; Xiaoran, S.; Zhen, L. Cost and energy synergy optimization model for ironmaking processes: Hybrid knowledge and data driven. *J. Clean. Prod.* **2025**, *486*, 144420. [CrossRef]
7. Lou, S.; Yang, C.; Zhang, X.; Zhang, H.; Wu, P. Data/mechanism hybrid-driven modeling of blast furnace smelting system and global sequential optimization. *J. Process Control* **2024**, *139*, 103235. [CrossRef]
8. Zheng, R.; Bao, Y.; Zhao, L.; Xing, L.-D. Enhanced steelmaking cost optimization and real-time alloying element yield prediction: A ferroalloy model based on machine learning and linear programming. *J. Iron Steel Res. Int.* **2024**, 1–16. [CrossRef]
9. Zhang, Y.; Liu, X.; Liu, R.; Li, H.; Duan, Y.; Li, X.; Sun, Y.; Li, L. A new method for prediction and decision-making of silicon content in hot metal: A hybrid approach based on machine learning and process optimization. *Preprints* **2024**, 1–16. [CrossRef]
10. Du, S.; Huang, C.; Ma, X.; Fan, H. A review of data-driven intelligent monitoring for geological drilling processes. *Processes* **2024**, *12*, 2478. [CrossRef]
11. Du, S.; Ma, X.; Wu, M.; Cao, W.; Pedrycz, W. Time series anomaly detection via rectangular information granulation for sintering process. *IEEE Trans. Fuzzy Syst.* **2024**, *32*, 4799–4804. [CrossRef]
12. Du, S.; Ma, X.; Fan, H.; Hu, J.; Cao, W.; Wu, M.; Pedrycz, W. Intelligent prediction and soft-sensing of comprehensive production indicators for iron ore sintering: A review. *Comput. Ind.* **2025**, *165*, 104215. [CrossRef]
13. Özdemir, M.E.; Beşkardeş, A.; Hameş, Y. Intelligent sinter machine speed control system using optimized fuzzy logic controller: An experimental study in iron and steel plant. *Arab. J. Sci. Eng.* **2024**, *49*, 16391–16406. [CrossRef]
14. Zhang, Y.; Zhang, P.; Du, S.; Dong, H. Economic optimal scheduling of integrated energy system considering wind–solar uncertainty and power to gas and carbon capture and storage. *Energies* **2024**, *17*, 2770. [CrossRef]
15. Zhang, Y.; Zhou, Y.; Wang, B.; Song, J. MMD-TSC: An adaptive multi-objective traffic signal control for energy saving with traffic efficiency. *Energies* **2024**, *17*, 5015. [CrossRef]
16. Meng, L.; Zhong, J.; Luo, S.; Zhu, X.; Wang, Y.; Zhang, S. Multimodal operation data mining for grid operation violation risk prediction. *Energies* **2024**, *17*, 5424. [CrossRef]
17. Wang, X.; Huang, Z.; Zhang, D.; Yuan, H.; Cai, B.; Liu, H.; Wang, C.; Cao, Y.; Zhou, X.; Dong, Y. Dynamic prediction of proton-exchange membrane fuel cell degradation based on gated recurrent unit and grey wolf optimization. *Energies* **2024**, *17*, 5855. [CrossRef]
18. Chen, F.; Wang, L.; Zheng, Z.; Pan, B.; Hu, Y.; Zhang, K. Data-driven-based full recovery technology and system for transformer insulating oil. *Energies* **2024**, *17*, 6345. [CrossRef]
19. Yang, Q.; Chen, G.; Guo, M.; Chen, T.; Luo, L.; Sun, L. Model predictive hybrid PID control and energy-saving performance analysis of supercritical unit. *Energies* **2024**, *17*, 6356. [CrossRef]
20. Li, F.; Shi, Z.; Ding, W.; Gan, Y. Intelligent optimization scheduling strategy for energy consumption reduction for equipment in open-pit mines based on enhanced genetic algorithm. *Energies* **2025**, *18*, 60. [CrossRef]
21. Zhu, R.; Li, N.; Duan, Y.; Li, G.; Liu, G.; Qu, F.; Long, C.; Wang, X.; Liao, Q.; Li, G. Well-production forecasting using machine learning with feature selection and automatic hyperparameter optimization. *Energies* **2025**, *18*, 99. [CrossRef]

22. Li, F.; Shi, Z.; Zhu, Z.; Gan, Y. Energy management strategy for direct current microgrids with consideration of photovoltaic power tracking optimization. *Energies* **2025**, *18*, 252. [CrossRef]
23. Kang, K.; Shi, N.; Cai, S.; Zhang, L.; Shao, X.; Cao, H.; Fei, M.; Zhou, S.; Wan, X. Distributed model predictive load frequency control for virtual power plants with novel event-based low-delay technique under cloud-edge-terminal framework. *Energies* **2025**, *18*, 1380. [CrossRef]

energies

MDPI

Article

Economic Optimal Scheduling of Integrated Energy System Considering Wind–Solar Uncertainty and Power to Gas and Carbon Capture and Storage

Yunlong Zhang [1], Panhong Zhang [2,*], Sheng Du [3] and Hanlin Dong [1]

[1] Power China Hubei Electric Engineering, Wuhan 430040, China
[2] School of Finance, Hubei University of Economics, Wuhan 430205, China
[3] School of Automation, China University of Geosciences, Wuhan 430074, China; dusheng@cug.edu.cn
* Correspondence: zhangphhbjj@sina.com

Abstract: With the shortage of fossil energy and the increasingly serious environmental problems, renewable energy based on wind and solar power generation has been gradually developed. For the problem of wind power uncertainty and the low-carbon economic optimization problem of an integrated energy system with power to gas (P2G) and carbon capture and storage (CCS), this paper proposes an economic optimization scheduling strategy of an integrated energy system considering wind power uncertainty and P2G-CCS technology. Firstly, the mathematical model of the park integrated energy system with P2G-CCS technology is established. Secondly, to address the wind power uncertainty problem, Latin hypercube sampling (LHS) is used to generate a large number of wind power scenarios, and the fast antecedent elimination technique is used to reduce the scenarios. Then, to establish a mixed integer linear programming model, the branch and bound algorithm is employed to develop an economic optimal scheduling model with the lowest operating cost of the system as the optimization objective, taking into account the ladder-type carbon trading mechanism, and the sensitivity of the scale parameters of P2G-CCS construction is analyzed. Finally, the scheduling scheme is introduced into a typical industrial park model for simulation. The simulation result shows that the consideration of the wind uncertainty problem can further reduce the system's operating cost, and the introduction of P2G-CCS can effectively help the park's integrated energy system to reduce carbon emissions and solve the problem of wind and solar power consumption. Moreover, it can more effectively reduce the system's operating costs and improve the economic benefits of the park.

Keywords: integrated energy systems; economically optimized dispatch; landscape uncertainty; carbon trading; P2G-CCS

Citation: Zhang, Y.; Zhang, P.; Du, S.; Dong, H. Economic Optimal Scheduling of Integrated Energy System Considering Wind–Solar Uncertainty and Power to Gas and Carbon Capture and Storage. *Energies* **2024**, *17*, 2770. https://doi.org/10.3390/en17112770

Academic Editor: Gabrial Anandarajah

Received: 11 May 2024
Revised: 30 May 2024
Accepted: 3 June 2024
Published: 5 June 2024

1. Introduction

In recent years, renewable new energy, mainly wind and solar power generation, has been gradually developed and applied to society. However, due to the wind and solar power generation system output having a certain degree of randomness and volatility, there has been a tendency for wind and solar power to be abandoned. This has resulted in the new energy utilization rate being relatively low, and the variability in and uncertainty of the energy source potentially affecting the stability and economy of the power system scheduling [1]. Concurrently, the energy industry, particularly the electric power industry, accounts for a considerable proportion of greenhouse gas emissions, and thus bears a significant responsibility for the achievement of green and low-carbon development. Consequently, the park integrated energy system (PIES), which couples a variety of energy sources through energy conversion equipment and exploits the complementary advantages of multiple energy sources, will play an important role in energy conservation and emission reduction in the energy industry [2,3]. This is due to the fact that it is supported by renewable energy generation, electricity-to-gas conversion, energy storage and other technologies.

The uncertainty of wind and solar output can be better managed to enhance the reliability, stability and economy of integrated energy systems. Currently, the mainstream research approach to the wind–solar power output uncertainty problem is to use statistical methods to model wind and solar resources, such as time series analysis and probability distribution fitting, to quantify the uncertainty of wind energy [4]. Nayak et al. [5] used a stochastic approach to model uncertainty and performed bi-objective optimization in a fuzzy domain using a 'fuzzy max-min' satisfaction criterion. Abdelmalak et al. [6] pointed out that Monte Carlo simulation and perturbation techniques are computationally intensive and cannot be directly embedded in the power system model, and proposed an approach based on generalized polynomial chaos (GPC). Cao et al. [7] proposed a combined LHS and K-means clustering method to solve the uncertainty problem of wind-glow output. Each of the aforementioned studies addresses uncertainty through extensive random sampling and parameter perturbation, or the use of polynomial approximation of the uncertainty distribution. However, constrained by the computational cost and parameter space, the LHS and fast prior scenario reduction methods appear to be more widely applicable and more efficient in sampling. When compared to other methods like Monte Carlo or random sampling, LHS offers better coverage of the parameter space and reduced sampling variance. It ensures a more even distribution of samples, which can lead to more accurate and reliable results, especially in high-dimensional problems. Many studies have used LHS to generate different scenarios and then reduced the simulation options using the fast forward selection algorithm (FSA), which is a time-consuming method of scenario simplification but more accurate compared to the backward reduction algorithm [8].

In addressing the issue of the low-carbon economic optimization of integrated energy systems (IESs), Duan et al. [9] employs the reward–punishment laddering carbon trading mechanism to quantify the carbon trading cost, thereby constructing a low-carbon economic optimization model of IESs. Wang et al. [10] utilizes the carbon emission factor and reward–punishment stepped carbon trading mechanism to calculate the carbon trading cost and evaluates the economic benefits of carbon capture power plants. By formulating an optimized stepped carbon trading mechanism, the internal carbon balance of the IES can be achieved and carbon emissions can be curbed to prevent the IES from overshooting the annual carbon emission settlement [11,12]. Li et al. [13] proposed an IES scheduling model based on full life cycle assessment and carbon capture and utilization (CCU) technology, and analyzed the correlation between different carbon trading prices and the low-carbon economic transport of the IES. Furthermore, the impact of CCU technology inputs on the low-carbonity of the IES was investigated. In a further contribution to the field, Li et al. [14] put forth an IES source–load coordinated optimal scheduling model with the objective of achieving comprehensive economic optimality for the system, encompassing system–system economy and low amounts of carbon. This model employs a reward and punishment ladder-type carbon trading mechanism to minimize the waste of wind energy, stimulate consumer participation, and reduce system operating costs and carbon emissions.

Power to gas (P2G) plants enhance the capacity of wind and solar power by converting excess electricity into natural gas in an IES [15]. The underlying principle is that H_2 and O_2 are initially produced by electrolysis of water in an electrolysis system (EL), and then natural gas is synthesized from H_2 and CO_2 by a methane reactor (MR). Carbon capture and storage (CCS) technology captures, sequesters and utilizes carbon, and the carbon captured by CCS can be made available to P2G. Introducing P2G-CCS from the equipment side is one of the important ways to help an IES reduce carbon emissions and achieve low-carbon operation. The P2G-CCS carbon cycle flow diagram is shown in Figure 1, in which the CCS captures and utilizes the CO_2 emitted from the operation of natural gas,

gas-fired internal combustion engines and gas-fired boilers, and transports a part of it to the methane reactor equipment to realize the coupling with the P2G.

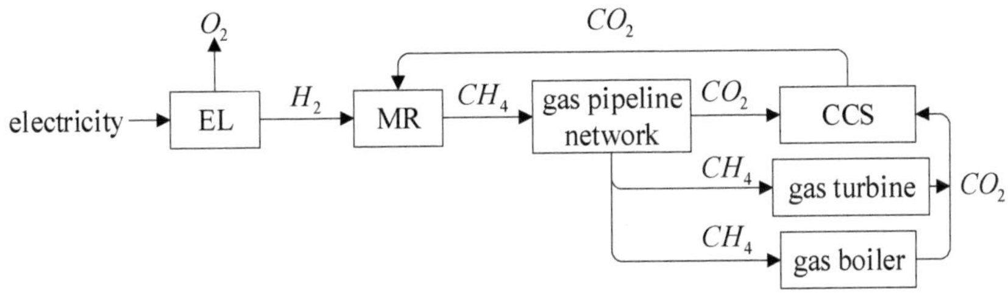

Figure 1. P2G-CCS flowchart.

Zhang et al. [16] proposed a two-tier optimal scheduling model for an integrated energy system with a carbon capture system and a P2G facility, which improves the economy of the IES and the capacity of wind and solar energy consumption. Furthermore, Li et al. [17] extended the P2G-CCS technology to a hybrid concentrated solar power and combined heat and power (CSP-CHP) IES with the aim of consuming more wind power and further reducing CO_2 emissions. In a further development, Liang et al. [18] created an Integrated Electricity–Gas System (IEGS) optimal dispatch model that couples gas-fired units and P2G links. This model was then used to investigate the impact of different constraints, such as power system constraints and natural gas constraints, on the model's wind power consumption. Zhang et al. [19] integrated P2G, carbon capture and the supercritical carbon dioxide ($S-CO_2$) cycle systems into a carbon cycle system. This system was then used to investigate the role of the carbon cycle system in an electric–thermal IES.

The literature above demonstrates the efficacy of P2G-CCS for the optimal low-carbon economic dispatch of the IES. However, it does present certain challenges to the economic dispatch due to the stochasticity of wind and solar power not being considered. Zhu et al. [20] proposed a low-carbon economic dispatch model of an IES containing P2G and hydrogen-doped gas units (HGT) coupling under the stepped carbon trading mechanism. This model addresses the wind abandonment problem of the IES by utilizing the operating characteristics of P2G in the system while considering the inherent uncertainty of wind power output. Furthermore, it enhances the system's renewable energy consumption capacity. In a further contribution to the field, Zhang et al. [21] employed conditional value-at-risk (CvaR) theory to analyze the uncertainty of wind power, utilizing P2G to reduce the system wind abandonment rate and energy storage devices to mitigate the operational risk of wind power fluctuation on the system. Both Sun et al. and Pan et al. [22,23] have proposed an optimal scheduling model for IESs considering flexible loads and P2G participation in the carbon trading market, and verified the importance and economics of P2G for system reduction of operating costs and the feasibility of wind power adaptability. In a further contribution to the field, Yang et al. [24] put forth a P2G multi-objective cooperative operation optimization model for an interconnected integrated gas-to-electric energy system, taking into account the inherent uncertainty of wind power. Through an arithmetic analysis, it was demonstrated that connecting the P2G equipment can reduce the total operating cost of the natural gas system and can improve the system wind energy consumption.

In light of the aforementioned background, it can be seen that the consideration of carbon trading mechanisms and wind and solar uncertainty is conducive to achieving energy savings, emission reductions and an economically viable integrated energy system. Furthermore, the role of P2G-CCS equipment in the system's wind and solar consumption capacity and low-carbon economy requires further investigation. However, the current study has the following shortcomings:

1. The low-carbon economy and wind-scenery consumption effect of integrated energy systems have not been adequately explored, despite the necessity of effectively quantifying wind-scenery uncertainty.
2. The sensitivity analysis of the construction scale parameters of P2G-CCS equipment has yet to be studied.

In light of the aforementioned considerations, this paper initially examines the potential for wind and solar output uncertainty, preprocesses the wind and solar resource data, generates a classical scenario set utilizing the Latin hypercube sampling and reduction technique, and then reduces the scenarios using the fast antecedent elimination technique. This process enables the scenario set to comprise a number of representative scenarios, which can then be used to estimate the actual wind and solar output effect and to reduce the influence of wind and PV output fluctuations.

Secondly, an integrated energy system mathematical model considering P2G-CCS is established. This model is used to analyze the coupling relationship between multiple energy sources and to establish a system carbon emission calculation model and a step-type carbon trading mechanism.

Subsequently, the branch bounding algorithm is employed to transform the system's wind and solar consumption and carbon emissions into wind and solar abandonment costs and carbon trading costs, which are then aggregated into the system's operating cost. The objective of this process is to identify the lowest system operating cost, which serves as the benchmark for the economic optimal scheduling study. This study aims to identify the optimal scheduling strategy for the system.

Finally, a typical park is used as an example to verify the optimization model proposed in this paper. The sensitivity of the construction scale parameters of the P2G-CCS is analyzed, and the results demonstrate that the model is effective in solving the aforementioned issues. The issue of wind and solar energy consumption, and the reduction in system operating costs and carbon emissions, is addressed. It is demonstrated that as the construction scale of the P2G-CCS increases, the system's operating costs decrease. This verifies the economic and effective nature of the optimization scheduling model proposed in this paper.

2. PIES System Architecture

The architectural framework of the PIES system is illustrated in Figure 2, which depicts the principal coupled energy types within the system. These include electric energy, cooling energy, heat energy and natural gas. With regard to the electric energy aspect, the power input side is primarily constituted by wind and solar power generation systems and combustion turbine power generation systems. In this latter case, the combustion turbine system performs the function of power transmission and distribution by consuming natural gas. With regard to the interaction between the PIES system and the main grid, it is important to note that the prices for the purchase and sale of power are identical. These prices are based on the same-day step tariff, as specified in [25].

In terms of heat generation, the primary source is a gas boiler and a waste heat recovery device, which utilizes the waste heat generated by the combustion engine to provide heat. On the cooling side, the primary cooling system is a lithium bromide absorption chiller, which operates on the principle of generating cooling energy by consuming heat energy. In order to facilitate the flexible scheduling of heat and cold energy, this paper incorporates heat and cold energy storage systems for energy storage and release.

The supply of natural gas to the PIES is comprised of two primary sources: the gas grid and the P2G facility on the campus. The gas grid is the primary source, where CO_2 is captured and purified by CCS to obtain a high concentration of CO_2 for delivery to the P2G facility. The P2G facility generates A and B by electrolysis of water, combines H_2 and CO_2 into natural gas, and sends CH_4 to the gas pipeline for use in the combustion engine and gas boiler systems. Consequently, the utilization of P2G-CCS apparatus cannot only facilitate the system's ability to utilize wind and solar energy but also facilitate the capture of CO_2 to produce natural gas for the system. This reduces the system's natural gas purchase costs and carbon emissions, and thus enables the system to operate in a low-carbon manner.

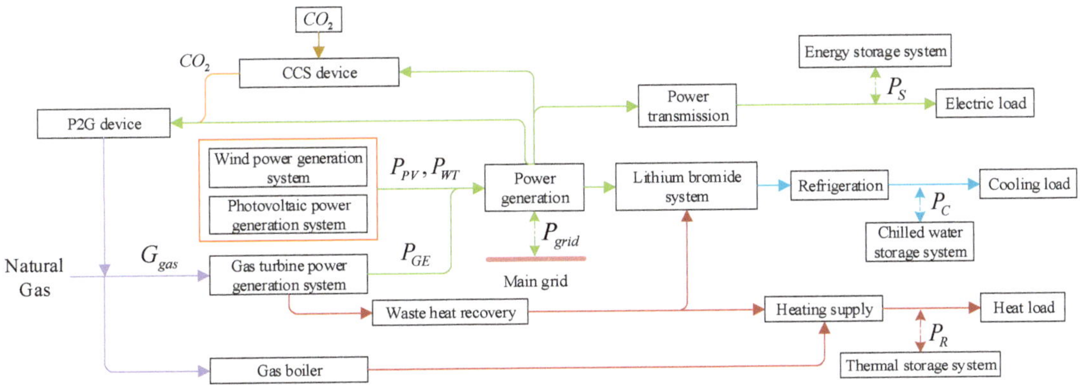

Figure 2. The architectural framework of the PIES system.

In this study, based on the information of the relevant price data in the past day, the optimal scheduling of each piece of equipment's output in the PIES in the past day is carried out with the optimization goal of minimizing the cost of one day's operation of the system, so as to achieve the economic and low-carbon operation of the system. Therefore, the mathematical model of each piece of equipment's output model in the PIES will be constructed according to the PIES system architecture, in which the natural gas supplier and the main network are the external energy suppliers of the system, and the remaining equipment is the internal equipment of the PIES. The specific mathematical model of each piece of equipment is shown in Section 3.

3. PIES Modelling

In the previous section, the system architecture of the PIES was established, and in this section, the mathematical modelling of the processing models of each device in the PIES will be continued. Among them, considering the spectral output uncertainty problem, this paper adopts the LHS method for scene generation, and then applies the fast antecedent elimination technique for scene reduction to deal with the wind and light output uncertainty problem. Each piece of energy supply equipment in the PIES sees the existence of coupling; therefore, this paper will establish a multi-energy coupling equipment power model in Section 2. At the same time, this paper takes the lowest cost of system operation as the optimization goal, and carries out the optimal scheduling of the output of each device before the day, and the scheduling process involves the relevant constraints to constrain the operation of the system. Therefore, this chapter will follow the construction of the mathematical model of the system optimization goal and constraints, so as to build up a mathematical model basis for the subsequent optimal scheduling research.

3.1. Wind Turbine Systems

Wind turbines convert the kinetic energy of the wind into mechanical energy through the wind turbine wheel. This wheel cuts the magnetic inductance to generate electric current after speeding up and then converts the mechanical energy into electric energy. Wind farms usually consist of several wind turbines, and the output power must be calculated according to factors such as wind velocity and wind direction. The power generation of wind turbines can be expressed as [26,27]:

$$P_{WT}(t) = \begin{cases} P_r \frac{v^2 - v_i^2}{v_r^2 - v_i^2}, & v_i < v < v_r \\ P_r, & v_r \leq v < v_f \\ 0, & \text{the rest} \end{cases} \tag{1}$$

where:

P_{WT}: Actual power of the WTG.
P_r: Rated power of the WTG.
v: Real-time wind velocity.
v_i: Cut-in wind velocity of the turbine during normal operation.
v_r: Rated wind velocity of the turbine.
v_f: Maximum wind velocity of the turbine during normal operation.
t: Current moment.

3.2. Photovoltaic System

The PV system, comprising photovoltaic panels, converts solar energy into direct current (DC) electrical energy [28]. The fundamental principle is based on the photovoltaic effect, whereby the semiconductor material on the panels absorbs solar energy to be converted into electrical energy. The photovoltaic array model is described in Equation (2) as the PIES. The PV system's active power output is primarily influenced by the intensity of light S on a given day, which can be expressed as:

$$P_{PV}(t) = P_{STC} \frac{L_C(t)}{L_{STC}} (1 + \mu(T_C - T_{STC})) \tag{2}$$

where:

P_{PV}: Actual power generated by the solar photovoltaic panel.
P_{STC}: Rated power of the solar photovoltaic panel.
L_C: Current ambient light intensity.
L_{STC}: Light intensity in the ideal environment.
μ: Power temperature coefficient.
T_C: Surface temperature of the solar photovoltaic panel at work.
T_{STC}: Surface temperature of the solar photovoltaic panel in the ideal environment.

3.3. Wind and Solar Scene Production and Weakening

The uncertainty of the system comes from two sources: turbine output and PV output. Wind turbine output is primarily affected by wind velocity and exhibits stochastic characteristics. PV output is predominantly contingent on outside weather conditions, particularly cloud cover. To effectively assess uncertainty, it is essential to propose probability density functions (PDFs) for wind and solar output [29]. It has been demonstrated that the forecast errors of wind turbine and PV output exhibit a Weibull distribution and a Beta distribution, respectively [30].

The PDF of wind turbine output:

$$f(P_{WT}) = (\frac{k_{WT}}{r}) \times (\frac{P_{WT}}{r})^{k_{WT}^{-1}} \exp[-(\frac{P_{WT}}{r})^{k_{WT}}] \tag{3}$$

where:

P_{WT}: The actual operating power of the turbine.
k_{WT}: The shape factor of the turbine output.
r: The scale factor.

The PDF of PV output:

$$f(P_{PV}) = P_{PV} \eta_m A_{PV} \eta_{PV} \cos\theta A \tag{4}$$

where:

P_{PV}: The actual operating power of the PV.
η_m: The tracking rate at the maximum power point.
A_{PV}: The radiation area of the PV.
η_{PV}: The conversion efficiency.
θ: The solar incidence angle.

3.3.1. Latin Hypercube Sampling Based on Probability Distributions

The LHS method is employed to generate a multitude of scenarios, which may be considered a form of stratified sampling with an inverse function transformation [31]. This is utilized to simulate the potential wind and solar power generation, in the context of a probability distribution function of a random variable. LHS may be stratified by dividing the sampling probability intervals in equal parts, thereby ensuring the completeness of the sampling information and enriching the information of the samples.

In this paper, the PIES is made to interact with the smart grid for information interaction, and through the previous day's wind velocity data and light information transmitted by the meteorological forecast central, it obtains the previous day's power generation data for preprocessing, determines its probabilistic model parameters, and generates the correlation sample matrix of wind velocity and light intensity, respectively, and then transforms the data in the matrix into wind and solar power data to form the base scenario according to the wind and solar system model.

3.3.2. Scene Weakening

To reduce the computational burden and improve the overall effectiveness and response speed, the fast antecedent elimination technique is applied to reduce the scenarios with similar characteristics, and the typical scenarios are selected while retaining the characteristics of the original set of scenarios. The simplified typical wind and PV scenarios are combined into a new scenario set and their corresponding probabilities are calculated [32].

The probability corresponding to each scene generated by LHS is $P_s(S = 1,2,\ldots,N)$ and satisfies $P_s \geq 0, \sum_S P_s = 1$. Let each scene generation probability $P_s = 1/N, \xi_S(S = 1,2,\ldots,N)$ denote the scene in the sample space, and the specific steps of the fast antecedent elimination technique are as follows:

Step 1: Calculate the two-paradigm number between each scene. This can be achieved by calculating the Euclidean distance, which allows for the determination of the elements that are more important according to their size. This enables the analysis of the characteristics and structure of the vector. The Euclidean distance formula is as follows:

$$d(\xi_i, \xi_j) = \sqrt{\sum_{t=1}^{24} (\xi_{t,i} - \xi_{t,j})^2} \tag{5}$$

where:

d: The Euclidean distance.
ξ_i/ξ_j: Two scenes in space.

Step 2: Identify the set of the minimum value of the product of the number of paradigms of each pair of scenes (ξ_i, ξ_j) and the probability P_s of the occurrence of scene ξ_i in all the scenes. This set is denoted as set D and its expression is as follows:

$$D = \left\{ \min_{\substack{i \in \{1,2,\cdots,N\} \\ i \neq j}} P_s d(\xi_i, \xi_j) \right\} \tag{6}$$

Replace ξ_i with ξ_j, the closest Euclidean distance to ξ_i in the scene set D. The probability of D_1 should then be added to the probability of D. ξ_i should be eliminated, a new scene set D_1 formed and the number of elements in the scene set updated.

Step 3: Repeat steps 1 and 2 until the reduction requirements are met and the number of remaining scenes is reached. At this point, the run should be stopped. Otherwise, the reduction process should be continued.

3.4. Energy Storage Equipment

In light of the uncertainty of wind and solar output, this paper proposes the introduction of energy storage systems in the PIES to enhance the stability of the system and increase the wind and solar consumption rate. The energy storage equipment can be categorized into three main types: battery storage, thermal energy storage and cold energy storage [33]. The storage systems can help the PIES to improve the stability of the system's function and increase the wind and solar consumption rate. Additionally, the incorporation of energy storage systems can increase the flexibility of the PIES, facilitate the optimal selection of energy sources, facilitate enhanced interaction between the system and the main grid, and facilitate the generation of higher economic returns.

(1) Accumulators' energy storage

In this paper, the lithium iron phosphate battery energy storage system with better safety and high-rate discharge performance is selected. With the breakthrough of battery formation and application technology, the technology of a lithium-ion battery in large-capacity, large-scale power storage system applications will be gradually improved. Its mathematical model is as follows:

$$E_t^{es} = E_{t-1}^{es} + (\eta_{es,in} P_t^{es,in} \mu_t^{es,in} - \frac{P_t^{es,out} \mu_t^{es,out}}{\eta_{es,out}})\Delta t \tag{7}$$

where:

E_t^{es}/E_{t-1}^{es}: The storage power of the battery storage equipment at t, $t-1$, kW.
$\eta_{es,in}/\eta_{es,out}$: The charging and discharging efficiency of the battery.
$P_t^{es,in}/P_t^{es,out}$: The charging and discharging power of the storage battery at t, kW.
$\mu_t^{es,in}/\mu_t^{es,out}$: The starting and stopping working state of the energy storage equipment.

(2) Heat and cold storage equipment

This paper presents a mathematical model for a heat and cold energy storage system that can be used to achieve peak shaving and valley filling of heat and cold energy. This model effectively improves the energy utilization and economy of the system. Its mathematical model is as follows:

$$E_t^{hs} = E_{t-1}^{hs} + (\eta_{hs,in} H_t^{hs,in} \mu_t^{hs,in} - \frac{H_t^{hs,out} \mu_t^{hs,out}}{\eta_{hs,out}})\Delta t \tag{8}$$

where:

E_t^{hs}/E_{t-1}^{hs}: The stored thermal power of heat storage equipment at t, $t-1$, kW.
$\eta_{hs,in}/\eta_{hs,out}$: The charging and discharging efficiency of heat storage equipment.
$H_t^{hs,in}/H_t^{hs,out}$: The charging and discharging power of thermal energy storage at t, kW.
$\mu_t^{hs,in}/\mu_t^{hs,out}$: The starting and stopping working state of energy storage equipment.

$$E_t^{cs} = E_{t-1}^{cs} + (\eta_{cs,in} Q_t^{cs,in} \mu_t^{cs,in} - \frac{Q_t^{cs,out} \mu_t^{cs,out}}{\eta_{cs,out}})\Delta t \tag{9}$$

where:

E_t^{cs}/E_{t-1}^{cs}: The stored thermal power of cold storage equipment at t, $t-1$, kW.

$\eta_{cs,in}/\eta_{cs,out}$: The charging and discharging efficiency of cold storage equipment.
$Q_t^{cs,in}/Q_t^{cs,out}$: The charging and discharging power of thermal energy storage at t, kW.
$\mu_t^{cs,in}/\mu_t^{cs,out}$: The starting and stopping working state of energy storage equipment.

3.5. Gas Internal Combustion Engine Systems

The internal combustion engine system generates electricity by consuming natural gas, while the waste heat generated by the internal combustion engine is utilized twice through a waste heat recovery unit. The mathematical model of the internal combustion engine system is as follow:

$$P_t^{GE} = G_t^{gas}\beta_{GE} \quad (10)$$

$$Q_t^{GE} = G_t^{gas}\beta_{RE} \quad (11)$$

where:

P_t^{GE}: The electrical energy generated by the combustion engine, kW.
G_t^{gas}: The flow rate of natural gas, m^3/h.
β_{GE}: The electrical energy conversion efficiency of the combustion engine.
Q_t^{GE}: The available waste heat generated in the process of generating electricity by the combustion engine.
β_{RE}: The waste heat recovery efficiency of the combustion engine.

3.6. Lithium Bromide Absorption Refrigeration System

The flue gas type lithium bromide absorption chiller utilizes a waste heat recovery device to absorb the waste heat generated by a gas-fired internal combustion engine for refrigeration. The mathematical model is as follows:

$$Q_t^{BL} = Q_t^{bl}\beta_{BL} \quad (12)$$

where:

Q_t^{BL}: The refrigeration capacity of the lithium bromide unit.
Q_t^{bl}: The heat absorbed and consumed by the lithium bromide unit for refrigeration.
β_{BL}: The conversion efficiency of the lithium bromide refrigeration unit.

3.7. The Waste Heat Recovery Device

The waste heat recovery device is mainly used to absorb the heat released during the operation of a gas internal combustion engine. Its mathematical model is as follows:

$$Q_t^{back} = Q_t^{ge}\beta_{back} \quad (13)$$

where:

Q_t^{back}: The heat generated by the waste heat recovery device.
Q_t^{ge}: The heat absorbed by the waste heat recovery device from the internal combustion engine.
β_{back}: The energy recovery efficiency of the waste heat recovery device.

3.8. P2G Device

The introduction of P2G technology into the PIES system, utilizing natural gas production, can effectively increase the wind and solar power integration rate and enhance the system's economic benefits. The mathematical model of the P2G device is as follows:

$$G_t^{P2G} = \frac{P_t^{P2G}\beta_{P2G}}{R_t} \quad (14)$$

where:

G_{P2G}: The natural gas flow rate produced by P2G equipment.

P_t^{P2G}: The electrical power consumed by the P2G equipment.
β_{P2G}: The electric-to-gas efficiency of the P2G equipment.
R_t: The lower heating value.

3.9. Objective Function

The integrated energy system constructed in this study achieves coordinated planning, optimal operation and interactive correspondence between energy subsystems by integrating multiple energy sources such as natural gas, electricity, heat and cold in the region, and introduces P2G-CCS equipment to efficiently improve energy use efficiency while improving the diversified energy demand in the system. When considering the effect of scenery uncertainty on the economic efficiency of the system, the costs under typical scenarios are calculated separately and then summed up to obtain the final operating costs. The system operation optimization cost is determined by the sum of the integrated energy system's cost of purchasing electricity from the main grid in the day-ahead market, the cost of purchasing gas at the natural gas end, the cost of operating and constructing each unit's equipment, the cost of carbon emission constraints, and the cost of the penalty for wind and solar energy system abandonment, which is expressed as shown in the following equation:

$$\min F_{PIES} = \sum_{s=1}^{n} P_s(F_s^{YW} + F_s^{JS} + F_{s,T,l}^{CO_2} + F_s^{ZW} + F_s^{Q} + F_s^{qfg}) \tag{15}$$

where

F_{PIES}: The minimum operating cost of the integrated energy system for one day.
n: The number of generated scenarios.
P_s: The probability of the scenario s occurring.
$F_{s,T,l}^{CO_2}$: The carbon emission cost under scenario s.

$$\begin{aligned} F_s^{YW} = \sum_{t=1}^{24} (& k_{PV}P_{s,t}^{PV} + k_{WT}P_{s,t}^{WT} + k_{MR}P_{s,t}^{MR} + k_{GE}P_{s,t}^{GE} + k_{GB}P_{s,t}^{GB} + k_{LB}P_{s,t}^{LB} \\ & + k_{GB}P_{s,t}^{GB} + k_{P2G}P_{s,t}^{P2G} + k_{CCS}P_{s,t}^{CCS} + \sum_{y\in[ES,CS,HS]} k_y(P_{y,s,t}^{c} + P_{y,s,t}^{f})) \end{aligned} \tag{16}$$

where
F_s^{YW}: The maintenance cost of the system's basic equipment under scenario s.
$k_{PV}/k_{WT}/k_{MR}/k_{GE}/k_{GB}/k_{LB}/k_{P2G}/k_{CCS}/k_y$: The maintenance cost factors for PV, WT, MR, GE, GB, LB, $P2G$, CCS, and energy storage units (ES, CS, HS), respectively.
$P_{s,t}^{WT}/P_{s,t}^{MR}/P_{s,t}^{GE}/P_{s,t}^{GB}/P_{s,t}^{LB}/P_{s,t}^{P2G}/P_{s,t}^{CCS}$: The output power of each device at time t.
$P_{y,s,t}^{c}$: The energy storage charging power at time t.
$P_{y,s,t}^{f}$: The energy storage discharging power at time t (energy storage cannot be charged and discharged simultaneously at the same time).

$$F_s^{JS} = \sum_{i=1}^{m} (h_i a_{i,s,t}) \tag{17}$$

where
F_s^{JS}: The system construction cost of scenario s.
m: The number of equipment types in the system.
h_i: The construction cost factor for equipment i.
a_i: The installed capacity of equipment i.

$$F_s^{ZW} = \sum_{t=1}^{24} (G_{price1} \cdot P_{s,t}^{grid,c} + G_{price2} \cdot P_{s,t}^{grid,f}) \tag{18}$$

where

F_s^{ZW}: The system power purchase costs of scenario s.
G_{price1}: The main grid tariff, which is sold to the system at time-of-day tariffs.
G_{price2}: The main grid power purchase price from the system.
$P_{s,t}^{grid}$: The electricity purchased by the system from the main grid.
$P_{s,t}^{grid,f}$: The electricity sold by the system to the main grid.

$$F_s^Q = \sum_{t=1}^{24} G_{Gasprice}(P_{s,t}^{GB} \cdot X_{GB} + P_{s,t}^{GE} \cdot X_{MT} - P_{s,t}^{P2G} \cdot X_{MR}) \tag{19}$$

where
F_s^Q: The cost of gas purchased by the system from the natural gas side of scenario s.
$G_{Gasprice}$: The price of natural gas.
$P_{s,t}^{GB}$: The exchange power of gas boiler at time t.
$P_{s,t}^{GE}$: The electrical power generated by the combustion engine at time t.
$P_{s,t}^{P2G}$: The exchange power of P2G at time t.
X_{GB}/X_{MT}: The efficiency values of natural gas consumption in gas boilers and combustion engines, respectively.
X_{MR}: The efficiency value of MR's consumption of hydrogen to make natural gas.

$$F_s^{qfg} = \sum_{t=1}^{24} c_{qfg}(P_{s,t}^{Zf} - P_{s,t}^{Z}) \tag{20}$$

where
F_s^{qfg}: The wind and light abandonment costs of the system under scenario s.
c_{qfg}: The abandonment penalty cost factor.
$P_{s,t}^{Zf}$: The WT and PV power generation at moment t.
$P_{s,t}^{Z}$: The local consumption of WT and PV energy at time t.

3.10. Constraints

The PIES introduces constraint conditions to regulate the rational, efficient, and stable scheduling and utilization of various system devices, assisting the PIES in its objective of economic optimization through multi-period, multi-objective system day-ahead scheduling research. Specific constraints include:

(1) Electric power balance constraint

$$\begin{aligned} P_t^{load} = P_t^{grid} + P_t^Z + P_t^{GE} + P_t^{es,out} - P_t^{P2G,e} - \\ P_t^{CCS} - P_t^{es,in} \end{aligned} \tag{21}$$

$$P_t^Z = P_t^{PV} + P_t^{WT} \tag{22}$$

where:

P_t^{load}: The system electrical load demand.
P_t^{grid}: At time t, the exchange power between the integrated energy system and the main grid, where a positive value indicates the system purchasing electricity from the main grid, and a negative value indicates the system selling electricity to the main grid.
P_t^Z: At time t, the sum of the output power from photovoltaic panels and wind turbines.

(2) Cooling power balance constraint

$$Q_t^{load} = Q_t^{BL} + Q_t^{cs,out} - Q_t^{cs,in} \tag{23}$$

where:

Q_t^{BL}: The cooling energy produced by lithium bromide.

Q_t^{cs}: The cooling energy exchange of the energy storage system.
Q_t^{load}: The system cooling energy demand.

(3) Thermal power balance constraint

$$H_t^{load} = H_t^{back} + H_t^{GB} + H_t^{hs,out} - H_t^{hs,in} - H_t^{BL} \tag{24}$$

where:

H_t^{load}: The thermal load demand of the system.

(4) The safety constraint on power exchange of the main grid transmission lines

$$-P_t^{grid,\max} \leq P_t^{grid} \leq P_t^{grid,\max} \tag{25}$$

where:

$P_t^{grid,\max}$: The maximum allowable power exchange of the transmission lines.

(5) The constraints of the electricity storage equipment

$$\begin{cases} P_t^{es,\min} \leq P_t^{es,in} \leq P_t^{es,\max} \\ P_t^{es,\min} \leq P_t^{es,out} \leq P_t^{es,\max} \\ \mu_t^{es,in} + \mu_t^{es,out} \leq 1 \\ \sum\limits_{t=1}^{24} (\mu_t^{es,in} + \mu_t^{es,out}) \leq R_{es} \\ SOC_{es,\min} \leq E_t^{es} \leq SOC_{es,\max} \\ SOC_{es,\min} = 0.1SOC \\ SOC_{es,\max} = 0.9SOC \end{cases} \tag{26}$$

where:

$P_t^{es,\max}$: The maximum charging and discharging power of the electricity storage equipment at time t.
$P_t^{es,\min}$: The minimum charging and discharging power of the electricity storage equipment at time t.
R_{es}: The maximum number of charge and discharge cycles allowed for the energy storage equipment within one day.
SOC: The installed capacity of the electricity storage system equipment, kWh.
$SOC_{es,\min}$: The minimum state of charge threshold for the energy storage system.
$SOC_{es,\max}$: The maximum state of charge threshold for the energy storage system.

(6) The constraints of the thermal storage equipment

$$\begin{cases} Q_t^{cs,\min} \leq Q_t^{cs,in} \leq Q_t^{cs,\max} \\ Q_t^{cs,\min} \leq Q_t^{cs,out} \leq Q_t^{cs,\max} \\ \mu_t^{cs,in} + \mu_t^{cs,out} \leq 1 \\ \sum\limits_{t=1}^{24} (\mu_t^{cs,in} + \mu_t^{cs,out}) \leq R_{cs} \\ SOC_{cs,\min} \leq E_t^{cs} \leq SOC_{cs,\max} \\ SOC_{cs,\min} = 0.1SOC_c \\ SOC_{cs,\max} = 0.9SOC_c \end{cases} \tag{27}$$

where:

$Q_t^{cs,\max}$: At time t, the maximum charging and discharging cooling power of the thermal energy storage equipment.
$Q_t^{cs,\min}$: At time t, the minimum charging and discharging cooling power of the thermal energy storage equipment.

R_{cs}: The maximum charging and discharging cycles allowed for thermal energy storage equipment within one day.
SOC_c: The installed capacity of the cold energy storage system equipment, kWh.
$SOC_{cs,\min}$: The minimum capacity state threshold of the energy storage system.
$SOC_{cs,\max}$: The maximum capacity state threshold of the energy storage system.

(7) Thermal energy storage device operating power constraint

$$\begin{cases} H_t^{hs,\min} \leq H_t^{hs,in} \leq H_t^{hs,\max} \\ H_t^{hs,\min} \leq H_t^{hs,out} \leq H_t^{hs,\max} \\ \mu_t^{hs,in} + \mu_t^{hs,out} \leq 1 \\ \sum\limits_{t=1}^{24} (\mu_t^{hs,in} + \mu_t^{hs,out}) \leq R_{hs} \\ SOC_{hs,\min} \leq E_t^{hs} \leq SOC_{hs,\max} \\ SOC_{hs,\min} = 0.1SOC_h \\ SOC_{hs,\max} = 0.9SOC_h \end{cases} \quad (28)$$

where:
$H_t^{hs,\max}$: At time t, the maximum charging and discharging heat power of the thermal energy storage equipment.
$H_t^{hs,\min}$: At time t, the minimum charging and discharging heat power of the thermal energy storage equipment.
R_{hs}: The maximum number of heat charge and discharge cycles allowed for the thermal storage equipment within one day.
SOC_h: The installed capacity of the thermal energy storage system equipment, kWh.
$SOC_{hs,\min}$: The minimum capacity state threshold of the energy storage system.
$SOC_{hs,\max}$: The maximum capacity state threshold of the energy storage system.

(8) The operational power constraints of the internal combustion engine equipment

$$\begin{cases} P_t^{GE,\min} \leq P_t^{GE} \leq P_t^{GE,\max} \\ P_t^{GE,down}\Delta t \leq P_t^{GE} - P_{t-1}^{GE} \leq P_t^{GE,up}\Delta t \end{cases} \quad (29)$$

where:
$P_t^{GE,\max}$: The maximum operating power of the gas engine.
$P_t^{GE,\min}$: The minimum operating power of the gas engine.
$P_t^{GE,up}$: The maximum ramping-up power of the gas engine unit.
$P_t^{GE,down}$: The maximum ramping-down power of the gas engine unit [34].

(9) Lithium bromide equipment operating power constraint

$$Q_t^{BL,\min} \leq Q_t^{BL} \leq Q_t^{BL,\max} \quad (30)$$

where:
$Q_t^{BL,\max}$: The maximum operating power of the lithium bromide system.
$Q_t^{BL,\min}$: The minimum operating power of the lithium bromide system.

(10) Power to gas (P2G) equipment operating power constraint

$$0 \leq P_t^{P2G} \leq P_t^{P2G,\max} \quad (31)$$

where:
$P_t^{P2G,\max}$: The maximum operating power allowed for P2G equipment.

3.11. Stepped Carbon Trading Mechanism

(1) Carbon emission calculation

The carbon emission modeling calculation of the carbon emission equipment in the system is carried out to further improve the calculation accuracy of the system's carbon transaction cost. The actual carbon emission calculation model is as follows:

$$\begin{cases} E_t^{CO_2,a} = E_t^{grid} + E_t^Z + E_t^{GE} + E_t^{GB} + E_t^{ES} \\ E_t^{grid} = \rho_g P_t^{grid} \\ E_t^Z = (\rho_p P_t^{pv} + \rho_w P_t^w) \\ E_t^{GE} = \rho_m(\kappa_h P_t^{GE} + H_t^{GE}) \\ E_t^{GB} = \rho_b H_t^{GB} \\ E_t^{ES} = \rho_e H_t^{ES} \end{cases} \quad (32)$$

where:

$E_t^{CO_2,a}$: The actual carbon emission of the equipment.
$\rho_g/\rho_p/\rho_w/\rho_m/\rho_b$: The carbon emission coefficients of power grid purchase, wind and solar power generation, gas turbine, gas boiler and energy storage, respectively.

(2) Carbon quota calculation

$$\begin{cases} E_t^{CO_2,b} = \sum_{t=1}^{T} (E_t^{grid} + E_t^Z + E_t^{GE} + E_t^{GB} + E_t^{ES}) \\ E_t^{grid} = \varpi_g P_t^{grid} \\ E_t^Z = (\varpi_p P_t^{pv} + \varpi_w P_t^w) \\ E_t^{GE} = \varpi_m(\kappa_h P_t^{GE} + H_t^{GE}) \\ E_t^{GB} = \varpi_b H_t^{GB} \\ E_t^{ES} = \varpi_e H_t^{ES} \end{cases} \quad (33)$$

where:

$E_t^{CO_2,b}$: The carbon emission quota of equipment.
$\varpi_g/\varpi_p/\varpi_w/\varpi_m/\varpi_b$: Initial carbon emission coefficient of power grid purchase, wind and solar power generation, gas turbine, gas boiler and energy storage, respectively.
T: One operating cycle, taking the value of 24 h a day.

Initial carbon allowances are the maximum amount of carbon dioxide that can be legally emitted in the carbon market. By limiting the total amount of carbon emissions, the government encourages companies to reduce their emissions through internal optimization or external trading.

Carbon emission allowances for each carbon source are shown in Table 1.

Table 1. Initial carbon emission allowances for each carbon emitting device.

Source of Carbon Emissions	Carbon Allowance/(kg/kWh)
Wind turbine	0.078
Photovoltaic generator	0.078
Gas boiler	0.152
Energy storage	0
Grids	0.798

(3) Carbon trading mechanism

The stepped carbon trading mechanism is a system that sets different carbon trading prices according to the different carbon emission intensities of enterprises. The difference between actual carbon emissions and carbon emission quotas is divided into multiple intervals. The higher the difference is, the higher the unit carbon trading price is, and the higher the additional cost that enterprises need to pay [35]. The cost mathematical model is as follows:

$$E_{tr,t} = \sum_{t=1}^{T} E_t^{CO_2,a} - E_t^{CO_2,b} \tag{34}$$

$$F_{T,l}^{CO_2} = \begin{cases} \lambda(1+2\partial)(E_{tr,t}+l), E_{tr,t} \le -l \\ -\lambda(1+2\partial)l + \lambda(1+\partial)E_{tr,t}, -l \le E_{tr,t} \le 0 \\ \lambda E_{tr,t}, E_{tr,t} < l \\ \lambda(1+\alpha)(E_{tr,t}-l) + \lambda l, l \le E_{tr,t} \le 2l \\ \lambda(1+2\alpha)(E_{tr,t}-2l) + \lambda(2+\alpha)l, 2l \le E_{tr,t} \le 3l \\ \lambda(1+3\alpha)(E_{tr,t}-3l) + \lambda(3+3\alpha)l, 3l \le E_{tr,t} \le 4l \\ \lambda(1+4\alpha)(E_{tr,t}-4l) + \lambda(4+6\alpha)l, E_{tr,t} \ge 4l \end{cases} \tag{35}$$

where:

$F_{T,l}^{CO_2}$: Stepped carbon transaction costs.
λ: Carbon trading base price.
$E_{tr,t}$: Actual carbon emissions minus carbon quotas.
l: The length of carbon emissions interval.
α: Trading price growth rate.
∂: The growth rate of transaction base price.

4. Multi-Objective Optimization Solution

Taking Equation (15) as the objective function and Equations (21) to (31) as the constraints, the system model can be summarized in the following form by optimizing the day-ahead power scheduling for each device of the system to achieve the intra-day economic operation of the system:

$$\min F_{PIES} = \sum_{s=1}^{n} P_s(F_s^{YW} + F_s^{JS} + F_{s,T,l}^{CO_2} + F_s^{ZW} + F_s^{Q} + F_s^{qfg})$$

$$\begin{cases} F_s^{YW} = \sum_{t=1}^{24} (k_{PV}P_{s,t}^{PV} + k_{WT}P_{s,t}^{WT} + k_{MR}P_{s,t}^{MR} + k_{GE}P_{s,t}^{GE} + k_{GB}P_{s,t}^{GB} + k_{LB}P_{s,t}^{LB} \\ +k_{P2G}P_{s,t}^{P2G} + k_{CCS}P_{s,t}^{CCS} + \sum_{y\in[ES,CS,HS]} k_y(P_{y,s,t}^{c} + P_{y,s,t}^{f})) \\ F_s^{JS} = \sum_{i=1}^{m} (h_i a_{i,s,t}) \\ F_s^{ZW} = \sum_{t=1}^{24} (G_{price1} \cdot P_{s,t}^{grid,c} + G_{price2} \cdot P_{s,t}^{grid,f}) \\ F_s^{Q} = \sum_{t=1}^{24} G_{Gasprice}(P_{s,t}^{GB} \cdot X_{GB} + P_{s,t}^{GE} \cdot X_{MT} - P_{s,t}^{P2G} \cdot X_{MR}) \\ F_s^{qfg} = \sum_{t=1}^{24} c_{qfg}(P_{s,t}^{Zf} - P_{s,t}^{Z}) \\ E.q.(35) \\ s.t. E.q.(21) \sim (31) \end{cases} \tag{36}$$

In the PIES, a day is considered to be divided into 24 time segments, with each hour as a unit, and it is assumed that the operation data of each system module are constant in the unit time.

Its system structure and algorithm correlation diagram is shown in Figure 3. The algorithm takes the minimum system operation cost as the goal, which in addition to the operation and maintenance cost of the integrated energy system in the park itself includes the cost of wind and light abandonment and the cost of carbon trading, and the constraints include the system energy supply and balance constraints, the equipment climbing constraints, and the equipment starting and stopping constraints. Since the model built in this study is a mixed integer linear programming model, this study considers the branch bounding algorithm and uses CPLEX for solving the problem. The optimal scheduling is shown in Figure 4, and the specific steps are shown below:

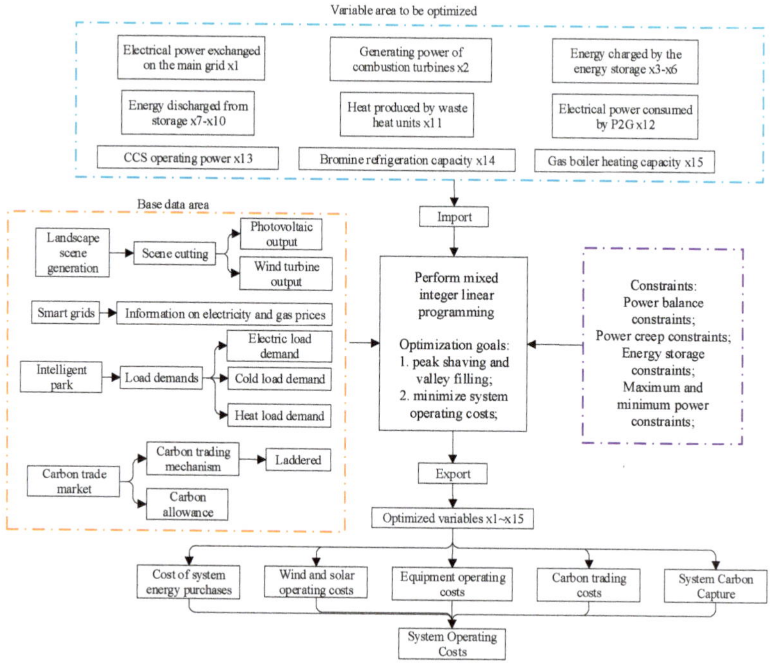

Figure 3. System architecture and algorithm correlation diagram.

Step 1: The wind and solar output scene is generated by Latin hypercube sampling, and then the scene reduction is carried out by using the fast previous generation elimination technology. Five typical wind and solar output scenes and the corresponding probabilities of each scene are obtained, and the wind and solar output data are transmitted to the model to be optimized.

Step 2: Obtain day-ahead information such as the ladder electricity price and gas price.

Step 3: Model each piece of equipment in the integrated energy system, construct the system carbon emission calculation model and the carbon trading mechanism model, and determine the optimization variables of each piece of equipment's output and their relevant parameters.

Step 4: Perform mixed integer linear programming, add conditional constraints, use the branch and bound algorithm, and use CPLEX for the optimal solution.

Step 5: Determine whether there is an optimal solution, and whether it is an integer; if not, select a non-integer solution variable to generate a branch. If yes, it is judged to search all branches, and the optimal solution is obtained by circular solution comparison.

Step 6: Obtain the solution results, output the output data of each piece of terminal equipment at each time, the system operation cost, carbon transaction cost and other information.

The optimization algorithm used by CPLEX can provide an accurate optimal solution; therefore, in this paper, for the integrated energy system model, and for the mixed integer linear programming problem, the CPLEX solver is used to find the best optimal scheduling solution [36].

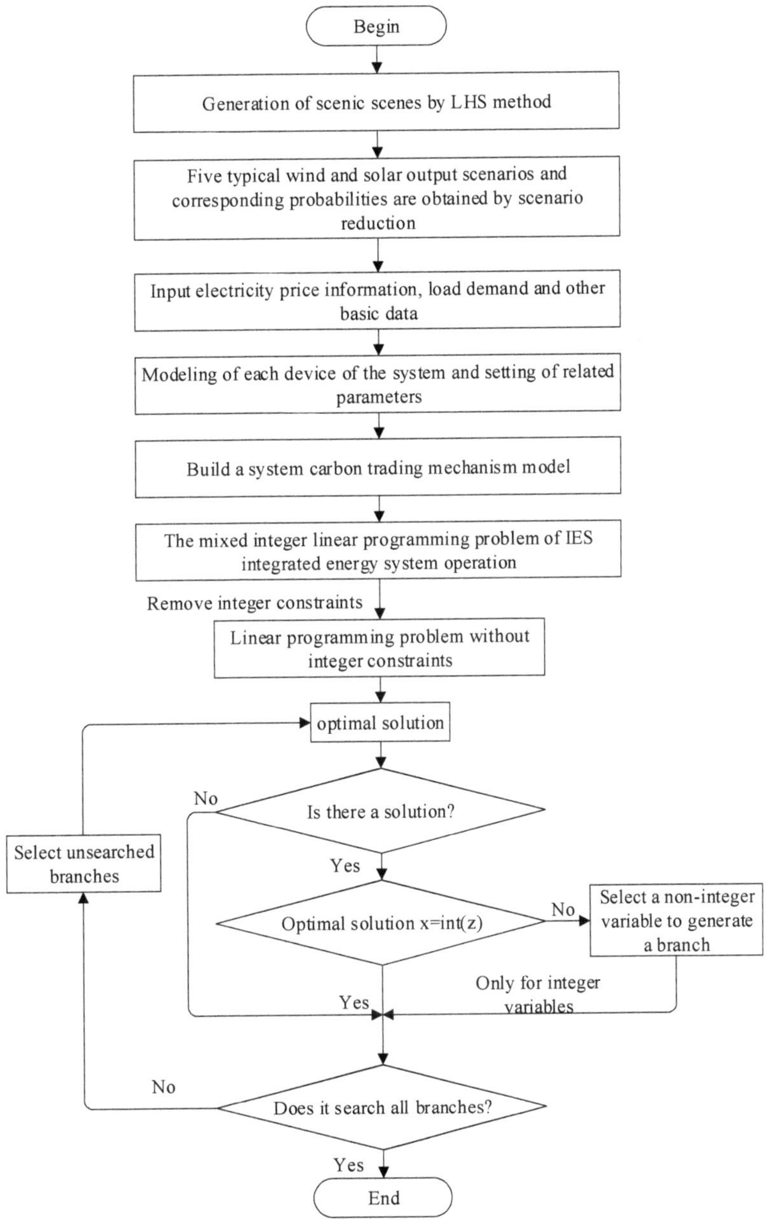

Figure 4. Flow chart of system optimized scheduling.

5. Example Analysis and Simulation

5.1. Basic Data

In this paper, the purchase and sale transactions between the PIES and the main grid are carried out based on time-sharing tariffs, and the tariff data are shown in Table 2, where the system purchases power from the grid according to the corresponding peak and valley time-sharing tariffs, and sells power to the main grid at a price of 0.36 yuan to save costs and improve economic efficiency in the process of optimal scheduling.

Table 2. Time division tariff schedule.

Time	Electricity Price/yuan
00:00–07:00	0.62
08:00–11:00	1.14
12:00–16:00	0.83
17:00–20:00	1.14
21:00–23:00	0.62
Main grid electricity tariff	0.36

In order to assess and compare the carbon emission levels of different equipment so as to take corresponding emission reduction measures, carbon emission coefficients are used to express the carbon emissions per unit of production, and the specific values are shown in Table 3.

Table 3. System carbon emission factor.

Source of Carbon Emissions	Carbon Emission Factor (kg/kWh)
Wind turbine	0.105
Photovoltaic generator	0.111
Natural gas (gas boiler)	0.236
Energy storage	0.112
Grids	1.303

Table 4 shows the values of the parameters such as power size, capacity range and number of times of charging and discharging energy for each device in the model of this paper.

Table 4. Partial equipment parameters of the system.

Parameter Name	Value	Unit
$P_t^{grid,\max}$	3000	kW
$P_t^{GE,\max}$	4000	kW
$P_t^{GE,\min}$	2830	kW
$P_t^{GB,\max}$	4000	kW
$P_t^{P2G,\max}$	3000	kW
$P_t^{CCS,\max}$	3000	kW
$P_t^{BL,\max}$	5000	kW
$P_t^{es,\max}$	2000	kW
$SOC_{es,\max}$	10,000	kWh
$H_t^{hs,\max}$	2000	kW
$SOC_{hs,\max}$	10,000	kWh
$Q_t^{cs,\max}$	2000	kW
$SOC_{cs,\max}$	10,000	kWh
R_{es}	10	times
R_{cs}	10	times
R_{hs}	10	times
X_{GB}	1.07	
X_{MT}	0.25	
X_{MR}	0.55	

5.2. Generation and Reduction of Landscape Scenes

In order to solve the wind power uncertainty problem, this paper adopts the previously described method to process the wind power uncertainty problem based on the day-ahead prediction data of wind power, as shown in Figure 5. Firstly, 1000 wind power scenarios are generated by using the Latin hypercubic sampling method, and then, five typical wind power scenarios are obtained by using the probability distance-based fast antecedent elimination technique to reduce them, The probability corresponding to each scenario is shown in Table 5.

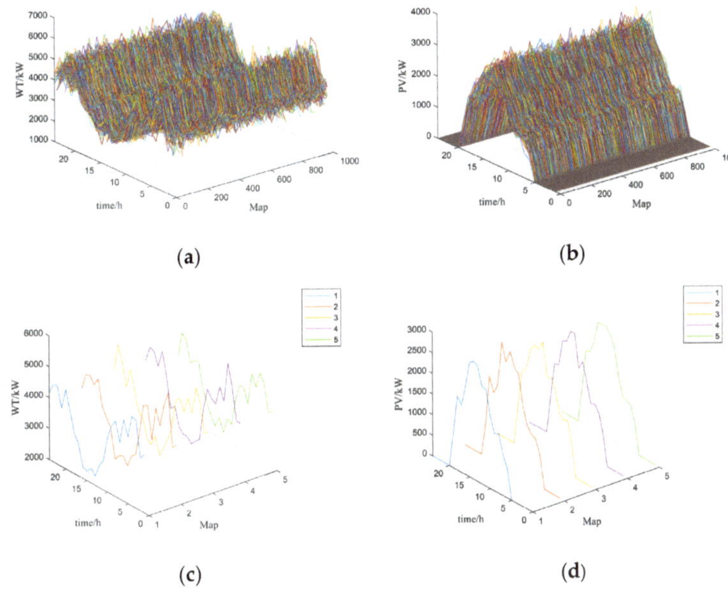

Figure 5. Generation and reduction of landscape scenes. Note: (**a**) generation of WT output scenarios; (**b**) generation of PV output scenarios; (**c**) scenario reduction diagram for WT output; (**d**) scenario reduction diagram for PV output.

Table 5. Corresponding probability for each scenario.

Scenario	Probability Value
1	0.158
2	0.258
3	0.133
4	0.195
5	0.256

The method used brings five typical scenarios of wind power output into the system optimal scheduling model for optimal operation to obtain the system's operation cost under different scenarios, then probabilistically sum up the operation costs under the five typical scenarios to obtain the final system's operation cost, and then carry out the optimal scheduling strategy of the system with the minimum operation cost through the solver to obtain the optimal operation scheduling strategy of the system considering the uncertainty of the wind power output.

5.3. Example Validation Analysis

To verify the effectiveness of the economic optimal scheduling strategy of the integrated energy system proposed in this paper considering wind power uncertainty and P2G-CCS, this paper takes the data of a typical integrated energy system park in Hubei as an example, as shown in Figure 6, which gives the initial data of the system's electric, heat and cold energy load demand, where *Pload* is the electric load, *Cload* is the cold load, and

Hload is the heat load. This paper also adopts the stepped carbon trading mechanism to carry out a comparative analysis of the system's operating costs under different scenarios, and the selected scenarios are shown as follows, with Scenario 4 being the scenario model proposed in this paper:

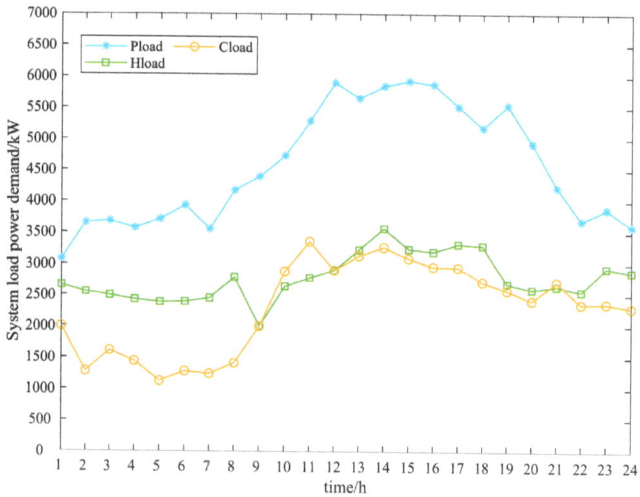

Figure 6. Initial system data diagram.

Scenario 1: Without considering the uncertainty of wind and solar processing, P2G-CCS is not introduced;

Scenario 2: Without considering the uncertainty of wind and solar power output, the introduction of P2G-CCS is considered;

Scenario 3: Considering the uncertainty of wind and solar output, the introduction of P2G-CCS is not considered;

Scenario 4: Considering the uncertainty of wind and solar power output, the introduction of P2G-CCS is considered.

Table 6 shows the comparative analysis results of the PIES's operating costs and other parameters in different scenarios. From Scenario 2 and Scenario 4, it can be seen that after the introduction of P2G-CCS in the system architecture of this paper, due to the scheduling flexibility of P2G-CCS, it can effectively help the system reduce the wind and light curtailment rate, reduce the carbon emissions of the system and thus reduce the operating costs of the system. In scenario 4, when considering the uncertainty of wind and solar output, on the basis of meeting the demand of various energy loads, the rate of wind and solar abandonment is basically 0, and the carbon emission has a downward trend. It not only solves the problem of the wind and solar consumption of the system, and reduces the carbon emissions of the system, but also further realizes the goal of the low-carbon economic operation of the PIES more effectively and saves economic costs.

Table 6. Comparative analysis table of system operating costs under different scenarios.

Scenario	Total System Operating Cost/million	Carbon Trading Costs/USD million	Carbon Emissions/ton	Cost of Wind and Light Abandonment/yuan	Wind and Light Rejection Rate
1	20.31	1.04	118.68	8339	12.07
2	13.12	−1.83	61.37	2096	3.24
3	18.74	0.93	106.07	6231	10.64
4	11.61	−2.25	57.42	0	0

In the Scenario 4 model, the energy supply optimization scheduling results of the system's equipment are shown in Figures 7–9. In the diagram, the positive value of each energy storage device indicates that the energy storage is discharged, and the negative value indicates that the energy storage system is charged. The positive value of the main grid indicates that the park buys electricity from the main grid, and the negative value indicates that the park sells electricity to the main grid.

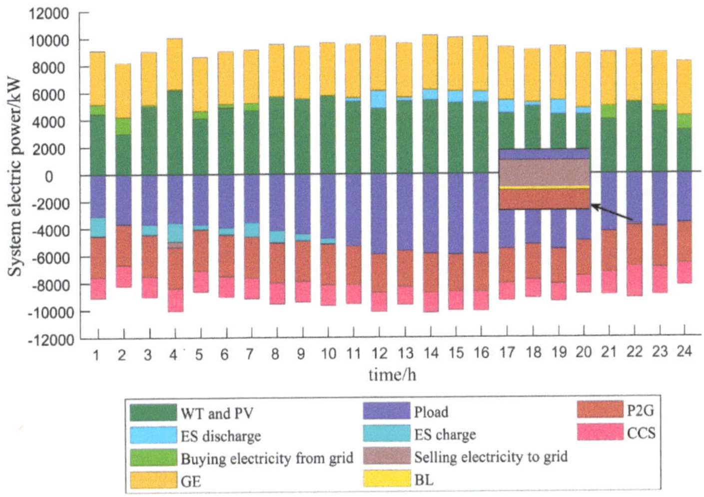

Figure 7. Optimized scheduling diagram for system power supply.

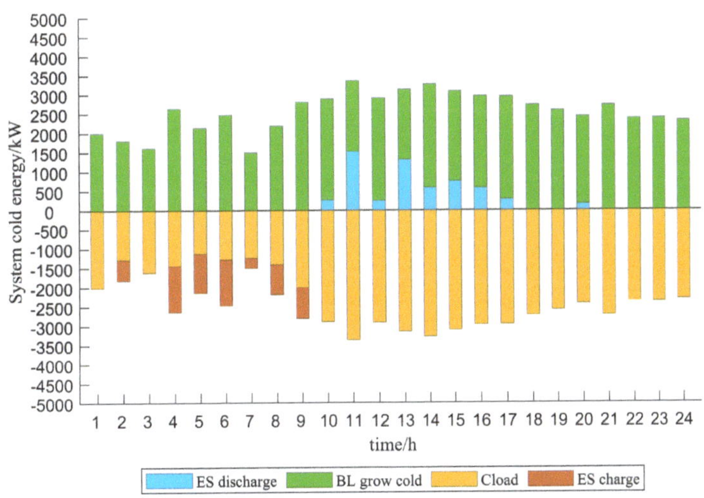

Figure 8. Optimized scheduling diagram for system cooling energy.

Figure 10 shows the carbon emissions diagram of the system. Figure 11 shows the distribution diagram of carbon dioxide utilization and storage after CCS captures carbon dioxide. After CCS captures carbon dioxide, a part of carbon dioxide is sent to P2G, and the MR in P2G consumes carbon dioxide and hydrogen to produce natural gas.

Combined with the chart, it can be observed that after the introduction of P2G-CCS in the system, since both P2G and CCS operate by consuming electric energy, the PIES flexibly schedules P2G equipment after the introduction of P2G-CCS, which solves the problem of wind and light consumption and makes the power supply of the system stable and balanced in each period. Moreover, P2G-CCS consumes carbon dioxide to produce natural gas for gas turbines and gas boilers, which reduces the cost of gas purchase of the PIES's natural gas to a certain extent. At the same time, it reduces the carbon emissions of the system, improves the economic benefits of the PIES participating in the carbon trading market and further reduces the operating cost of the system.

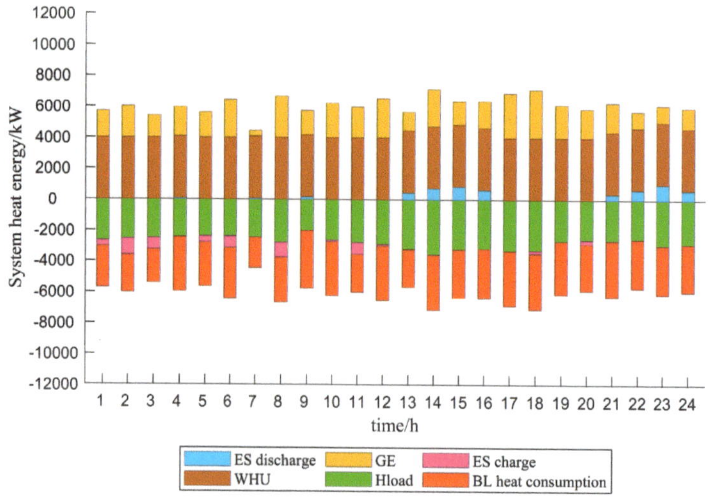

Figure 9. Optimized scheduling diagram for system heating energy.

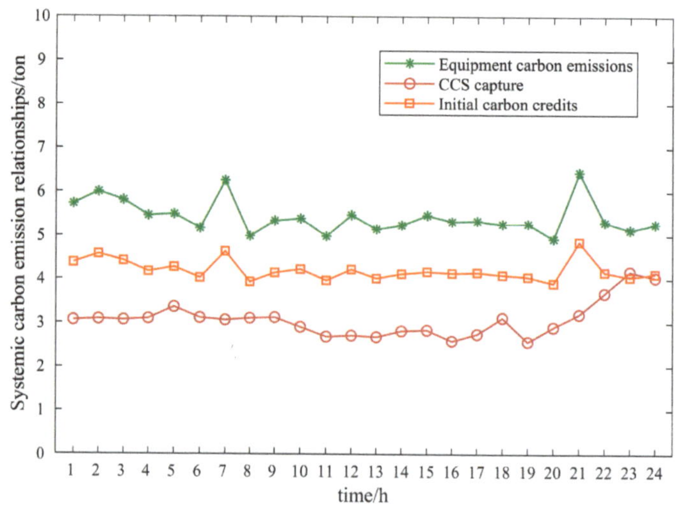

Figure 10. Systematic carbon emission relationships.

In order to analyze the impact of changing the construction scale of P2G-CCS on the operating costs and carbon emissions of the PIES, this paper proceeds with the sensitivity analysis of the P2G-CCS construction scale parameter in Scenario 4, as shown in Figure 12, which shows that, with the increase in the P2G-CCS construction scale parameter, the operating cost of the system is gradually reduced, and the total carbon emissions of the system decrease first and then increase. Therefore, this paper suggests that in the system architecture of this paper, the value of the P2G-CCS construction scale parameter between 3 MWh and 4 MWh can better achieve the goal of the system's low-carbon and economic operation, and this paper chooses to introduce the P2G-CCS construction scale of 3 MW in consideration of the environmental and economic aspects of the system's operation.

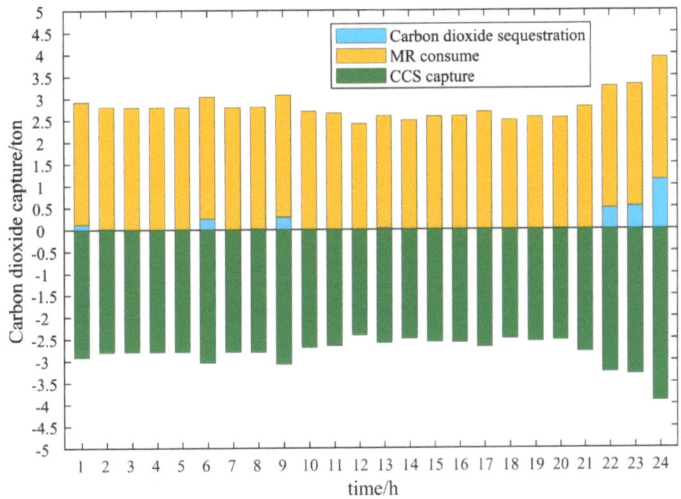

Figure 11. Utilization map between carbon capture and storage in the system.

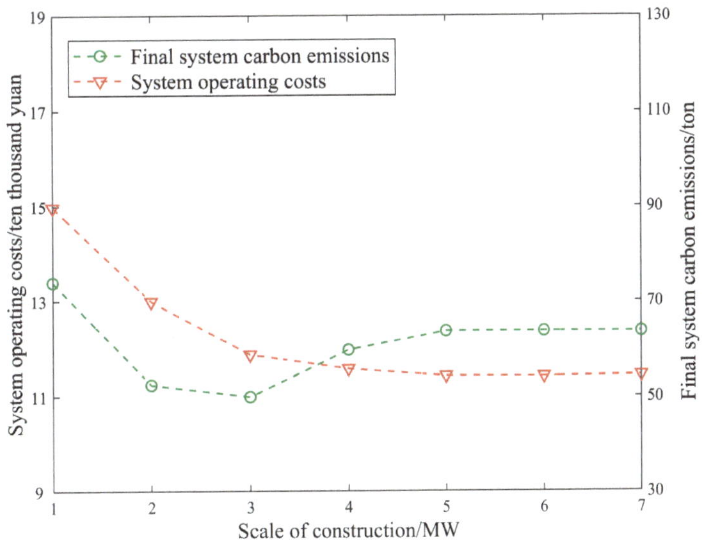

Figure 12. P2G construction scale sensitivity analysis chart.

6. Conclusions

Aimed at solving the problem of the low-carbon economic operation of a multi-energy coupling industrial park system with wind–solar new energy, this paper proposes the economic optimization research of an integrated energy system considering the uncertainty of wind–solar output and P2G-CCS. Through the simulation analysis of the example, the following conclusions can be obtained:

1. Considering the uncertainty of wind and solar output, the proposed model can more effectively improve the reliability, stability and economy of the IES, and reduce the operating costs of the system.
2. The introduction of P2G-CCS can more effectively help the system to solve the problem of wind and solar consumption, achieve the balance of energy supply, reduce the carbon emissions of the system and achieve the goal of the low-carbon economic operation of the system.
3. With the increase in P2G-CCS construction scale parameters, the system's operating costs gradually decrease. However, due to the influence of the scale of other equipment architecture of the system, the carbon emissions of the system show a trend of decreasing first and then increasing. Therefore, in the system architecture of this paper, it is suggested that the value of the P2G-CCS construction scale parameters should be between 3 MWh and 4 MWh.
4. The integrated energy system optimization model proposed in this paper, which takes into account the wind and light uncertainty and introduces P2G-CCS, can greatly improve the wind and light consumption rate, reduce the system's operating costs and carbon emissions, and achieve stable, efficient and environmentally friendly operation of the system.

Since the natural gas produced by P2G is incorporated into the natural gas pipeline to supply energy to the system's equipment, the impact of the dynamic response of the natural gas pipeline network on the low-carbon economic operation of the system can be considered in future research. The hot and cold pipe network in the system has certain delay characteristics. In future research, the optimal scheduling of the IES's hot and cold pipe network can be studied based on user-side comfort.

Author Contributions: Conceptualization, Y.Z. and P.Z.; methodology, Y.Z.; software, H.D.; validation, S.D.; formal analysis, P.Z.; investigation, H.D.; resources, S.D.; data curation, P.Z.; writing—original draft preparation, Y.Z.; writing—review and editing, P.Z.; visualization, H.D.; supervision, S.D.; project administration, Y.Z.; funding acquisition, Y.Z. All authors have read and agreed to the published version of the manuscript.

Funding: This research was funded by Power China Hubei Electric Engineering Co., Ltd., grant number DJ-ZDXM-2021-23.

Data Availability Statement: The original contributions presented in the study are included in the article, further inquiries can be directed to the corresponding author.

Conflicts of Interest: Authors Yunlong Zhang and Hanlin Dong were employed by the Power China Hubei Electric Engineering. The remaining authors declare that the research was conducted in the absence of any commercial or financial relationships that could be construed as a potential conflict of interest. The funder was not involved in the study design, collection, analysis, interpretation of data, the writing of this article or the decision to submit it for publication.

Abbreviations

CS	cold energy storage
CCS	carbon capture and storage
CCU	carbon capture and utilization
CSP-CHP	concentrated solar power and combined heat and power
CvaR	conditional value-at-risk
EL	electrolysis system
ES	electrical energy storage
GB	gas boiler
GE	gas engine
GPC	generalized polynomial chaos
HGT	hydrogen-doped gas units
HS	thermal energy storage

IEGS	Integrated Electricity–Gas System
IES	integrated energy systems
LB	lithium bromide
LHS	Latin hypercube sampling
MR	methane reactor
PIES	park integrated energy system
P2G	power to gas
PDF	probability density functions
PV	photovoltaic power generation
S-CO_2	supercritical carbon dioxide
WT	wind turbine power generation

References

1. Fan, M.; Li, Z.; Ding, T.; Huang, L.; Dong, F.; Ren, Z.; Liu, C. Uncertainty Evaluation Algorithm in Power System Dynamic Analysis with Correlated Renewable Energy Sources. *IEEE Trans. Power Syst.* **2021**, *36*, 5602–5611. [CrossRef]
2. Tan, Z.; De, G.; Li, M.; Lin, H.; Yang, S.; Huang, L.; Tan, Q. Combined Electricity-Heat-Cooling-Gas Load Forecasting Model for Integrated Energy System Based on Multi-Task Learning and Least Square Support Vector Machine. *J. Clean. Prod.* **2020**, *248*, 119252. [CrossRef]
3. Gilani, M.A.; Kazemi, A.; Ghasemi, M. Distribution System Resilience Enhancement by Microgrid Formation Considering Distributed Energy Resources. *Energy* **2020**, *191*, 116442. [CrossRef]
4. Li, F.; Chen, S.; Ju, C.; Zhang, X.; Ma, G.; Huang, W. Research on Short-Term Joint Optimization Scheduling Strategy for Hydro-Wind-Solar Hybrid Systems Considering Uncertainty in Renewable Energy Generation. *Energy Strategy Rev.* **2023**, *50*, 101242. [CrossRef]
5. Nayak, A.; Maulik, A.; Das, D. An Integrated Optimal Operating Strategy for a Grid-Connected AC Microgrid under Load and Renewable Generation Uncertainty Considering Demand Response. *Sustain. Energy Technol. Assess.* **2021**, *45*, 101169. [CrossRef]
6. Abdelmalak, M.; Benidris, M. A Polynomial Chaos-Based Approach to Quantify Uncertainties of Correlated Renewable Energy Sources in Voltage Regulation. *IEEE Trans. Ind. Appl.* **2021**, *57*, 2089–2097. [CrossRef]
7. Cao, W.; Yu, J.; Xu, M. Optimization Scheduling of Virtual Power Plants Considering Source-Load Coordinated Operation and Wind–Solar Uncertainty. *Processes* **2023**, *12*, 11. [CrossRef]
8. Dupačová, J.; Gröwe-Kuska, N.; Römisch, W. Scenario Reduction in Stochastic Programming. *Math. Program.* **2003**, *95*, 493–511. [CrossRef]
9. Duan, J.; Xia, Y.; Cheng, R.; Gao, Q.; Liu, F. Low Carbon and Economic Optimal Operation of Electricity-Gas Integrated Energy System Considering Demand Response. Sustain. *Energy Grids Netw.* **2024**, *38*, 101290.
10. Wang, R.; Wen, X.; Wang, X.; Fu, Y.; Zhang, Y. Low Carbon Optimal Operation of Integrated Energy System Based on Carbon Capture Technology, LCA Carbon Emissions and Ladder-Type Carbon Trading. *Appl. Energy* **2022**, *311*, 118664. [CrossRef]
11. Yan, N.; Ma, G.; Li, X.; Guerrero, J.M. Low-Carbon Economic Dispatch Method for Integrated Energy System Considering Seasonal Carbon Flow Dynamic Balance. *IEEE Trans. Sustain. Energy* **2023**, *14*, 576–586. [CrossRef]
12. Chen, H.; Wu, H.; Kan, T.; Zhang, J.; Li, H. Low-Carbon Economic Dispatch of Integrated Energy System Containing Electric Hydrogen Production Based on VMD-GRU Short-Term Wind Power Prediction. *Int. J. Electr. Power Energy Syst.* **2023**, *154*, 109420. [CrossRef]
13. Li, M.; Qin, J.; Han, Z.; Niu, Q. Low-carbon Economic Optimization Method for Integrated Energy Systems Based on Life Cycle Assessment and Carbon Capture Utilization Technologies. *Energy Sci. Eng.* **2023**, *11*, 4238–4255. [CrossRef]
14. Li, W.; He, M.; Cai, T. Low-Carbon Economic Dispatch of Hydrogen-Containing Integrated Energy System Considering Stepped Demand Response. *Energy Rep.* **2024**, *11*, 4222–4232. [CrossRef]
15. Wu, M.; Wu, Z.; Shi, Z. Low Carbon Economic Dispatch of Integrated Energy Systems Considering Utilization of Hydrogen and Oxygen Energy. *Int. J. Electr. Power Energy Syst.* **2024**, *158*, 109923. [CrossRef]
16. Zhang, Z.; Du, J.; Li, M.; Guo, J.; Xu, Z.; Li, W. Bi-Level Optimization Dispatch of Integrated-Energy Systems with P2G and Carbon Capture. *Front. Energy Res.* **2022**, *9*, 784703. [CrossRef]
17. Li, X.; Li, T.; Liu, L.; Wang, Z.; Li, X.; Huang, J.; Huang, J.; Guo, P.; Xiong, W. Operation Optimization for Integrated Energy System Based on Hybrid CSP-CHP Considering Power-to-Gas Technology and Carbon Capture System. *J. Clean. Prod.* **2023**, *391*, 136119. [CrossRef]
18. Liang, Y.-L.; Zhang, H.; Yang, C.-T.; Li, K.-J. Research on Optimization Scheduling of Integrated Electricity-Gas System Considering Carbon Trading and P2G Operation Characteristics. *Electr. Power Syst. Res.* **2023**, *225*, 109797. [CrossRef]
19. Zhang, G.; Wang, W.; Chen, Z.; Li, R.; Niu, Y. Modeling and Optimal Dispatch of a Carbon-Cycle Integrated Energy System for Low-Carbon and Economic Operation. *Energy* **2022**, *240*, 122795. [CrossRef]
20. Zhu, X.; Xue, J.; Hu, M.; Liu, Z.; Gao, X.; Huang, W. Low-Carbon Economy Dispatching of Integrated Energy System with P2G-HGT Coupling Wind Power Absorption Based on Stepped Carbon Emission Trading. *Energy Rep.* **2023**, *10*, 1753–1764. [CrossRef]

21. Zhang, X.; Zhang, Y.; Ji, X.; Ye, P.; Li, J. Unit Commitment of Integrated Energy System Considering Conditional Value-at-Risk and P2G. *Electr. Power Syst. Res.* **2023**, *221*, 109398. [CrossRef]
22. Sun, H.; Sun, X.; Kou, L.; Ke, W. Low-Carbon Economic Operation Optimization of Park-Level Integrated Energy Systems with Flexible Loads and P2G under the Carbon Trading Mechanism. *Sustainability* **2023**, *15*, 15203. [CrossRef]
23. Pan, C.; Jin, T.; Li, N.; Wang, G.; Hou, X.; Gu, Y. Multi-Objective and Two-Stage Optimization Study of Integrated Energy Systems Considering P2G and Integrated Demand Responses. *Energy* **2023**, *270*, 126846. [CrossRef]
24. Yang, B. Multi-objective Optimization of Integrated Gas–Electricity Energy System Based on Improved Multi-object Cuckoo Algorithm. *Energy Sci. Eng.* **2021**, *9*, 1839–1857. [CrossRef]
25. Wu, D.; Han, S.; Wang, L.; Li, G.; Guo, J. Multi-Parameter Optimization Design Method for Energy System in Low-Carbon Park with Integrated Hybrid Energy Storage. *Energy Convers. Manag.* **2023**, *291*, 117265. [CrossRef]
26. Qiao, Y.; Hu, F.; Xiong, W.; Li, Y. Energy Hub-based Configuration Optimization Method of Integrated Energy System. *Int. J. Energy Res.* **2022**, *46*, 23287–23309. [CrossRef]
27. Sun, H.; Zou, H.; Wen, J.; Ke, W.; Kou, L. Optimal Scheduling Considering Carbon Capture and Demand Response under Uncertain Output Scenarios for Wind Energy. *Sustainability* **2024**, *16*, 970. [CrossRef]
28. Haghighat Mamaghani, A.; Avella Escandon, S.A.; Najafi, B.; Shirazi, A.; Rinaldi, F. Techno-Economic Feasibility of Photovoltaic, Wind, Diesel and Hybrid Electrification Systems for off-Grid Rural Electrification in Colombia. *Renew. Energy* **2016**, *97*, 293–305. [CrossRef]
29. Ju, L.; Tan, Z.; Yuan, J.; Tan, Q.; Li, H.; Dong, F. A Bi-Level Stochastic Scheduling Optimization Model for a Virtual Power Plant Connected to a Wind–Photovoltaic–Energy Storage System Considering the Uncertainty and Demand Response. *Appl. Energy* **2016**, *171*, 184–199. [CrossRef]
30. Li, Y.; Yang, Z.; Li, G.; Zhao, D.; Tian, W. Optimal Scheduling of an Isolated Microgrid With Battery Storage Considering Load and Renewable Generation Uncertainties. *IEEE Trans. Ind. Electron.* **2019**, *66*, 1565–1575. [CrossRef]
31. Mckay, M.D.; Beckman, R.J.; Conover, W.J. A Comparison of Three Methods for Selecting Values of Input Variables in the Analysis of Output From a Computer Code. *Technometrics* **2000**, *42*, 55–61. [CrossRef]
32. Lei, K.; Chang, J.; Wang, X.; Guo, A.; Wang, Y.; Ren, C. Peak Shaving and Short-Term Economic Operation of Hydro-Wind-PV Hybrid System Considering the Uncertainty of Wind and PV Power. *Renew. Energy* **2023**, *215*, 118903. [CrossRef]
33. Cho, J.; Jeong, S.; Kim, Y. Commercial and Research Battery Technologies for Electrical Energy Storage Applications. *Prog. Energy Combust. Sci.* **2015**, *48*, 84–101. [CrossRef]
34. Zhang, W.; Wang, W.; Fan, X.; He, S.; Wang, H.; Wu, J.; Shi, R. Low-Carbon Optimal Operation Strategy of Multi-Park Integrated Energy System Considering Multi-Energy Sharing Trading Mechanism and Asymmetric Nash Bargaining. *Energy Rep.* **2023**, *10*, 255–284. [CrossRef]
35. Chen, J.; Tang, Z.; Huang, Y.; Qiao, A.; Liu, J. Asymmetric Nash Bargaining-Based Cooperative Energy Trading of Multi-Park Integrated Energy System under Carbon Trading Mechanism. *Electr. Power Syst. Res.* **2024**, *228*, 110033. [CrossRef]
36. Amrollahi, M.H.; Bathaee, S.M.T. Techno-Economic Optimization of Hybrid Photovoltaic/Wind Generation Together with Energy Storage System in a Stand-Alone Micro-Grid Subjected to Demand Response. *Appl. Energy* **2017**, *202*, 66–77. [CrossRef]

energies

Article

MMD-TSC: An Adaptive Multi-Objective Traffic Signal Control for Energy Saving with Traffic Efficiency

Yuqi Zhang, Yingying Zhou, Beilei Wang and Jie Song *

Software College, Northeastern University, Shenyang 112000, China; 2190088@stu.neu.edu.cn (Y.Z.); 20193043@stu.neu.edu.cn (Y.Z.); wangbl@swc.neu.edu.cn (B.W.)
* Correspondence: songjie@mail.neu.edu.cn

Abstract: Reducing traffic energy consumption is crucial for smart cities, and vehicle carbon emissions are a key energy indicator. Traffic signal control (TSC) is a useful method because it can affect the energy consumption of vehicles on the road by controlling the stop-and-go of vehicles at traffic intersections. However, setting traffic signals to reduce energy consumption will affect traffic efficiency and this is not in line with traffic management objectives. Current studies adopt multi-objective optimization methods with high traffic efficiency and low carbon emissions to solve this problem. However, most methods use static weights, which cannot adapt to complex and dynamic traffic states, resulting in non-optimal performance. Current energy indicators for urban transportation often fail to consider passenger fairness. This fairness is significant because the purpose of urban transportation is to serve people's mobility needs not vehicles. Therefore, this paper proposes **M**ulti-objective Adaptive **M**eta-**D**QN **TSC** (MMD-TSC), which introduces a dynamic weight adaptation mechanism to simultaneously optimize traffic efficiency and energy saving, and incorporates the per capita carbon emissions as the energy indicator. Firstly, this paper integrates traffic state data such as vehicle positions, velocities, vehicle types, and the number of passengers and incorporates fairness into the energy indicators, using per capita carbon emissions as the target for reducing energy consumption. Then, it proposes MMD-TSC with dynamic weights between energy consumption and traffic efficiency as reward functions. The MMD-TSC model includes two agents, the TSC agent and the weight agent, which are responsible for traffic signal adjustment and weight calculation, respectively. The weights are calculated by a function of traffic states. Finally, the paper describes the design of the MMD-TSC model learning algorithm and uses a SUMO (Simulation of Urban Mobility) v.1.20.0 for traffic simulation. The results show that in non-highly congested traffic states, the MMD-TSC model has higher traffic efficiency and lower energy consumption compared to static multi-objective TSC models and single-objective TSC models, and can adaptively achieve traffic management objectives. Compared with using vehicle average carbon emissions as the energy consumption indicator, using per capita carbon emissions achieves Pareto improvements in traffic efficiency and energy consumption indicators. The energy utilization efficiency of the MMD-TSC model is improved by 35% compared to the fixed-time TSC.

Keywords: sustainable transition; energy saving; reinforcement learning; meta-learning

Citation: Zhang, Y.; Zhou, Y.; Wang, B.; Song, J. MMD-TSC: An Adaptive Multi-Objective Traffic Signal Control for Energy Saving with Traffic Efficiency. *Energies* **2024**, *17*, 5015. https://doi.org/10.3390/en17195015

Academic Editors: Sheng Du, Xiongbo Wan, Zixin Huang and Li Jin

Received: 3 September 2024
Revised: 5 October 2024
Accepted: 6 October 2024
Published: 9 October 2024

1. Introduction

Severe global warming has motivated people to increase their awareness of environmental protection. The development trend of smart cities is gradually shifting to-wards a sustainable transition [1]. Sustainable transition aims to improve traffic efficiency and focuses on saving energy and ensuring fairness. This includes reducing carbon emissions during driving and ensuring that all passengers have equal rights to road usage [2]. Therefore, fairness and carbon emissions have become key indicators to evaluate the development level of smart cities and the effectiveness of their sustainable transition initiatives.

Traffic signal control (TSC) provides a way to manage mixed traffic flow effectively. TSC can optimize traffic flow by reasonably setting each signal's phases and duration [3]. Therefore, TSC has become a potentially effective method to promote energy saving. However, traffic efficiency will be decreased if energy saving is the only goal. Traffic efficiency is one of the most important goals of a smart city [4]. An efficient transportation system can alleviate congestion and save travel time. TSC is a crucial method to improve traffic efficiency by regulating traffic flow in different directions at intersections based on traffic states. Therefore, the research object is a TSC method that can simultaneously improve traffic efficiency and reduce per capita carbon emissions under the traffic states of mixed traffic flow and vehicle types.

Reinforcement learning has become a popular research technology for TSC due to its ability to model and make decisions in complex environments [5]. The agents can control the phases of traffic signals based on traffic states. TSC models the traffic state at intersections based on the spatial characteristics of single or multiple intersections and vehicle positions from current or previous times. Most reward functions maximize traffic efficiency [6]. Traffic efficiency is often measured by traffic velocity and average queue length [7]. Few papers consider sustainability, such as harmful gas emissions and safety [8]. Dynamic sequence and fixed sequence are two types of actions. In the dynamic sequence mode, the order of phases is dynamically adjusted based on real-time traffic states and optimization algorithms to seek the optimal control strategy. The fixed sequence mode executes the phases in a predetermined order [9].

The current reinforcement learning methods for TSC primarily focus on ensuring smooth traffic flow [6]. While the results of these studies often include improvements in both traffic flow and energy saving, the main objective is typically to optimize traffic efficiency [10]. However, even when traffic efficiency is maximized, energy saving is not optimal, as additional factors beyond traffic efficiency, such as vehicle types and the number of passengers, influence carbon emissions. Consequently, per capita carbon emissions are not minimized when traffic efficiency is maximized. Traffic efficiency and per capita carbon emissions are distinct objectives that cannot be directly substituted for each other. Furthermore, most methods employ a fixed weighting approach to balance these two objectives, which fails to adapt to dynamic traffic environments.

Therefore, the existing TSC methods have the following limitations: First, they mainly focus on optimizing traffic efficiency, lacking a consideration of energy indicators such as per capita carbon emissions, and cannot directly use energy indicators as optimization objectives, making it difficult for the optimization results to meet the requirements of a sustainable transition. Second, when considering the sustainable transition, existing methods lack the consideration of fairness factors and fail to adequately address the travel needs and experiments of different passenger groups, while fairness is an essential component of the sustainable transition [11]. Moreover, most methods use fixed weights to balance traffic efficiency and energy saving, which cannot adapt to dynamically changing traffic environments. These limitations prevent existing methods from achieving the sustainable transition and going beyond only traffic efficiency optimization.

The research question is: How can a TSC model simultaneously optimize traffic efficiency and energy saving, incorporate fairness factors into the energy indicator, and dynamically balance these objectives through adaptive weights based on complex and varying traffic states?

This paper proposes the Multi-objective Adaptive Meta-DQN TSC model (MMD-TSC). DQN is a value-based deep reinforcement learning algorithm that uses deep neural networks to approximate the optimal Q-function (state-action value function). It can handle large-scale, high-dimensional state spaces and is very suitable for solving complex sequential decision-making problems. A Multi-objective DQN for TSC (TSC-Agent) was built by integrating multiple influencing factors into the state and designing the reward function. Based on the spatial distribution of vehicles as the state, the state adds the dimensions of vehicle types and the number of passengers. These are vital factors that influence traffic efficiency and energy saving. The multi-objective reward function adopts the weighting method. One of the objectives is queue length, and the other is per capita carbon emission. The weights of the two indicators range from zero to one, and the sum of these is one. An adaptive weight calculation method for traffic efficiency and per capita carbon emissions based on the state is designed to solve the problem of static weight design. A neural network is built to calculate the weight. It takes the state as input and weight as output.

The challenge lies in achieving multi-objective weight adaptation; namely, how does the algorithm train the neural network parameters for calculating the weight? Weights are often static values. The objective will be improved if weights are dynamic values obtained from the current traffic state. The dynamic weight is like a teacher, and the DQN is like a student. The student learns how to control traffic signals. The teacher determines the student's learning objectives and adjusts the teaching progress according to the student's learning progress. The two cooperate to make the student's learning better. The teacher will adjust the objectives in time, when the student's current learning objectives are well-learned. This paper adopts the meta-learning approach to construct an agent that dynamically controls weights based on traffic states. The dynamic weight agent is designed as the actor–critic model (weight agent). The actor of the AC model is the weight generator, and the PPO algorithm is applied to train the weight agent. Applying the reward of the TSC agent, the weight is adjusted dynamically according to the traffic states. When the training time of the TSC agent is greater than the threshold, the weight agent starts to train. Then, according to the new weights, the TSC-Agent continues to be trained, realizing cooperative learning. Figure 1 illustrates the differences between TSC and MMD-TSC. Firstly, MMD-TSC focuses more on the passengers inside the vehicle and optimizes based on per capita carbon emissions. Secondly, MMD-TSC constructs a weight agent to implement a dynamic and adaptive multi-objective weight calculation.

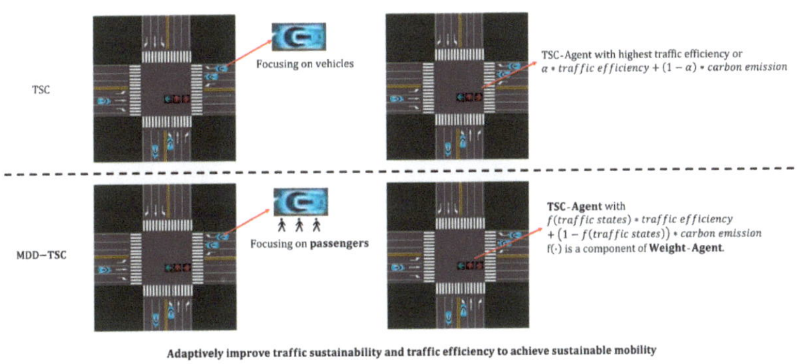

Figure 1. Overview of MDD-TSC.

This paper makes the following main contributions:

- We propose an MMD-TSC model. The MMD-TSC model is a dynamic multi-objective optimization model capable of simultaneously optimizing traffic efficiency and energy saving, surpassing the limitations of existing methods that only focus on traffic efficiency.
- We introduce a dynamic weight adaptation mechanism in the MMD-TSC model through a meta-learning weight agent. The model adaptively adjusts the optimization weights of traffic efficiency and energy saving based on traffic states.
- We incorporate the per capita carbon emissions indicator as the energy indicator in the reward function design of the MMD-TSC model to show the development level of the sustainable transition.

The structure of the paper is as follows. Section 2 summarizes related works. In Section 3, it introduces the MMD-TSC background. Section 4 elaborates on the model structure and training algorithms in detail. Section 5 describes the experimental design, shows experiment results, and discusses the results. Section 6 draws the conclusions and proposes future works.

2. Related Work

Deep reinforcement learning (DRL), with its ability to recognize complex environments, can be applied to decision-making problems in dynamic environments. TSC aims to optimize traffic flow and improve efficiency by decreasing queue length and increasing velocity. The DRL's action is to determine which phases should be green. When selecting the action, it is necessary to analyze the complex road environment and the dynamic changes in traffic flow. Therefore, DRL can be applied to TSC. Chu et al. [3] constructed a 3D model of the traffic intersection for state acquisition, using the vehicle position images as the current state. They designed an end-to-end model, and the result is superior to other TSC methods. Lin et al. [12] constructed a lane-level graph to represent the traffic intersection to identify which lane had special situations. This microscopic design method helps represent the relationships between lanes and discover problematic lanes. Razack et al. [7] extracted the spatial and temporal characteristics. The results were better than other TSC models in terms of queue length and the number of waiting vehicles. Antes et al. proposed a hierarchical multiagent reinforcement learning approach for traffic signal control, where regional agents aggregated information and generated recommendations. The proposed method outperforms fixed-time and non-hierarchical RL-based approaches on a synthetic traffic grid [13]. Abdoos et al. proposed a hierarchical reinforcement learning approach to control traffic lights. The upper layer is responsible for predicting traffic flow based on global information, while the lower layer makes traffic light action decisions based on the prediction results and the conditions at the intersection. Experimental results show that this approach is feasible in a scenario involving multiple traffic lights at 16 intersections [14]. Traffic carbon emissions account for one-tenth of the total carbon emissions in China, so low carbon emissions is a goal of smart cities. Low carbon should have been the goal of TSC [15], but Koch et al. [10] found that maximum traffic flow and minimum carbon emissions are not necessarily conflicting goals. That means the two goals can be improved simultaneously. Most of the research considers carbon emissions as a secondary goal while focusing on traffic efficiency. Guo et al. [16] focused on the sustainability of Connected Autonomous Vehicles (CAVs) and designed a multi-agent system that controls both traffic signals and vehicles. The model can balance travel time and fuel use in a large-scale complex road network. It essentially focuses on vehicles rather than traffic flow. Reddy et al. [6] focused on potential route conflicts of lanes at intersections and designed lane-level intelligent TSC to improve throughput and low carbon performance. Górka et al. [17] the control model with priority, adjusts the traffic lights at intersections containing tram to reduce energy consumption, and verifies it on a real dataset using simulation methods. Kolat et al. [18] designed the standard deviation of the number of in-lane parking as a reward function to improve traffic efficiency while also improving sustainability. The electric vehicles with high velocity will have excessive energy consumption. Chen et al. [19] designed a centralized coordinated control method connecting vehicles and traffic signals to achieve a balance between energy consumption and traffic efficiency. Only one study focuses on optimizing carbon emissions through TSC. Kang et al. [8] built a dueling DQN to train TSC agents to minimize harmful gas emissions. The result shows the agent can reduce carbon emissions and improve traffic efficiency.

Fairness for passengers is a core concept of sustainability, so in addition to carbon emissions, fairness is an important component of sustainability. Research also focuses on fairness from perspectives of vehicle and road priority. Sun et al. [20] designed a reward function to avoid excessive waiting for passengers and vehicles on non-main roads, and a priority scheduling TSC method for emergency vehicles. The results increased the throughput of intersections, effectively solved traffic congestion, and ensured fairness. To ensure fairness, a cost function was used to prevent the traffic signal of a lane from being always red. Ye et al. [21] designed a hierarchical TSC model with vehicle fairness. Vehicle fairness is the ratio of waiting time to passing time. Kumar et al. [22] designed three TSC modes: fair mode, priority mode, and emergency mode. Each mode set vehicle priorities according to importance. Cui et al. [23] designed traffic-network fairness from the perspective of phase actuation fairness and network resource utilization fairness.

Traffic efficiency and energy saving are two different goals. For such multi-objective TSC, current research focuses on three methods: (1) Pareto, (2) weighted sum (with static weights, as well as dynamic weight represented by entropy according to traffic state), and (3) other algorithms to balance multiple objectives. Akyol et al. [24] used heuristic algorithms to calculate the Pareto solutions that minimize pedestrian waiting time and carbon emissions. The results showed that MOABC (Multi-objective Artificial Bee Colony) is better. Gong et al. [25] used a multi-objective reinforcement learning method to optimize TSC, aiming to ensure both efficiency and safety. The solution for multi-objective models is to use different models and parameter learning methods for different value functions. Saiki et al. [26] proposed a multi-strategy multi-objective reinforcement learning method. Zhang et al. [27] used NSGA-III (Nondominated Sorting Genetic Algorithm-III) to solve multi-objective TSC with goals of signal control delay and traffic capacity indices. Zhang et al. [28] designed a dynamic multi-objective reinforcement learning model to optimize traffic signals in terms of low-carbon, efficiency, etc. The entropy weight was used to calculate the weights under different states. Fang et al. [29] constructed a multi-agent multi-objective collaborative reinforcement model for traffic signals across the entire road network. The goals include traffic efficiency, safety, and network coordination. The weights are state coefficients. Reyad et al. [30] used a static weighting method to balance traffic efficiency and safety in TSC.

The current methods lack the process of weight optimization. This paper adopts meta learning to train the weight agent to calculate the weight based on traffic state. There are only two researchers studying meta-learning methods of TSC. Wang et al. [31] addressed the problem that TSC cannot effectively make use of the potential relations of spatial–temporal information based on a cooperative strategy. Meta learning is the Meta-Knowledge Learner in this paper. It is applied to identify traffic flow dynamics in the road network and adjust the parameter settings in the original spatio–temporal feature extraction. It is difficult for the intersection agent to fully capture the influence of the adjacent intersection, resulting in a poor agent effect. Zhu et al. [32] used meta-learning to replace the information of adjacent intersections and serve as an intrinsic motivation method to improve performance and stability.

3. Problem Definition

This section explains the existing approaches' limitations to promoting sustainable transition and proves two assumptions: (1) When defining a reward function, traffic efficiency differs from per capita carbon emissions. So, they cannot be directly substituted. (2) Dynamic weights for traffic efficiency and low carbon emission based on traffic state are necessary.

3.1. Limitations of Existing Approaches to Promote Sustainable Transition

There are some practical approaches to promote traffic transition, such as promoting shared mobility and public transportation. The goal of promoting shared mobility and public transportation is to increase the number of passengers per vehicle and use low-carbon vehicles [33]. Its essence is to reduce per capita carbon emissions by improving the utilization rate of carbon emissions through multi-passenger vehicles and low-carbon vehicles. However, this approach does not achieve its goal. On the one hand, ensuring punctuality on public transportation is challenging due to dynamic traffic efficiency. Low service levels will result in insufficient attractiveness for passengers. On the other hand, low traffic efficiency can lead to low-carbon vehicles not being low carbon.

Setting up dedicated lanes for specific vehicle types is an approach to promoting a sustainable traffic transition, such as setting bus lanes [33]. However, this method has limitations. First, there is insufficient road space because the additional lane allocation requires too many road resources. Second, allocating lanes and setting traffic regulations for different types of vehicles can only partially improve traffic efficiency. Finally, dividing lanes by vehicle type for all roads is unsuitable because of the complex traffic flow and traffic regulations.

In summary, promoting shared mobility, public transportation, and dedicated lanes has limitations and is not highly applicable in various scenarios. These methods fail to achieve a sustainable transition. Therefore, there is a need for a comprehensive traffic management solution that can effectively regulate mixed traffic flow, such as traffic signal control. The goal is to design a model that can adaptively balance traffic efficiency and traffic carbon emissions based on traffic states.

3.2. Traffic Efficiency vs. Per Capita Carbon Emissions

Most of the research on TSC to improve traffic efficiency often also reduces per capita carbon emissions [10]. However, traffic efficiency and low per capita carbon emissions are only partially interchangeable. Under different objectives, the actions taken by the TSC agent should be different for the same state. So, the optimal strategies targeting per capita carbon emissions and traffic efficiency are different. This means that as long as one action is different, the strategy is different.

Traffic state changes are Markov processes. The factors that influence traffic efficiency are vehicle distribution and velocity. The factor influencing carbon emissions needs to add vehicle type. The factor influencing per capita carbon emissions needs to add the number of people in each vehicle. According to the Bellman equation as Equation (1), actions with traffic efficiency, carbon emissions, and per capita carbon emissions as reward functions are expressed as Equation (2).

$$action = argmax_a(R(s,a) + \gamma \sum P(s' \mid s,a) V(s')) \tag{1}$$

$$action' = argmax_a(R(s+\Delta s,a) + \gamma \sum P(s'+\Delta s' \mid s+\Delta s,a) V(s'+\Delta s')) \tag{2}$$

V represents the state-value function, which estimates the sum of future rewards starting from a given state. S is the velocity value of the vehicle position at the traffic intersection, and Δs is a vehicle type and the number of people in the vehicle. *Action* equals *action'* only if Δs does not affect R and V. That means Δs is constant. However, because Δs changes at any time, *action* equals *action'* is unequal. Since s and Δs are independent of each other, it can be similarly concluded that the actions are not entirely the same. The conclusion indicates that when indicators are substituted for each other, different actions will be generated, leading to various strategies. Therefore, traffic efficiency and per capita carbon emissions are not interchangeable indicators.

3.3. Static Weights vs. Dynamic Weights

Static weights and dynamic weights refer to the fixed settings of static weights and dynamic adaptive settings of weights for traffic efficiency and per capita carbon emissions in the context of constantly changing traffic states [29,31]. If dynamic weights are unnecessary, static weights will be optimal in different states. Due to the large state space, this paper shows two typical scenarios. If the optimal weights in the two scenarios are different, it can be demonstrated that dynamic weights are essential.

Scenario 1: peak time. There are too many vehicles at the intersection. If the weight of sustainability is too large, it will increase congestion, contrary to traffic management objectives. Therefore, the weight should be appropriately reduced to alleviate congestion.

Scenario 2: off-peak time. There is very little traffic at the intersection. The traffic efficiency is very high. If the weight of traffic efficiency is too large, the objective becomes ineffective because the traffic velocity is already at its maximum. Therefore, the weight of sustainable mobility should be increased.

In summary, the above examples illustrate the limitations of static weights. Adaptive dynamic weights based on traffic states are necessary.

4. Models and Training Method

The TSC changes with the traffic state. DQN can recognize dynamic and complex traffic states, so this method is often applied to the TSC. In this paper, the agent for TSC is called the TSC agent, and its reward represents the long-term optimization of traffic efficiency and sustainability.

The weights of traffic efficiency and per capita carbon emissions are dynamic. For example, improving traffic efficiency in traffic congestion is more important, and when road loads are low, traffic sustainability is more important. The traffic state is the basis of dynamic weights. That is, the dynamic weight is based on the vehicle state at the intersection. It can be expressed by Equation (3).

$$\alpha = f(s) \tag{3}$$

The s is the vehicle state at the intersection, f is a function that translates the vehicle state to dynamic weights, and α is the dynamic weight. This paper designs and trains a neural network model to fit f. The model's input is the s, and the output is α. A is a critical component of the reward in the TSC agent and guides the training direction. Therefore, the TSC agent and f should be collaboratively trained to optimize the reward of the TSC agent. Since f is also trained by interacting with the environment, reinforcement learning algorithms should also be used as training. This agent is called a weight agent. Therefore, the model is a multi-agent model comprising the TSC agent and the weight agent. Two agents train together with the goal of an optimal TSC agent.

4.1. Model

4.1.1. TSC Agent

The TSC agent uses the DQN method to control traffic signals. The agent can control the traffic signal phases to maximize efficiency and per capita carbon emissions. The TSC agent is a DQN-based agent responsible for selecting the optimal traffic signal phase at each time step. The traffic signal is at a typical four-way intersection, where each road has four lanes: one left-turn lane, one right-turn lane, and two through lanes. The critical components of the TSC agent are as follows:

State: The state captures the vehicle position distribution and driving state at the intersection. Divide each lane into N small segments of fixed length.

For each segment i and each lane j, we record the following information:

- $v_{i,j}$: the average velocity of vehicles within the segment, normalized to [0, 1].
- $p_{i,j}$: the number of passengers in the vehicles within the segment.
- $c_{i,j}$: the one-hot encoded vehicle type distribution within the segment, where the number of vehicle types is C.

With M lanes and N segments, the state at time step t is a tensor s_t with shape (M, N, C + 2). Equation (4) represents the expression for s_t

$$s_t = \begin{bmatrix} [v_{0,0}, p_{0,0}, c_{0,0}] & \cdots \\ \cdots & [v_{M,N}, p_{M,N}, c_{M,N}] \end{bmatrix} \tag{4}$$

To incorporate temporal information, we stack the states from the past T time steps, resulting in a final state tensor s with shape $(T, M, N, C + 2)$. Equation (5) represents the expression for s.

$$s = [s_1, s_2, \ldots, s_T] \tag{5}$$

Action: The action space is discrete, representing the selection of traffic signal phases as Figure 2 shows.

- Phase 1: No—h-south through traffic.
- Phase 2: E—t-west through traffic.
- Phase 3: No—h-south left-turn traffic.
- Phase 4: E—t-west left-turn traffic.

At each time step, the TSC agent selects the optimal phase a_t based on the state s_t using an ϵ-greedy policy. The calculation method for a_t is given by Equation (6).

$$a_t = \begin{cases} argmax_a Q(s_t, a; \theta), & with\ probability\ 1 - \epsilon \\ random\ action, & with\ probability\ \epsilon \end{cases} \tag{6}$$

where $Q(s_t, a; \theta)$ is the Q-value of action in a given state s_t, parameterized by θ, and ϵ is the exploration rate.

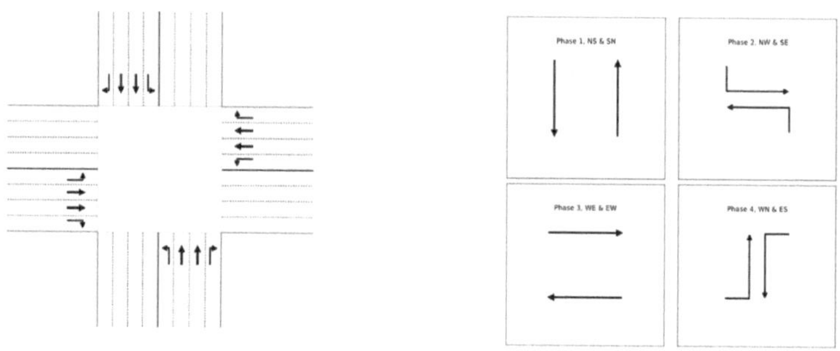

(a) (b)

Figure 2. Intersection layout and signal phase settings. (**a**) The intersection layout with through and left-turn lanes: One left-turn lane, two through lanes, and one right-turn lane. (**b**) Four signal phases: Phase 1 (NS and SN through traffic), Phase 2 (NW and SE left-turn traffic), Phase 3 (WE and EW left-turn traffic), and Phase 4 (WN and ES through traffic).

Reward: The reward function represents the long-term optimization of traffic efficiency and sustainability. Traffic efficiency is measured by the queue length, which is the ratio of vehicles with zero velocity at the intersection to the number of segments. A smaller queue length indicates higher traffic efficiency. Sustainability is measured by the average carbon emissions per person, reflecting fairness and low-carbon transportation. A smaller value indicates better sustainability. The reward r_t at time step t is defined as Equation (7).

$$r_t = -((1 - \alpha_t) \cdot queueLength_t + \alpha_t \cdot per\_capita\_carbon\_emissions_t) \tag{7}$$

where α_t is the dynamic weight for sustainability generated by the weight agent at time step t.

4.1.2. Weight Agent

The weight agent can generate the dynamic weight for balancing traffic efficiency and per capita carbon emissions in the TSC agent's reward function. The reason for choosing to use two agents instead of a single agent to complete this task is that the optimal balance between traffic efficiency and per capita carbon emissions depends on the current traffic state. Because the states are dynamic and complex, building another agent can focus on learning this mapping from traffic states to weights without interfering with the TSC agent's learning of the optimal control policy. By using different reinforcement learning algorithms that are more suitable for each task, weight generation is decoupled from control policy learning.

The collaborative training of the TSC agent and weight agent is a form of meta-learning. The weight agent learns to generate an adaptive reward function for the TSC agent based on the traffic state, helping the TSC agent learn a more effective control policy. The weight agent follows the actor–critic architecture. The critical components of the weight agent are as follows:

State: The state for the weight agent is the same as for the TSC agent, capturing the vehicle position, velocity, number of passengers, and vehicle type information at the intersection.

Action: The action of the weight agent is the dynamic weight α, which is a continuous value between zero and one.

Reward: The intuition behind the weight agent's reward design is that it should learn to generate weights that maximize the TSC agent's long-term reward. Therefore, we define the weight agent's reward as the TSC agent's reward.

This formulation encourages the weight agent to find the optimal balance between traffic efficiency and sustainability, minimizing the weighted sum of queue length and average carbon emissions. Figure 3 illustrates the computational process of the model. Both agents interact directly with the environment, and the weight agent provides the dynamic weight to the TSC agent.

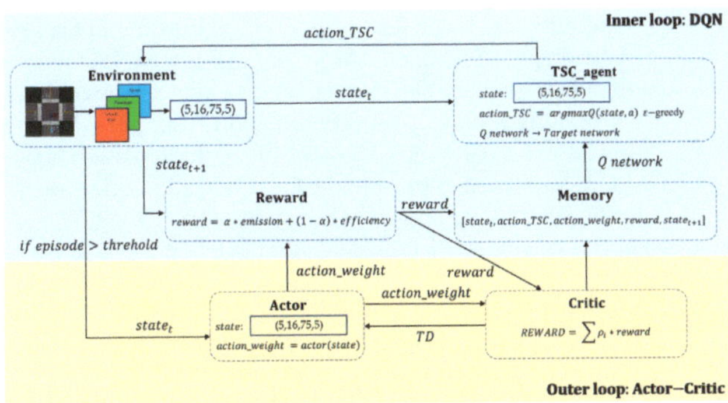

Figure 3. MDD-TSC model diagram. The model consists of two agents, the inner loop is the TSC agent and the outer loop is the weight agent. Two agents collaborate to calculate actions and rewards, and share memory information.

4.2. Training Process

The MMD-TSC model introduces a meta-learning approach to coordinate the training process of the TSC agent and the weight agent. The weight agent computes weights to guide the TSC agent towards a better control policy.

The training process involves the collaborative training of the TSC agent and the weight agent. At each time step, the TSC agent selects an action (traffic signal phase) based on the current state, and the weight agent generates the weight based on the same state. The environment transitions to the next state based on the selected action and returns the reward calculated using Equation (7). The transition tuple $(s_t, a_t, \alpha_t, r_t, s_{t+1})$ is stored in the replay buffer for both agents.

4.2.1. Training Algorithms

For the TSC agent, the training follows the DQN algorithm. The Q-network is updated by minimizing the temporal difference error between the predicted Q-values and the target Q-values. The target Q-values are calculated using the Bellman equation as Equation (8) shows.

$$Q_{target} = r_t + \gamma \max_{a'} Q(s_{t+1}, a'; \theta) \tag{8}$$

where γ is the discount factor and θ represents the parameters of the target Q-network.

For the weight agent, the training follows the Proximal Policy Optimization (PPO) algorithm. The actor network is updated by maximizing the clipped surrogate objective as Equation (9) shows.

$$L^{CLIP}(\phi) = \hat{E}_t\left[min\left(r_t(\phi)\hat{A}_t, clip(r_t(\phi), 1-\epsilon, 1+\epsilon)\hat{A}_t\right)\right] \tag{9}$$

where $r_t(\phi) = \frac{\pi_\phi(a_t|s_t)}{\pi_{\phi_{old}}(a_t|s_t)}$ is the probability ratio, $\hat{A}_t$ is the estimated advantage, and ε is the clipping threshold. The critic network is updated by minimizing the mean squared error (MSE) between the predicted state values and the target values.

The critic network $V_\psi(s)$ parameterized by ψ estimates the state-value function, which predicts the expected cumulative reward starting from state s. The critic is trained to minimize the mean squared error between the predicted state values and the target values as Equation (10) shows.

$$L(\psi) = \frac{1}{B}\sum_{i=1}^{B}\left(V_\psi(s_i) - V_i^{target}\right)^2 \tag{10}$$

where B is the batch size, and V_i^{target} is the target value calculated using the Generalized Advantage Estimation (GAE). The GAE is a method for estimating the advantage function in policy gradient algorithms, which balances the trade-off between bias and variance by combining Monte Carlo estimation and Temporal Difference (TD) estimation. The GAE introduces a parameter λ to control the interpolation between the two estimation methods. In the proposed approach, the GAE is used to calculate the target values for updating the critic network, as shown in Equation (11):

$$V_i^{target} = \sum(\gamma\lambda)^t \delta_{i+t}^V \tag{11}$$

where $\delta_t^V = r_t + \gamma V_\psi(s_{t+1}) - V_\psi(s_t)$ is the TD residual, γ is the discount factor, and λ is the GAE parameter.

In summary, the weight agent learns to generate dynamic weights for balancing traffic efficiency and sustainability in the TSC agent's reward function. The actor–critic architecture and the PPO algorithm enable the weight agent to adapt to the dynamic traffic state and provide informative rewards for the TSC agent's learning.

4.2.2. Multi-Agent Collaborative Training

(1) Information Exchange and Shared Replay Buffer

The collaborative training of the TSC agent and the weight agent involves a continuous exchange of information between the two agents. The TSC agent relies on the weights generated by the weight agent to calculate its reward, while the weight agent's reward is based on the performance of the TSC agent. The transition tuples $(s_t, a_t, \alpha_t, r_t, s_{t+1})$ are stored in a shared replay buffer, allowing both agents to learn from the same experiments.

(2) Training Trigger

To ensure stable training, a training trigger is implemented for the weight agent. In the current implementation, the weight agent is trained once for every *threshold* step of TSC agent training. This allows the TSC agent to adapt to the new weights before the weight agent updates its policy. The training trigger can be adjusted based on the convergence of the TSC agent's reward. For example, if the TSC agent's reward converges, the frequency of the weight agent's training can be increased to further fine-tune the weights.

The appropriate value of the threshold depends on the specific problem and should be determined through empirical experimentation. If the *threshold* is too small, the weight agent will update its policy too frequently, causing instability in the TSC agent's learning process as it struggles to adapt to the rapidly changing weights. On the other hand, if the *threshold* is too large, the weight agent's policy updates will be too infrequent, leading to slow convergence and suboptimal performance.

Algorithm 1 presents the pseudocode for the collaborative training process of the TSC agent and the weight agent. The training process consists of M episodes, each with T time steps. At each time step, the TSC agent selects an action a_t using an ε-greedy policy based on its Q-network, while the weight agent generates a weight α_t using its actor network. The transition tuple $(s_t,\ a_t,\ \alpha_t,\ r_t,\ s_{t+1})$ is then stored in the shared replay buffer D. Every threshold step, the weight agent is trained for K epochs using the PPO algorithm, updating its actor and critic networks. Meanwhile, the TSC agent is trained at each time step by sampling a batch of transitions from D, computing the target Q-values, and updating its Q-network parameters. This collaborative training process enables both agents to learn from each other and adapt to the dynamic traffic environment.

Algorithm 1 **Collaborative training of the TSC agent and the weight agent.**

```
Initialize Q-network parameters θ for TSC agent
Initialize target Q-network parameters θ_target = θ for TSC agent
Initialize actor network parameters φ for weight agent
Initialize critic network parameters ψ for weight agent
Initialize shared replay buffer D
for episode = 1 to M do the following
    Initialize state s
    for t = 1 to T do the following
        Select action a_t for TSC agent using ε − greedy policy based on Q(s, a; θ)
        Generate weight α_t for weight agent using actor network π(a|s; φ)
        Execute action a_t, observe reward r_t and next state s_{t+1}
        Store transition (s_t, a_t, α_t, r_t, s_{t+1}) in the shared replay buffer D
        if t % threshold == 0 then
            for epoch = 1 to K do the following
                Sample a batch of transitions (s_i, a_i, α_i, r_i, s_{i+1}) from D
                Compute probability ratio r_t(φ) and estimated advantage _t
                Compute clipped surrogate objective L^CLIP(ϕ) for the Weight agent
                Compute critic loss L(ψ) for weight agent
                Update actor network parameters φ for weight agent
                Update critic network parameters ψ for weight agent
            end for
        end if
        Sample a batch of transitions (s_i, a_i, α_i, r_i, s_{i+1}) from D
        Compute target Q-values using Equation (8) for TSC agent
        Compute Q-network loss for TSC agent Update Q-network parameters θ for TSC agent
        Update target Q-network parameters θ_target = θ for TSC agent
        s_t = s_{t+1}
    end for
end for
```

5. Experiments

This section shows the results of training the MMD-TSC model and controlling the traffic flow at intersections, and evaluates the traffic conditions in the given traffic flow using the Simulation of Urban Mobility (SUMO) v.1.20.0.

5.1. Experiments Goals

In this section, we present an experimental evaluation of the model proposed in this paper. The goals of our study are as follows:

(1) To demonstrate the improvement of the sustainable transition, and to compare the performance of MAM-TSC with other weight design methods, such as single-objective TSC and multi-objective TSC with static weights.
(2) To evaluate the influence of different traffic state segments on dynamic weights.

5.2. Experimental Setting

5.2.1. Baseline

In this section, MMD-TSC is compared with four other TSC methods. One comparison is with single-objective TSC methods, including maximum traffic efficiency and minimum per capita carbon emissions. The other comparison is with state weighting methods.

- FT-TSC (fixed time): Traffic lights control traffic flow by setting periodic signal changes. This is a common baseline.
- CEP-TSC (carbon emissions priority): The reward function is to achieve maximum cumulative traffic efficiency. Research aimed at energy saving generally targets the minimization of harmful gas emissions [2,8]. In this paper, minimizing total carbon dioxide emissions represents this research.
- CEQB-TSC (carbon emissions and queue length balance): Traffic efficiency and per capita carbon emissions weights are static parameters in the reward function. Set this parameter to equal, that is, 1:1. Research on the multi-objective optimization of traffic efficiency and energy saving typically uses fixed weights [25,26,29,30]. In this paper, a 1:1 goal ratio serves as a representation of this research.
- QLP-TSC (queue length priority): The reward function is to achieve the minimum queue length. Research targeting maximum traffic efficiency often focuses on minimizing queue length or travel time [1,7]. In this paper, minimizing queue length is used as a representative goal for this research.

5.2.2. Simulation Parameters

The experiments in this study are conducted using the SUMO, an open-source, microscopic, and multi-modal traffic simulation tool. The SUMO allows for the creation of realistic traffic scenarios, including road networks, traffic demand, and vehicle behavior. It supports the definition of traffic signal systems, which can be controlled using the TraCI (Traffic Control Interface) API. This API enables real-time interaction between the simulation and external control algorithms, making the SUMO suitable for evaluating the proposed multi-agent reinforcement learning approach for adaptive traffic signal control.

The corresponding SUMO simulation parameters are shown in Table 1.

Table 1. Sumo simulation parameters in experiments.

Title 1	Title 2
Steps	800
The number of vehicles	1000
Green light duration	10
Yellow light duration	4

These parameters constitute a representative set of simulation settings that effectively capture the dynamic evolution of traffic conditions and the formation of congestion. The simulation is conducted on an empty road network with a continuous arrival of vehicles, resulting in a gradual increase in traffic flow throughout the simulation period.

The simulation duration is set to 800 steps, which provides sufficient time to observe the transition of traffic conditions from free-flow to congested states. This extended simulation period allows for a comprehensive analysis of the traffic dynamics and the impact of increasing traffic volume on road performance. With a total of 1000 vehicles, the simulation scale is representative of a realistic traffic scenario, enabling the evaluation of traffic management strategies under conditions that closely resemble real-world traffic patterns. The 1000 vehicles consist of cars, buses, and EVs. The number of passengers in each vehicle is a random number up to its capacity, with cars carrying 1–4 people and buses carrying 1–10 people. In addition to the difference in passenger capacity, the type of vehicle also affects its speed and carbon emissions.

The traffic light settings, with a green light duration of ten and a yellow light duration of four, are consistent with commonly observed values in real-world traffic signal configurations. These settings play a crucial role in the simulation, as they contribute to the accumulation of vehicles at intersections, leading to the formation of congestion in the latter half of the simulation period.

5.3. Results

5.3.1. Experiment Results

Figure 4 illustrates the relationship between crucial traffic indicators and time during the traffic simulation process. The critical traffic indicators are average queue length and vehicle waiting time. Shorter queue lengths and waiting times indicate that vehicles can pass through intersections more quickly, improving overall traffic efficiency and reflecting better traffic signal regulation.

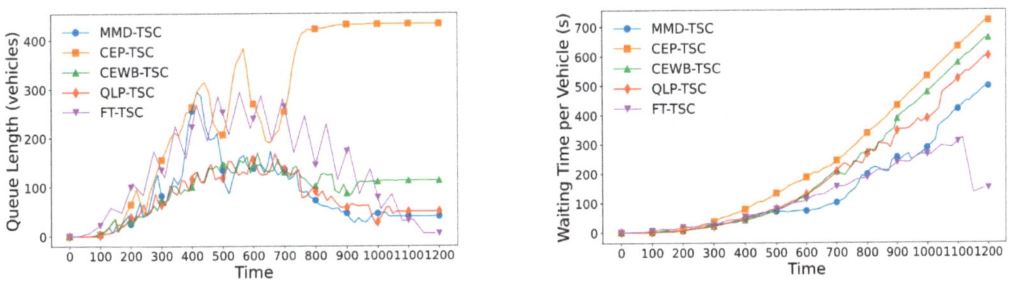

(**a**) (**b**)

Figure 4. Comparison of sustainability indicators among different TSC models over time. (**a**) Relationship between per capita carbon emissions and simulation time for each model. (**b**) Relationship between cumulative carbon emissions and simulation time for each model.

Figure 4a shows that MMD-TSC performs the best at most time points. At the end of the simulation, the queue length of MMD-TSC can be reduced to a minimal value, indicating that this signal light regulation method can effectively manage vehicle traffic. Compared to QLP-TSC and FT-TSC, MMD-TSC outperforms QLP-TSC at most time points, except for the time range of 300–500. Compared to CEWB-TSC, MMD-TSC performs better in the later stages because dynamic weights can better adapt to various traffic states to efficiently control traffic flow through intersections.

In Figure 4b, the performance of other models varies based on the weight assigned to traffic efficiency, with higher weights generally leading to greater efficiency. FT-TSC outperforms QLP-TSC. Although QLP-TSC aims to maximize traffic efficiency, the waiting time indicator may not be the best since the model's objective is queue length. MMD-TSC has the best overall performance, with its line at the bottom. Adaptive weight adjustment enables MMD-TSC to improve not only on a single objective but also to comprehensively enhance traffic efficiency.

In summary, MMD-TSC performs better than other models in reducing queue lengths and waiting times throughout the simulation process, achieving higher overall traffic efficiency. This highlights the advantage of MMD-TSC in dynamically adjusting signal lights to adapt to changes in traffic states.

Figure 5 illustrates the relationship between sustainability indicators and simulation time throughout the traffic simulation process. The critical sustainability indicators are per capita carbon emissions and cumulative carbon emissions. Per capita carbon emissions are a microscopic indicator that reflects an individual's sustainability. In contrast, cumulative carbon emissions are a macroscopic indicator reflecting all vehicles' overall carbon emissions effect.

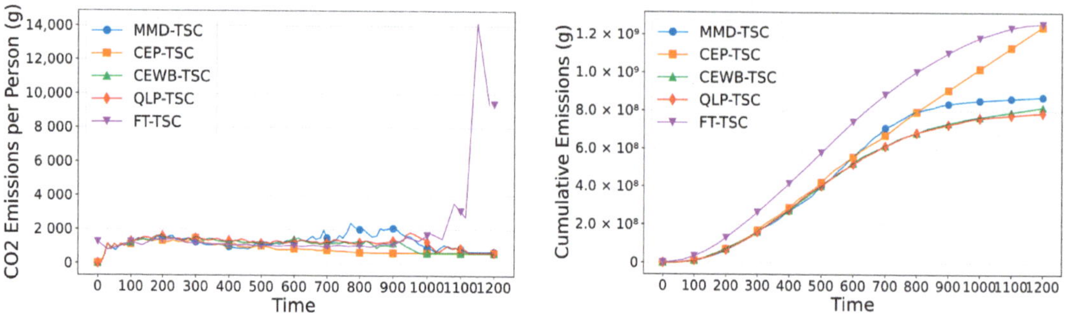

(a) (b)

Figure 5. Comparison of traffic indicators among different TSC models over time. (**a**) Relationship between average queue length and simulation time for each model. (**b**) Relationship between average vehicle waiting time and simulation time for each model.

In Figure 5a, the CEP-TSC model performs the best because it aims to minimize per capita carbon emissions. The per capita carbon emissions of the MMD-TSC model are slightly higher than those of the CEP-TSC model only in the time range of 650–800.

In Figure 5b, QLP-TSC and CEWB-TSC are the best models for total carbon emissions. Although CEP is the model aimed at optimal sustainability, per capita carbon emissions are a microscopic indicator and cannot represent the macroscopic total carbon emissions. The cumulative emissions of MMD-TSC are slightly higher in the latter half compared to QLP-TSC and CEWB-TSC models.

From the above analysis, MMD-TSC can better promote a sustainable transition. In most cases, MMD-TSC outperforms other models in terms of traffic efficiency and is also close to the best model in terms of sustainability.

5.3.2. Spatial Influence Analysis on Actor of Weight Agent

To investigate which spatial locations in the traffic state have a more significant influence on the weights and to better explain the actor model, permutation importance is used to examine the impact of traffic environment values in different regions on the weights.

Divide the road length into proportions of 2:3:4:6, and examine the influence of each section's values on the results when they vary randomly. The data used are a sample of traffic states, including uniformly collected data from the entire simulation process, totaling 100 samples.

This set of figures shows the influence of different regions at the traffic intersection on the weights. Figure 6a–c are graphs for three time periods, with lighter colors indicating a greater degree of influence. For time periods one, two, and three, the regions with a high degree of influence shift backward successively, which aligns with the common sense of vehicle flow approaching the intersection and demonstrates the reasonableness of the model. Furthermore, with each lane approaching the intersection, only one region usually has a significantly high degree of influence. This indicates that, even though vehicles are in motion, only one region has a notable impact.

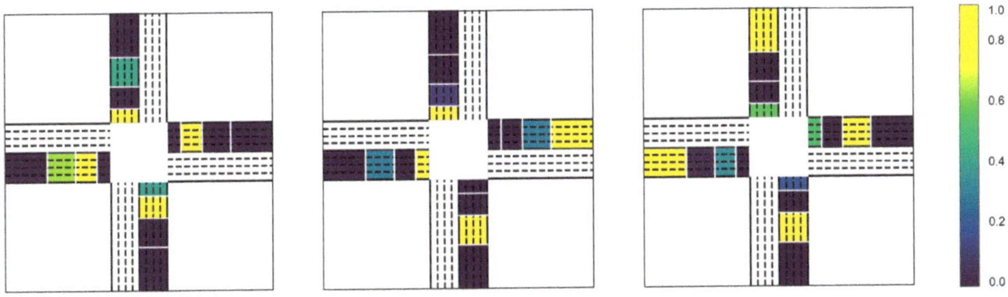

(a) (b) (c)

Figure 6. Heatmaps of the influence of different spatial locations on weights at three time periods: (**a**) Time period 1, (**b**) Time period 2, and (**c**) Time period 3. The influence is standardized to a range of 0–1, and the larger the value, the greater the degree of influence. Yellow has the greatest impact here.

5.3.3. Comparative Analysis of Per Capita and Per Vehicle Carbon Emissions

Most studies focus on vehicle carbon emissions without passengers. To demonstrate the necessity of incorporating vehicle occupancy into the model, compare the goals of per capita carbon emissions with per vehicle carbon emissions. Then, compare the simulation results of the minimum per capita and minimum per vehicle carbon emissions as sustainability indicators. Queue length is the traffic state evaluation indicator, and cumulative carbon emissions are the sustainability evaluation indicator.

Figure 7 is a three-dimensional graph with dimensions of time, cumulative carbon emissions, and queue length. MMD-TSC (per vehicle) represents the model with per vehicle carbon emissions as the sustainability objective, while MMD-TSC (per capita) represents the model with per capita carbon emissions. As simulation time increases, cumulative carbon emissions continuously increase. Along the time axis, the green part indicates that MMD-TSC (per capita) performs better. At the beginning of the simulation, the cumulative carbon emissions indicator is red, but from Figure 7c, it can be seen that there is little difference between the two indicators. Only during the time period of 600–800, MMD-TSC (per vehicle) has slightly better traffic efficiency. Therefore, except for the time when the number of passing vehicles is at its maximum, MMD-TSC (per capita) is Pareto superior to MMD-TSC (per vehicle). This suggests that, except for the case of excessive vehicle congestion, MMD-TSC (per capita) can more effectively regulate traffic intersections for sustainable transition. The reason is that when there are too many vehicles, excessively weakening vehicle attributes will affect the improvement of traffic efficiency.

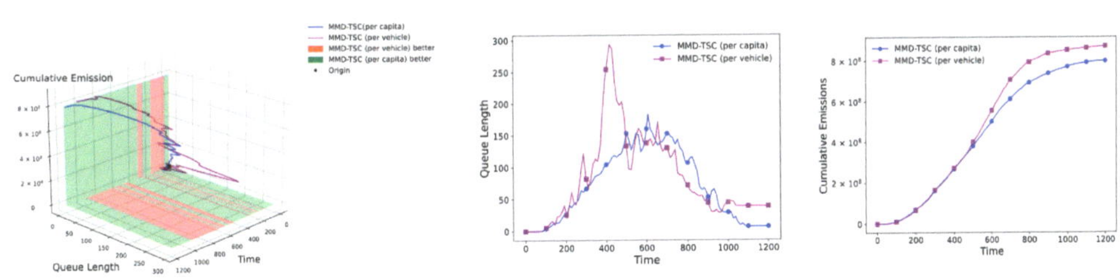

(a) (b) (c)

Figure 7. The degree of influence of different locations in the intersection on the dynamic weights is standardized to a range of 0–1. Larger values indicate a greater degree of influence. Yellow indicates the greatest influence. (**a**) A 3D plot shows the relationship between simulation time, queue length, and cumulative carbon emissions in a single graph. To provide a clearer view of the relationship between the two indicators and simulation time, (**b**,**c**) are line graphs showing the relationship between queue length, cumulative carbon emissions, and simulation time.

5.3.4. Comparative Analysis of Carbon Emission Utilization

Figure 8 illustrates the gasoline consumption utilization of different TSC models. The horizontal axis represents the simulation time, while the vertical axis represents the number of vehicles and passengers that can drive per liter of gasoline. As the simulation time increases, the gasoline consumption utilization of all models shows a trend of rising first and then stabilizing. It can be seen that MMD-TSC outperforms FT-TSC by about 35% in terms of gasoline consumption utilization. This indicates that MMD-TSC is more effective in optimizing traffic signal control strategies, allowing more vehicles and passengers to pass through with the same amount of gasoline consumption.

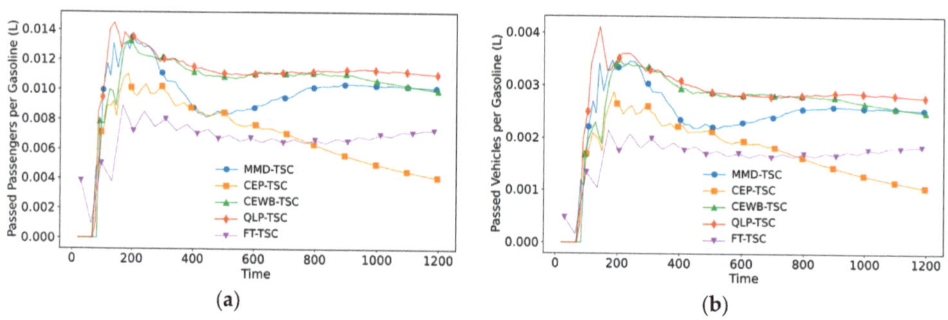

Figure 8. Comparative analysis of energy consumption utilization among different TSC models. (**a**) Relationship between passed passengers per gasoline for each model. (**b**) Relationship between passed vehicles per gasoline for each model.

6. Discussion

Reducing carbon emissions can be achieved while improving traffic efficiency. The coordinated training mode of MDD-TSC has a better effect. In the first half of the simulation, when traffic flow increases, but congestion has not yet occurred, the total carbon emissions of MMD-TSC are lower. In the second half of the simulation, when vehicle congestion occurs, MMD-TSC has better traffic efficiency and can effectively alleviate congestion pressure. Therefore, MMD-TSC can better adapt to complex traffic environments and provide solutions that meet traffic management requirements.

Compared to FT-TSC, the energy consumption utilization of the MMD-TSC model has improved by 35%. With the same amount of carbon emissions, more vehicles and passengers can pass through an intersection. However, MMD-TSC is slightly less efficient in terms of gasoline consumption compared to QLP-TSC. On the one hand, the number of passengers, vehicle type, and vehicle velocity can enhance the analysis of traffic states. This enhancement positively impacts traffic efficiency. On the other hand, the QLP-TSC model focuses on reducing queue length to improve traffic efficiency. This approach will result in frequent changes to traffic signals, which can slightly increase gasoline consumption in vehicles. Furthermore, the waiting time under MMD-TSC is consistently better than that of other models. The conclusion is that, compared to QLP-TSC, the slightly lower gasoline consumption of MMD-TSC contributes to achieving more effective traffic management.

7. Conclusions

This paper investigates the problem of current TSC models having single objectives and multi-objective models only having static weights, making it difficult for the models to achieve a sustainable transition. The challenge lies in constructing a multi-objective learning model with dynamic weights and how to train the parameters. We propose the MMD-TSC model, which realizes its functionality through the interaction of the TSC agent and the weight agent, and alternately trains the two agents using the meta-learning approach. Experimental results show that the MMD-agent model achieves higher traffic efficiency; in non-congested states, it has lower per capita carbon emission, and in congested states, it can better ensure traffic efficiency and dynamically meet traffic management requirements.

TSC models that aim to optimize traffic efficiency can reduce carbon emissions while improving traffic efficiency, but they cannot lower emissions to the lowest possible level. Models with static parameters as multi-objective reward functions can better reduce carbon emissions, but the improvement in traffic efficiency is insufficient. Moreover, they cannot identify traffic states, making traffic efficiency not the primary objective even when the road is congested, which does not align with traffic management requirements. The MMD-TSC model proposed in this paper can dynamically identify traffic states, ensuring both traffic efficiency and energy saving, thereby achieving sustainable transition. Additionally, we find that the main factor influencing the weights is a specific region of the road.

Furthermore, this paper analyzes the gasoline consumption utilization of different TSC models. The results show that the MMD-TSC model reduces gasoline consumption by approximately 35% compared to the FT-TSC model. Although the gasoline consumption of the MMD-TSC model is slightly greater than that of the QLP-TSC model, it achieves more effective traffic management objectives.

Our work can be extended in multiple directions. On one hand, the TSC optimization problem for single intersections can be extended to multi-intersection problems and can be refined to lane-level navigation. The influence of pedestrians or bicyclists would also be important. On the other hand, the application of meta-learning in TSC can be explored to improve control precision. Future research can further investigate optimal TSC under various variable traffic parameters, such as green light duration and signal light sequence.

Author Contributions: Conceptualization, B.W. and J.S.; Methodology, Y.Z. (Yuqi Zhang), Y.Z. (Yingying Zhou), and J.S.; Software, Y.Z. (Yuqi Zhang) and Y.Z. (Yingying Zhou); Validation, J.S. and Y.Z. (Yuqi Zhang); Formal analysis, Y.Z. (Yuqi Zhang); Data curation, Y.Z. (Yuqi Zhang); Writing—original draft, Y.Z. (Yuqi Zhang); Writing—review and editing, J.S.; Visualization, Y.Z. (Yingying Zhou); Supervision, B.W. and J.S. All authors have read and agreed to the published version of the manuscript.

Funding: This research received no external funding.

Data Availability Statement: The original contributions presented in the study are included in the article, further inquiries can be directed to the corresponding author.

Conflicts of Interest: The authors declare no conflicts of interest.

References

1. Mora, L.; Deakin, M.; Zhang, X.; Batty, M.; de Jong, M.; Santi, P.; Appio, F.P. Assembling Sustainable Smart City Transitions: An Interdisciplinary Theoretical Perspective. *J. Urban Technol.* **2020**, *28*, 1–27. [CrossRef]
2. Zhang, Y.; Wang, H.; Wang, X. Towards Fairness-Aware Crowd Management System and Surge Prevention in Smart Cities. In Proceedings of the 2024 IEEE Workshop on Design Automation for CPS and IoT (DESTION), Hong Kong, China, 13–14 May 2023; pp. 46–54.
3. Chu, K.-F.; Lam, A.Y.S.; Li, V.O.K. Traffic Signal Control Using End-to-End Off-Policy Deep Reinforcement Learning. *IEEE Trans. Intell. Transp. Syst.* **2022**, *23*, 7184–7195. [CrossRef]
4. Zhang, Y.; Wang, H.; Wang, X. Research on the improvement of transportation efficiency of smart city by traffic visualization based on pattern recognition. *Neural Comput. Appl.* **2022**, *35*, 2211–2224. [CrossRef]
5. Farazi, N.P.; Zou, B.; Ahamed, T.; Barua, L. Deep reinforcement learning in transportation research: A review. *Transp. Res. Interdiscip. Perspect.* **2021**, *11*, 100425. [CrossRef]
6. Reddy, R.; Almeida, L.; Gaitán, M.G.; Santos, P.M.; Tovar, E. Synchronous Management of Mixed Traffic at Signalized Intersections Toward Sustainable Road Transportation. *IEEE Access* **2023**, *11*, 64928–64940. [CrossRef]
7. Razack, A.J.; Ajith, V.; Gupta, R. A Deep Reinforcement Learning Approach to Traffic Signal Control. In Proceedings of the 2021 IEEE Conference on Technologies for Sustainability (SusTech), Kollam, India, 8–9 August 2021.
8. Kang, L.; Huang, H.; Lu, W.; Liu, L. A Dueling Deep Q-Network method for low-carbon traffic signal control. *Appl. Soft Comput.* **2023**, *141*, 110304. [CrossRef]
9. Ma, D.; Li, S.; Han, B.; Zhang, Y. A decentralized model predictive traffic signal control method with fixed phase sequence for urban networks. *J. Intell. Transp. Syst.* **2020**, *25*, 455–468. [CrossRef]

10. Koch, L.; Brinkmann, T.; Wegener, M.; Badalian, K.; Andert, J. Adaptive Traffic Light Control with Deep Reinforcement Learning: An Evaluation of Traffic Flow and Energy Consumption. *IEEE Trans. Intell. Transp. Syst.* **2023**, *24*, 15066–15076. [CrossRef]
11. Huttunen, S.; Tykkyläinen, R.; Kaljonen, M.; Kortetmäki, T.; Paloviita, A. Framing just transition: The case of sustainable food system transition in Finland. *Environ. Policy Gov.* **2024**, *34*, 463–475. [CrossRef]
12. Lin, W.-Y.; Song, Y.-Z.; Ruan, B.-K.; Shuai, H.-H.; Shen, C.-Y.; Wang, L.-C.; Li, Y.-H. Temporal Difference-Aware Graph Convolutional Reinforcement Learning for Multi-Intersection Traffic Signal Control. *IEEE Trans. Intell. Transp. Syst.* **2024**, *25*, 327–337. [CrossRef]
13. Antes, T.d.O.; Bazzan, A.L.; Tavares, A.R. Information upwards, recommendation downwards: Reinforcement learning with hierarchy for traffic signal control. *Procedia Comput. Sci.* **2022**, *201*, 24–31. [CrossRef]
14. Abdoos, M.; Bazzan, A.L. Hierarchical Traffic Signal Optimization Using Reinforcement Learning and Traffic Prediction with Long-Short Term Memory. *Expert Syst. Appl.* **2021**, *171*, 114580. [CrossRef]
15. Alshayeb, S.; Stevanovic, A.; Mitrovic, N.; Espino, E. Traffic Signal Optimization to Improve Sustainability: A Literature Review. *Energies* **2022**, *15*, 8452. [CrossRef]
16. Guo, J.; Cheng, L.; Wang, S. CoTV: Cooperative control for traffic light signals and connected autonomous vehicles using deep reinforcement learning. *IEEE Trans. Intell. Transp. Syst.* **2023**, *24*, 10501–10512. [CrossRef]
17. Górka, A.; Czerepicki, A.; Krukowicz, T. The Impact of Priority in Coordinated Traffic Lights on Tram Energy Consumption. *Energies* **2024**, *17*, 520. [CrossRef]
18. Kolat, M.; Kővári, B.; Bécsi, T.; Aradi, S. Multi-Agent Reinforcement Learning for Traffic Signal Control: A Cooperative Approach. *Sustainability* **2023**, *15*, 3479. [CrossRef]
19. Chen, H.; Wu, F.; Qiu, T.Z. Achieving Energy-Efficient and Travel Time-Optimized Trajectory and Signal Control for CAEVs. *IEEE Trans. Intell. Transp. Syst.* **2024**, *25*, 10246–10259. [CrossRef]
20. Sun, G.; Qi, R.; Liu, Y.; Xu, F. A dynamic traffic signal scheduling system based on improved greedy algorithm. *PLoS ONE* **2024**, *19*, e0298417. [CrossRef]
21. Ye, Y.; Ding, J.; Wang, T.; Zhou, J.; Wei, X.; Chen, M. FairLight: Fairness-Aware Autonomous Traffic Signal Control With Hierarchical Action Space. *IEEE Trans. Comput. Des. Integr. Circuits Syst.* **2023**, *42*, 2434–2446. [CrossRef]
22. Kumar, N.; Rahman, S.S.; Dhakad, N. Fuzzy Inference Enabled Deep Reinforcement Learning-Based Traffic Light Control for Intelligent Transportation System. *IEEE Trans. Intell. Transp. Syst.* **2020**, *22*, 4919–4928. [CrossRef]
23. Cui, S.; Xue, Y.; Gao, K.; Wang, K.; Yu, B.; Qu, X. Delay-throughput tradeoffs for signalized networks with finite queue capacity. *Transp. Res. Part B Methodol.* **2024**, *180*, 102876. [CrossRef]
24. Akyol, G.; Göncü, S.; Silgu, M.A. Multi-objective Optimization Framework for Trade-Off Among Pedestrian Delays and Vehicular Emissions at Signal-Controlled Intersections. *Arab. J. Sci. Eng.* **2024**, *49*, 14117–14130. [CrossRef]
25. Gong, Y.; Abdel-Aty, M.; Yuan, J.; Cai, Q. Multi-Objective reinforcement learning approach for improving safety at intersections with adaptive traffic signal control. *Accid. Anal. Prev.* **2020**, *144*, 105655. [CrossRef] [PubMed]
26. Saiki, T.; Arai, S. Flexible Traffic Signal Control via Multi-Objective Reinforcement Learning. *IEEE Access* **2023**, *11*, 75875–75883. [CrossRef]
27. Zhang, X.; Fan, X.; Yu, S.; Shan, A.; Men, R. Multi-Objective Optimization Method for Signalized Intersections in Intelligent Traffic Network. *Sensors* **2023**, *23*, 6303. [CrossRef]
28. Zhang, G.; Chang, F.; Jin, J.; Yang, F.; Huang, H. Multi-objective deep reinforcement learning approach for adaptive traffic signal control system with concurrent optimization of safety, efficiency, and decarbonization at intersections. *Accid. Anal. Prev.* **2024**, *199*, 107451. [CrossRef]
29. Fang, J.; You, Y.; Xu, M.; Wang, J.; Cai, S. Multi-Objective Traffic Signal Control Using Network-Wide Agent Coordinated Reinforcement Learning. *Expert Syst. Appl.* **2023**, *229*, 120535. [CrossRef]
30. Reyad, P.; Sayed, T. Real-Time multi-objective optimization of safety and mobility at signalized intersections. *Transp. B Transp. Dyn.* **2022**, *11*, 847–868. [CrossRef]
31. Wang, M.; Wu, L.; Li, M.; Wu, D.; Shi, X.; Ma, C. Meta-learning based spatial-temporal graph attention network for traffic signal control. *Knowl.-Based Syst.* **2022**, *250*, 109166. [CrossRef]
32. Zhu, L.; Peng, P.; Lu, Z.; Tian, Y. MetaVIM: Meta Variationally Intrinsic Motivated Reinforcement Learning for Decentralized Traffic Signal Control. *IEEE Trans. Knowl. Data Eng.* **2023**, *35*, 11570–11584. [CrossRef]
33. Zhang, X.; Zhong, S.; Ling, S.; Jia, N.; Qi, H.; He, Z. How to promote the transition from solo driving to mobility services delivery? An empirical study focusing on ridesharing. *Transp. Policy* **2022**, *129*, 176–187. [CrossRef]

energies

MDPI

Article

Multimodal Operation Data Mining for Grid Operation Violation Risk Prediction

Lingwen Meng [1], Jingliang Zhong [1], Shasha Luo [1], Xinshan Zhu [2], Yulin Wang [2] and Shumei Zhang [2,*]

1 Electric Power Scientific Research Institute of Guizhou Power Grid Guizhou, Guiyang 550000, China; menglw000@163.com (L.M.); 13985469075@139.com (J.Z.); luoshasha931@163.com (S.L.)
2 School of Electrical Automation and Information Engineering, Tianjin University, Tianjin 300072, China; xszhu@tju.edu.cn (X.Z.); yulin_2000@tju.edu.cn (Y.W.)
* Correspondence: shumeizhang@tju.edu.cn

Abstract: With the continuous expansion of the power grid, the issue of operational safety has attracted increasing attention. In power grid operation control, unauthorized operations are one of the primary causes of personal accidents. Therefore, preventing and monitoring unauthorized actions by power grid operators is of critical importance. First, multimodal violation data are integrated through information systems, such as the power grid management platform, to construct a historical case database. Next, word vectors for three types of operation-related factors are generated using natural language processing techniques, and key vectors are selected based on generalized correlation coefficients using mutual information, enabling effective dimensionality reduction. Independent component analysis is then employed for feature extraction and further dimensionality reduction, allowing for the effective characterization of operational scenarios. For each historical case, a risk score is derived from a violation risk prediction model constructed using the Random Forests (RF) algorithm. When a high-risk score is identified, the K-Nearest Neighbor (KNN) algorithm is applied to locate similar scenarios in the historical case database where violations may have occurred. Real-time violation risk assessment is performed for each operation, providing early warnings to operators, thereby reducing the likelihood of violations, and enhancing the safety of power grid operations.

Keywords: violation dictionary construction; operation risk prediction; random forests; independent component analysis; mutual information

Citation: Meng, L.; Zhong, J.; Luo, S.; Zhu, X.; Wang, Y.; Zhang, S. Multimodal Operation Data Mining for Grid Operation Violation Risk Prediction. *Energies* **2024**, *17*, 5424. https://doi.org/10.3390/en17215424

Academic Editor: Tek Tjing Lie

Received: 20 September 2024
Revised: 21 October 2024
Accepted: 29 October 2024
Published: 30 October 2024

1. Introduction

The smart grid plays a crucial role as a foundational platform in the Chinese national strategy to promote energy transformation and drive innovation-led development [1]. With the ongoing expansion of electric power infrastructure, construction safety has garnered significant attention [2]. Enhancing risk resilience is a top priority, and strengthening grid security must be the primary goal of grid development [3]. Grid operations face numerous challenges, including a large number of operators, high personnel turnover, outdoor construction risks, and poor working conditions, all of which contribute to potential safety hazards. To implement the policy of "safety first, prevention foremost, and comprehensive management", power grid enterprises have undertaken extensive safety and anti-violation efforts [4]. Workplace accidents pose a significant threat to employees' health and safety. Despite advancements in safety technology across various industries, non-compliant behavior remains a major contributor to workplace accidents [5]. Memarian et al. [6] argue that the rational use of risk analysis and assessment methods is essential for effectively mitigating risk factors in power safety production. The Southern Power Grid Corporation embraces the safety concept that "all accidents can be prevented". However, the current management of power operation violations primarily focuses on site control, dispatching safety personnel to inspect work sites, or using cameras for video surveillance. Peng [7] proposed an object detection model, CORY-Net, combining contrastive self-supervised pretraining and YOLOv5 for intelligent safety monitoring on power grid construction sites. Nevertheless, unauthorized operations are still largely identified manually by monitoring personnel, which is inefficient [8]. Additionally, violation prevention mainly relies on improving management systems, with little focus on assessing operational violation risks before grid operations, making effective prevention challenging.

On-site power operations involve numerous risk factors, such as temperature, operating environment, voltage levels, and work modes, which are characterized by complexity, dynamic changes, randomness, and uncontrollability. A scientific analysis of these risk factors is essential for accurately assessing the risks associated with on-site power operations [9]. In the complex power grid system, the diversity of data sources and the complexity of data connotations result in multimodal and cross-modal data characteristics. This means that the data contain various modalities that correspond to or interact with one another, such as text, information tables, and more. For text data, converting textual information into numerical representations is often required. In this context, Qaiser et al. [10] utilized Term Frequency-Inverse Document Frequency (TF-IDF) to comprehensively evaluate document content, increasing the significance of important keywords and reducing the impact of high-frequency words. However, when the keyword "command" is included in text data, the TF-IDF formula assigns it a low weight, despite its critical role in operational information. Midya et al. [11] employed independent component analysis (ICA) to extract response features for automotive fault diagnosis, simplifying computation and aiding in system design for vehicle diagnostics and quality control. In the power sector, Zhou et al. [12] examined the deep relationship between noncompliance and violations using data from electric power companies, paving the way for improved workplace safety. Awasthi et al. [13] proposed a K-Nearest Neighbor (KNN) algorithm for accurately and efficiently classifying fault types in distribution networks with distributed generation (DG), thereby enhancing the reliability and stability of complex distribution systems. Mao [14] introduced a deep multi-task learning approach, addressing issues such as insufficient feature mining and error accumulation in multi-node load forecasting by incorporating multi-modal feature extraction and a soft-shared multi-task learning model with strong practical value. Chang et al. [15] combined dynamic information, such as operator images and locations, with static information derived from historical data to perform a comprehensive risk assessment and estimate safety probabilities. Maria et al. [16] proposed a new algorithm for managing unexpected events in smart grids, evaluating potential impacts to ensure smooth grid operation. Praveen [17] developed a CNN-based model for fault classification in power distribution networks integrated with distributed generation, achieving 99.52% accuracy without the need for pre-processing or feature engineering. While some research has focused on risk prediction and prevention mechanisms for violations in field power operations, these studies remain inadequate. There is currently no system that provides a future operational plan combined with relevant data to predict the likelihood of violations, key violation types, and potential issues prior to the operation. This information, if available, would offer operators valuable insights into preventing violations before they occur.

To address these challenges, this paper first constructs a historical case database by leveraging multimodal historical operation data from information systems such as power grid management platforms. Natural language processing techniques are then applied to generate word vectors for three types of operation-related factors. Generalized correlation coefficients based on mutual information are used to select key word vectors, enabling efficient approximation and dimensionality reduction of the word vectors. ICA is employed for feature extraction and dimensionality reduction, facilitating accurate representation of operational scenarios. For each historical case, a violation risk prediction model, built using the Random Forest (RF) algorithm, is utilized to calculate a risk score. If the risk score is high, the KNN algorithm is applied to identify potential violations under similar operational conditions from the historical case database. Real-time violation risk assessment is conducted with each issued operation, providing early warnings to operators.

This approach effectively mitigates the occurrence of violations and enhances the safety and security of power grid operations.

1.1. Jieba Segmentation

Segmentation is a crucial task in natural language processing, involving the division of continuous text into meaningful words or phrases based on specific rules. Unlike English, where spaces naturally separate words, Chinese word segmentation is more complex. The challenges arise from differing tasks, perspectives, and interpretations of what constitutes a "word", making it difficult to establish a universal standard for segmentation.

The Jieba segmentation tool in Python employs a comprehensive general-purpose lexicon, making it both powerful and broadly applicable across various domains. It also allows for the integration of a custom lexicon with the general-purpose lexicon, facilitating efficient segmentation in specialized fields [18]. Jieba's segmentation algorithm is based on Hidden Markov Models (HMMs), which encompasses three distinct segmentation models: the exact model, the complete model, and the search engine model. HMMs are statistical models that describe a Markov process with parameters that are not directly observable. The complete model scans all possible words in the text according to a Chinese thesaurus, which, while fast, does not effectively resolve ambiguities. In contrast, the exact model aims for precise segmentation, ensuring that no redundant words appear in the output. The search engine model builds upon the exact model by re-segmenting long words to enhance recall, albeit at the expense of introducing some redundancy in the results.

1.2. Independent Component Analysis

Independent Component Analysis (ICA) is a signal processing technique grounded in the principle of statistical independence of mixed signals [19]. It operates by separating observed signals into mutually independent source signals through a linear transformation. ICA assumes that the mixed signals are linear combinations of the underlying source signals and aims to maximize the statistical independence of the output signals by adaptively adjusting the transformation matrix. This method has extensive applications in addressing the challenge of separating independent components within multidimensional data.

FastICA, developed by Hyvärinen et al. [20], is a widely recognized and efficient blind source separation algorithm in the field of multidimensional signal processing, noted for its rapid convergence and superior performance. It aims to recover independent source signals from mixed signals. The fundamental principles of FastICA encompass several key processes: centering, whitening, initialization of the separation matrix, iterative updating, and termination condition testing. By iteratively updating the separation matrix, the non-Gaussianity of the output signal is progressively enhanced, allowing for the identification of a transformation matrix that facilitates the blind source separation of the signals. FastICA is particularly effective in processing non-Gaussian signals.

1.3. K-Nearest Neighbor Algorithm

The KNN algorithm is a basic machine learning algorithm for classification and regression problems [21]. It works on the principle of classification based on the votes of the K nearest neighbors in the feature space or regression based on the mean of these neighbors. For a given data point, the KNN determines the class to which it belongs by calculating the distance to its nearest neighbor training samples by majority voting. Specifically, it uses KNN in the feature space to perform classification or regression based on their labels. Among them, the sample distance is measured by Euclidean distance, Manhattan distance, Minkowski distance, and cosine similarity; the choice of K value is also particularly critical [22]. When the K value is too small, the model will be particularly sensitive to noise and is easily affected by local noise, which leads to poor generalization of the model to the overall data, and when the K value is too large, it easily leads to the phenomenon of overfitting of the model.

1.4. Random Forests Algorithm

The RF algorithm was introduced by Leo Breiman and Adele Cutler [23], who elaborated on the principles and applications of the Random Forests algorithm in their 2001 paper. The RF algorithm is an ensemble learning method that addresses classification and regression tasks by aggregating the outputs of multiple decision trees. Each decision tree is constructed using a randomly selected subset of features and a randomly selected subset of samples. The final prediction is then determined through voting or averaging the results from all the trees [24].

2. Construction of Grid Operation Violation Risk Prediction Model for Operation Data Mining

This paper presents a grid operation violation risk prediction method based on operational data mining, enabling real-time violation risk assessment for each issued operation and providing early warnings for operators. Specifically, as illustrated in Figure 1, this violation assessment method constructs a predictive model through the processing of multimodal unstructured data. The modeling process primarily involves the following steps: preprocessing of multimodal unstructured data, construction of an operational lexicon and word vector space, keyword vector filtering and dimensionality reduction, feature extraction via ICA, violation score prediction via RF, and subsequent violation risk assessment.

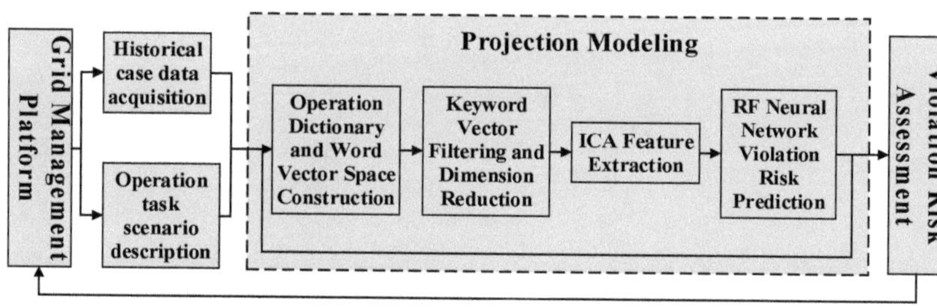

Figure 1. Flowchart for the proposed method.

2.1. Data Acquisition and Case Base Construction

Relying on the grid management platform and other information systems to obtain multimodal data on grid operation violations, recorded as $\widetilde{X} = [\widetilde{X}_1, \widetilde{X}_2, \widetilde{X}_3] \in \mathbb{R}^{n\times m}$, its corresponding violation label is noted as $Y \in \mathbb{R}^{n\times 1}$, where $\widetilde{X}_1 \in \mathbb{R}^{n\times m_1}$ denotes m_1 factors whose attributes are in numerical form—examples include weather temperature, latitude and longitude of the workplace, etc.; $\widetilde{X}_2 \in \mathbb{R}^{n\times m_2}$ denotes m_2 factors in the form of fixed text phrases; $Y \in \mathbb{R}^{n\times 1}$ is the violation label. n denotes the number of historical cases, and $m = m_1 + m_2 + m_3$ denotes the number of operational factors.

In order to construct a standardized database, the multimodal historical operation data are first preprocessed to construct word vectors of operation-related factors.

Step 2.1-1 Pre-processing of numerical form factors

For operation factors whose attributes are in numerical form $\widetilde{\mathbf{X}}_1^i (i = 1, 2, \ldots, m_1)$, normalization is performed to remove the effects of different magnitudes:

$$\mathbf{X}_1^{i,j} = \frac{\widetilde{\mathbf{X}}_1^{i,j} - \min\left(\widetilde{\mathbf{X}}_1^i\right)}{\max\left(\widetilde{\mathbf{X}}_1^i\right) - \min\left(\widetilde{\mathbf{X}}_1^i\right)} \tag{1}$$

where $\widetilde{\mathbf{X}}_1^{i,j}$ denotes the value of the attribute of the operational factor $\widetilde{\mathbf{X}}_1^i (i = 1, 2, \ldots, m_1)$ in the $jth(j = 1, 2, \ldots n)$ historical case; $\max\left(\widetilde{\mathbf{X}}_1^i\right)$ and $\min\left(\widetilde{\mathbf{X}}_1^i\right)$ denote the maximum and minimum values of the attributes of the operational factor $\widetilde{\mathbf{X}}_1^i (i = 1, 2, \ldots, m_1)$, respectively.

Step 2.1-2 Pre-processing of fixed text phrase form assignment factors

For factors that fix the form of textual phrases $\widetilde{\mathbf{X}}_2^i(i = 1,2,\ldots,m_2)$, their subfactor is used as a keyword in the operational violation dictionary. The attribute values in the factor are counted and coded sequentially, and these values are converted to numeric vector form.

Step 2.1-3 Pre-processing of descriptive linguistic form assignment factors

For descriptive linguistic forms with the factor $\widetilde{\mathbf{X}}_3^i(i = 1,2,\ldots,m_3)$, the Jieba segmentation engine is firstly employed for text segmentation. After removing low-frequency partitions by filtering out deactivated words with low information content, a lexicon is constructed. Subsequently, the bag-of-words (BoW) model is applied based on this filtered lexicon, resulting in a total of J words. The frequency of each word from the lexicon appearing in the segmented results of each operational case is recorded as the corresponding position in the J-dimensional vector. This process generates a J-dimensional vector for each operational case, thereby establishing the word vector space by applying the previously described procedures across all operational cases.

With the above steps, the word vector space of grid operations can be constructed, and n operation cases are represented as a structured data matrix $X \in \mathbb{R}^{n\times J}$ with variable dimension J and number of samples n:

$$X = \begin{pmatrix} x_1(1) & x_2(1) & \cdots & x_J(1) \\ x_1(2) & x_2(2) & \cdots & x_J(2) \\ \vdots & \vdots & \ddots & \vdots \\ x_1(n) & x_2(n) & \cdots & x_J(n) \end{pmatrix} \in \mathbb{R}^{n\times J} \tag{2}$$

where the value of $x_j(i)$ denotes the value of the attribute corresponding to the jth factor or sub-phrase in the ith historical case.

Step 2.1-4 Pre-processing of violation label

For the violation label, there are two elements: "violation" and "no violation". "No violation" is coded as 0, while "violation" is coded as 1.

2.2. Key Segmentation Extraction and Classical Word Vector Space Construction

To address the challenges posed by the high-dimensional sparse word vector matrix, this paper introduces a keyword vector extraction method grounded in mutual information. First, the mutual information between each keyword in the dictionary and the violation label is calculated. Subsequently, correlation coefficients based on this mutual information are computed, enabling the selection of word vectors with higher correlation coefficients as keywords. The specific steps involved in this process are outlined below.

Step 2.2-1 Mutual information between segmentation and violation label

For the high-dimensional sparse word vector matrices, the mutual information between each keyword in the dictionary and the violation label is first calculated:

$$I(X_j, Y) = H(X_j) + H(Y) - H(X_j, Y) \tag{3}$$

where $\mathbf{X}_j = [x_j(1), x_j(2), \ldots, x_j(n)]^T$ denotes the $jth(j = 1,2,\ldots,J)$ segmentation, $H(X_j)$ is the information entropy of the segmentation X_j:

$$H(X_j) = -\int_x p(X_j) \log p(X_j) dx \tag{4}$$

$H(Y)$ is the information entropy of the violation label Y:

$$H(Y) = -\int_y p(Y) \log p(Y) dx \tag{5}$$

$H(X_j, Y)$ is the joint information entropy of the partition X_j and Y:

$$H(X_j, Y) = -\iint_{x_j, y} p(X_j, Y) \log p(X_j, Y) dx_j dy \tag{6}$$

where $p(X_j)$ and $p(Y)$ denote the probability density functions of the variables X_j and Y, and $p(X_j, Y) = \iint_{x_j, y} p(x, y) dx_j dy$ is the joint probability density function.

Step 2.2-2 Generalized correlation coefficients for segmentation and violation Label

The generalized correlation coefficients of the segmentation X_j and segmentation Y are obtained based on the mutual information derived from Equation (6):

$$r_{X_j, Y} = r(X_j, Y) = \frac{I(X_j, Y)}{\min(H(X_j), H(Y))} \tag{7}$$

where $r_{X_j, Y} \in [0, 1]$.

Step 2.2-3 Filtering of keywords

If the correlation coefficient $r_{X_j, Y}$ of the jth segmentation with the label Y is greater than the threshold δ, it means that the segmentation is a keyword:

$$r_{X_j, Y} = r(X_j, Y) > \delta \tag{8}$$

Otherwise, the segmentation is considered to contain less information and discarded, where the threshold δ is selected in practice based on the mean value of the mutual information values between the calculated segmentation and the violation label.

Step 2.2-4 Construction of key segmentation lexicon and word vector matrices

After filtering out the keyword segments through the previous step, the key lexicon of historical cases is built and the word vector matrix is constructed as

$$\widehat{X} = \begin{pmatrix} \widehat{x}_1(1) & \widehat{x}_2(1) & \cdots & \widehat{x}_J(1) \\ \widehat{x}_1(2) & \widehat{x}_2(2) & \cdots & \widehat{x}_J(2) \\ \vdots & \vdots & \ddots & \vdots \\ \widehat{x}_1(n) & \widehat{x}_2(n) & \cdots & \widehat{x}_J(n) \end{pmatrix} \in \mathbb{R}^{n \times J} \tag{9}$$

where the value of $\widehat{x}_k(i)$ denotes the attribute value corresponding to the kth key segmentation in the ith historical case, and J is the number of key segmentations.

2.3. Feature Extraction and Operational Scenario Characterization

To further extract the underlying information from the word vector matrix, this paper employs ICA for feature extraction and dimensionality reduction. The identified potential features serve as characterizations of the operational scenarios for each historical case in the subsequent steps.

Step 2.3-1 Whitening process

Whiten the sparse high-dimensional word vector matrix $\widehat{X}$ obtained in Equation (9):

$$Z = \Lambda^{-1/2} U^T \widehat{X} = Q\widehat{X} \in \mathbb{R}^{n \times J} \tag{10}$$

where matrix Z is the whitened data, $Q = \Lambda^{-1/2} U^T$ is the whitened matrix, and U and Λ are the orthogonal eigenvector array and eigenvalue diagonal array obtained by eigenvalue decomposition of the covariance matrix $E(\widehat{X}\widehat{X}^T)$, respectively.

Step 2.3-2 Low-dimensional feature extraction

Mapping the high-dimensional word vector matrices to the low-dimensional space using FastICA yields the independent principal element matrices:

$$S = WZ^T = \begin{bmatrix} W_{1:p} \\ W_{p+1:J} \end{bmatrix} Z^T \in \mathbb{R}^{J\times n} \tag{11}$$

where S denotes the independent principal elements and p is the number of retained principal elements; W is the decomposition matrix, the solution of which can be solved iteratively according to the non-Gaussian maximization principle using the FastICA algorithm proposed by Hyvarine [17].

In this way, with minimal loss of information, the operational scenarios for each historical case are characterized as a few of the most representative independent principal elements:

$$\widetilde{S} = ZW_{1:p}^T = \begin{bmatrix} s_1(1) & s_2(1) & \cdots & s_p(1) \\ s_1(2) & s_2(2) & \cdots & s_p(2) \\ \vdots & \vdots & \ddots & \vdots \\ s_1(n) & s_2(n) & \cdots & s_p(n) \end{bmatrix} \tag{12}$$

2.4. Modeling of Violation Risk Estimation

To facilitate the estimation of violation risk within a specific operational scenario, it is essential to develop a prediction model that incorporates both the operational scenario representation vector and the violation risk indicator. The key steps involved in this process are described below.

Step 2.4-1 Violation risk indicators

For the *ith* historical case, the KNN method is used to find the operational cases under similar operational scenarios in the historical case database, and the distance between the two cases is calculated:

$$d_{ij} = \sqrt{\sum_{t=1}^{p} (s_i(t) - s_j(t))^2} \tag{13}$$

Select the K cases closest to $\widetilde{S}_i$, noted as $N(\widetilde{S}_i) = \left\{\widetilde{S}_1^i, \widetilde{S}_2^i, \cdots, \widetilde{S}_k^i\right\}$, and find the corresponding violation label $y(\widetilde{S}_i) = \left\{y_i^1, y_i^2 \cdots, y_i^k\right\}$.

The probability of violation is calculated as an indicator of the risk of violation under the operational scenario of the *ith* historical case:

$$p(\widetilde{S}_i) = \frac{sum(y(\widetilde{S}_i))}{k} = \frac{\sum_{t=1}^{k} y_i^t}{k} \tag{14}$$

where k can be determined using a cross-validation algorithm.

Step 2.4-2 Violation risk estimation modeling

The RF algorithm is utilized to construct the prediction model of operational key factors and violation risk indicators, the structure of which is shown in Figure 2. $\widetilde{S} = (\widetilde{S}_1, \widetilde{S}_2, \ldots, \widetilde{S}_n)$ is the input layer data, $Y_p = (p(\widetilde{S}_1), p(\widetilde{S}_2), \ldots, p(\widetilde{S}_n))$ is the output layer data, and the RF model consists of multiple decision trees, each of which is a type of base learner. The input layer data is transmitted to each decision tree, where each node represents a decision based on a feature of the input data. The RF algorithm aggregates the outputs of all decision trees through voting or averaging to ultimately produce a prediction of the target variable. The construction of each decision tree introduces diversity into the model by randomly selecting subsets of samples and features. The overall RF model enhances its predictive capability by adjusting model parameters and the number of trees while minimizing loss functions, such as the root mean square error, to achieve an accurate prediction model.

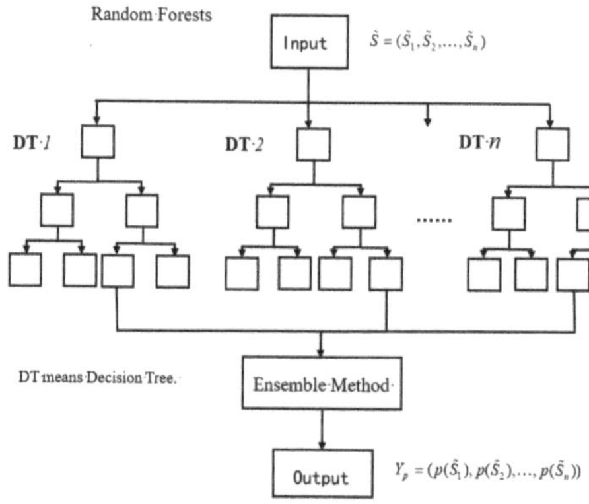

Figure 2. Model structure of RF.

2.5. Risk Prediction of Behavioral Violations in Operational Tasks

In order to realize the violation risk assessment of the grid operation, it is necessary to predict the violation risk in advance according to the operation scenario when the operation is issued, the specific steps for which are described below.

Step 2.5-1 Pre-processing of data for the online operations work plan

When a grid operation is planned, its operation factors can be represented as $\widetilde{x} = [\widetilde{x}_1, \widetilde{x}_2, \widetilde{x}_3] \in \mathbb{R}^{1\times m}$, where $\widetilde{x}_1 \in \mathbb{R}^{1\times m_1}$ denotes m_1 factors whose attributes are in numerical form; $\widetilde{x}_2 \in \mathbb{R}^{1\times m_2}$ denotes m_2 factors in the form of fixed text phrases; and $\widetilde{x}_3 \in \mathbb{R}^{1\times m_3}$ denotes m_3 groups of factors in descriptive language.

Step 2.5-2 Pre-processing of numerical type factors for the online operations work plan

For numerical type factors, normalization is performed using the minimum and maximum values of the operational factors in Step 2.1-1:

$$x_1^i = \frac{\widetilde{x}_1^i - \min(\widetilde{X}_1^i)}{\max(\widetilde{X}_1^i) - \min(\widetilde{X}_1^i)} \tag{15}$$

Step 2.5-3 Constructing keyword vectors for the online operations work plan

Using the key dictionary developed in Step 2.2-4, a keyword vector is constructed to represent the current grid operation task scenario.:

$$\widehat{x} = [\widehat{x}(1) \quad \widehat{x}(2) \quad \cdots \quad \widehat{x}(J)] \in \mathbb{R}^{1\times J} \tag{16}$$

Step 2.5-4 The whitening vector of the online operations work plan

Using the whitening transformation matrix obtained in Step 2.3-1, the whitening vector for the current grid operation task scenario is derived:

$$z = Q\widehat{x} \in \mathbb{R}^{1\times J} \tag{17}$$

Step 2.5-5 Independent principal element representation of the online operations work plan

Using the decomposition matrix obtained in Step 2.3-2, the independent principal element representation for the current grid operation task scenario is derived:

$$\widetilde{s} = zW_{1:p}^T = [s_1 \quad s_2 \quad \cdots \quad s_p] \in \mathbb{R}^{1\times p} \tag{18}$$

Step 2.5-6 Violation risk indicators for the online operations work plan

The independent principal element representation under the current grid operation task scenario is input into the RF prediction model obtained in Step 2.4-2 to calculate the violation risk indicator $p(\widetilde{s})$ for the current grid operation task scenario. When this indicator is relatively large, the violation situation of cases under similar operation scenarios in the historical case database is searched, and the corresponding violation dictionary is queried to provide a reference for the grid operator to effectively reduce the occurrence of violation behaviors and provide safety for the grid operation.

3. Application and Analysis of Results

3.1. Data Presentation and Pre-Processing

A total of 24,001 operational data points from January–July 2022 were obtained from a power grid company, and 21,715 historical cases were derived by removing data cases with incomplete information. A total of 90% of the data were randomly selected as historical cases to construct a case database. Therefore, 19,544 data points were used for model training and the remaining 2171 data points were used as testing data. As shown in Table 1, each piece of operational data contains 17 operational factors, $\widetilde{X} = [\widetilde{X}_1, \widetilde{X}_2, \widetilde{X}_3] \in \mathbb{R}^{19544\times 17}$ Among them, $\widetilde{X}_1 \in \mathbb{R}^{19544\times 5}$ denotes five factors whose attributes are in the form of numerical values, including { Longitude of the operating site, Latitude of the operating site, Maximum temperature on working days, Minimum temperature on working days, Voltage level}; $\widetilde{X}_2 \in \mathbb{R}^{19544\times 11}$ represents 11 factors in the form of fixed text phrases, including { municipal bureau, Supervisory levels, systematic classification, Project management units, Operating units (shifts), Main body of operation, Task origin, Type of operation, Professional fields, Risk levels and Operation methods}; $\widetilde{X}_3 \in \mathbb{R}^{19544\times 1}$ represents the descriptive language factor " Operational plan implementation contents"; and $Y \in \mathbb{R}^{19544\times 1}$ is the violation label, which is transformed into a numeric form with the attribute value {1, 0}, where 1 means "violation" and 0 means " no violation".

Table 1. Example of a numerical factor in an operational plan.

Operational Influencing Factors	Examples of Contents	Factors Format
Municipal bureaus	XX Power Supply Bureau	Fixed texts
Supervisory levels	Prefectural level	
systematic classification	Supervision system	
Project management units	XX Power Supply Bureau	
Operating units (shifts)	XX XX Development Ltd.	
Main body of operation	Home unit, outside unit	
Task origin	Infrastructure	
Type of operation	Infrastructure type	
Professional fields	Power transmission	
Risk levels	Medium	
Operation methods	Non-stop power supply	
Operating site longitude (°)	104.95764	Numeric data
Operating site latitude (°)	26.54658	
Maximum temperature (°C)	14	
Minimum temperature (°C)	5	
Voltage level	35 kV	
Operational plan implementation contents	35 kV Guan Steepening Line to 110 kV Stroke Field Substation 35 kV Project (P53 Pylon Establishment)	Descriptive Language Texts

In this case, relying on the grid management platform and other information systems to obtain the data from the database of grid operation violation behavior is noted as $\widetilde{X} = [\widetilde{X}_1, \widetilde{X}_2, \widetilde{X}_3] \in \mathbb{R}^{19544\times 17}$, and its corresponding violation label is noted as $Y \in \mathbb{R}^{19544\times 1}$. Following the data preprocessing method outlined in Section 2.1, a word vector space for grid operations is constructed. Attributes such as "temperature" and "longitude", represented as numerical data, are processed for their magnitudes. A factor dictionary is created to quantify these attribute values and convert them into numerical vector form; for instance, the factor "Task Origin" encompasses three textual attributes—{Infrastructure, Marketing, Production}—which are encoded as {1, 2, 3}. For the descriptive language of the factor "Operational Plan Implementation Contents", the Jieba segmentation engine is employed to segment the text. Words with low information content are filtered out by setting deactivation words, resulting in a total of 8675 filtered segments that include key descriptors relevant to grid operations, such as "Operations", "Site", "Switch", and "Ticket". After removing less frequent segments, 113 segmentation entries are retained. The bag-of-words (BoW) model is then utilized to establish the word vector space, leading to the formation of a structured data matrix that represents 19,544 operation cases across 113 variable dimensions and 19,544 samples, with the initial word cloud illustrated in Figure 3.

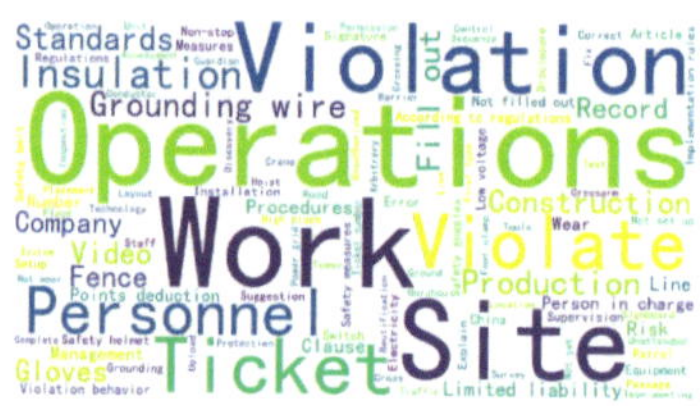

Figure 3. Word cloud map of initial segmentation.

3.2. *Keyword Dictionary Construction and Feature Extraction Results*

To extract fault-related key segments, the mutual information between each segment and the violation label in the dictionary is first calculated. Subsequently, the correlation coefficients based on the mutual information are computed, allowing for the selection of word vectors with larger correlation coefficients as key segments. Through practical analysis, it has been determined that a correlation coefficient $r_{X_j,Y} > 0.22$ between a segment and the violation label qualifies that segment as a key segment. Figure 4 illustrates the mutual information of each segment in the lexicon relative to the violation label, with the horizontal axis representing the lexicon keywords (such as "violation", "operation", etc.) and the vertical axis indicating the mutual information value between each corresponding keyword and the violation occurrence.

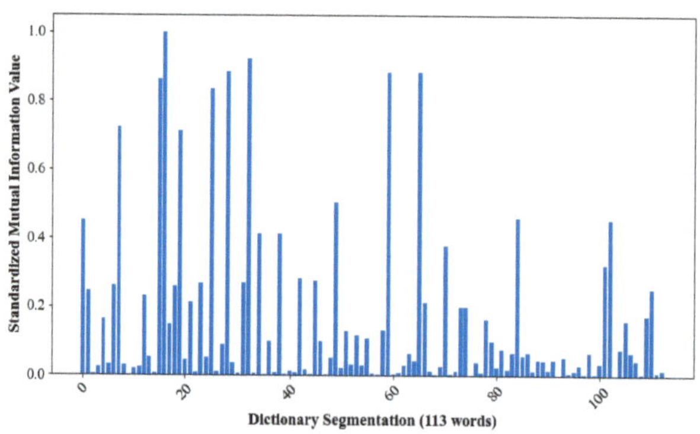

Figure 4. Mutual information between key segmentations and violation tags.

Finally, a total of 26 key sub-words, such as "Violation", "Personnel", "Site", " Work", "Violate", "Insulation", "Ticket", etc., are screened as shown in Figure 5 to establish the keyword dictionary for the historical cases and construct the word vector matrix.

Figure 5. Word cloud map of key segmentations (26 words.).

Subsequently, the 26-dimensional sparse high-dimensional word vector matrix is processed using the ICA algorithm. Through whitening and FastICA, the high-dimensional word vector matrix is mapped to a low-dimensional space, resulting in a 3-dimensional independent principal matrix. Figure 6 illustrates the independent principal components of the historical cases. The ICA algorithm is employed for feature extraction, effectively mapping the high-dimensional word vector matrix to the low-dimensional space, where the most representative features are selected as the operational scene representations for each historical case, ensuring minimal loss of information.

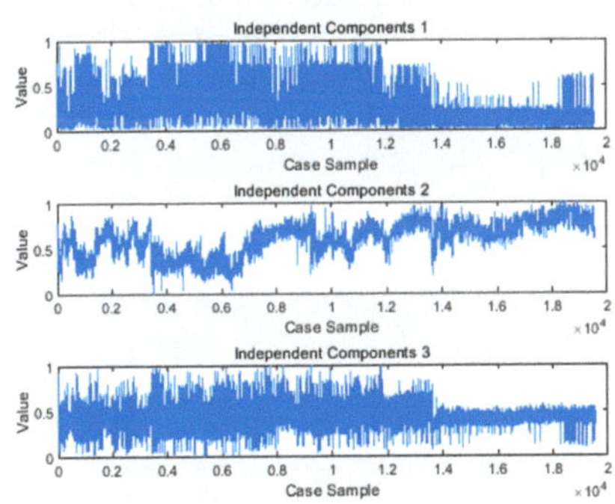

Figure 6. Independent feature representation of historical casework scenarios.

3.3. *Operational Risk Prediction Modeling and Online Assessment Results*

When establishing the operational risk prediction model, for each historical case, the KNN algorithm is used to find operational cases under similar operational scenarios in the historical case database, and the probability of violation of the k historical operational cases in the nearest neighbor is statistically analyzed and used as the violation risk indicator under the operational scenario. The RF algorithm is used to construct the estimation model of operation key factors and violation risk index, $\widetilde{S} = \left(\widetilde{S}_1, \widetilde{S}_2, \ldots, \widetilde{S}_{19544}\right)$ is the input data, $Y_p = \left(p(\widetilde{S}_1), p(\widetilde{S}_2), \ldots, p(\widetilde{S}_{19544})\right)$ is the output data, the number of decision trees is set to 100, the minimum number of leaf nodes is set to 5, the out-of-bag error estimation is turned on, and the importance of features is calculated, which is used for evaluating the performance of the model.

To validate the effectiveness of the proposed grid operation violation risk prediction algorithm, a total of 2171 grid operation tasks were evaluated. The RF algorithm employed in this study was compared against six other algorithms: Extreme Learning Machine (ELM), Radial Basis Function (RBF) neural network, Back Propagation (BP) neural network along with its variants—Genetic Algorithm-Back Propagation (GA-BP) and Particle Swarm Optimization-Back Propagation (PSO-BP)—as well as a Convolutional Neural Network (CNN). The differences between the predicted and actual values of violation risk for all models on the testing set are illustrated in Figure 7.

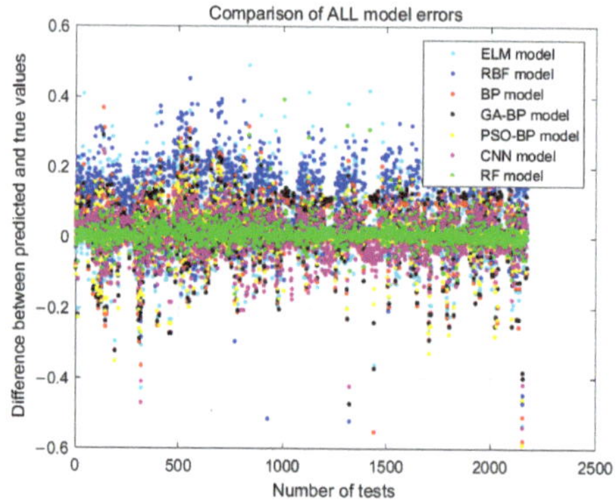

Figure 7. Predicted error on the testing set for different models.

To gain a more comprehensive understanding of the advantages of the proposed method, a comparative analysis was conducted with other algorithms based on three key metrics: R2 (i.e., coefficient of determination), MAE (mean absolute error), and MBE (mean bias error). The comparison of R^2 values for the training and testing sets across different algorithmic models is presented in Figure 8, while the comparisons of the MAE and MBE metrics are summarized in Tables 2–4.

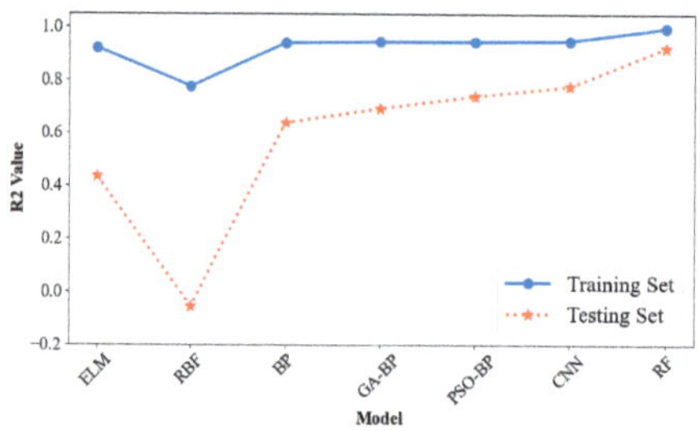

Figure 8. Comparison of R2 values for different models.

Table 2. Comparison of R2, MAE, and MBE metrics for initial word vector space.

Models Indicators		ELM	RBF	BP	GA-BP	PSO-BP	CNN	RF
R2	Training set	0.3165	0.4182	0.4861	0.5353	0.5390	0.6165	0.8232
	Testing set	0.2414	0.3851	0.4692	0.4791	0.4797	0.4847	0.5728
MAE	Training set	0.3501	0.3104	0.2691	0.2443	0.2314	0.2251	0.1228
	Testing set	0.3924	0.3472	0.2949	0.2768	0.2722	0.2741	0.2374
MBE	Training set	−0.0170	−0.0195	−0.0051	0.0259	−0.0192	0.0036	0.0019
	Testing set	0.0505	−0.0142	0.0296	0.0679	0.00378	0.0165	0.0130

First, the performance metrics considered are R^2, MAE, and MBE. The R^2 metric is a statistical measure used to evaluate the fit of a regression model to the data, assessing how well the predicted values align with the actual observed values. Its range is from 0 to 1, where a value of 0 indicates poor predictive accuracy, while values closer to 1 suggest higher accuracy, with 1 signifying perfect prediction. MAE is commonly employed to assess the accuracy of predictive models; it represents the average of the absolute differences between predicted and actual values. Similarly, MBE measures predictive accuracy and is calculated as the average of the differences between predicted and actual values, divided by the total number of samples. A more accurate model will yield a smaller absolute value of MBE, ideally approaching zero.

Table 3. Comparison of R2, MAE, and MBE metrics for Keywords vector space.

Models Indicators		ELM	RBF	BP	GA-BP	PSO-BP	CNN	RF
R2	Training set	0.3376	0.4031	0.5839	0.5745	0.5586	0.5888	0.8173
	Testing set	0.2955	0.3492	0.4797	0.4810	0.4816	0.5006	0.5757
MAE	Training set	0.3461	0.3199	0.2084	0.2088	0.2176	0.2368	0.1231
	Testing set	0.3830	0.3554	0.2561	0.2519	0.2510	0.2794	0.2272
MBE	Training set	−0.0168	−0.0182	0.0013	−0.0021	−0.0028	0.0028	0.0014
	Testing set	0.0185	−0.0191	0.0141	0.0087	−0.0030	0.0089	−0.0125

Table 4. Comparison of R2, MAE, and MBE metrics for dimensional vector space.

Models Indicators		ELM	RBF	BP	GA-BP	PSO-BP	CNN	RF
R2	Training set	0.9210	0.7758	0.9404	0.9442	0.9453	0.9494	0.9971
	Testing set	0.4348	−0.0551	0.6386	0.6941	0.7403	0.7776	0.9237
MAE	Training set	0.0653	0.1204	0.0548	0.0518	0.0497	0.0511	0.0107
	Testing set	0.0661	0.1062	0.0545	0.0499	0.0412	0.0492	0.0227
MBE	Training set	−0.0052	−0.0234	0.0005	−0.0025	−0.0008	−0.0021	0.0001
	Testing set	0.0230	0.0897	0.0206	0.0168	0.012	0.0087	0.0132

The analysis reveals that, for the performance indicator R2, the RF algorithm achieves values closer to 1 in both the training and testing sets. It indicates that the RF algorithm exhibits stronger explanatory power for the target variables compared to other algorithms, effectively capturing the relationship between independent and dependent variables. Notably, in the testing set, the RBF algorithm yields a negative R2 value when it processes 3-dimensionality-reduced data. This may be attributed to the limited information present in low-dimensional data, which restricts the model's fitting capability and hinders its generalization. Additionally, for the performance metric MAE, the RF algorithm model demonstrates relatively lower values in both the training and testing sets, indicating that the prediction results obtained through the RF algorithm are more accurate than those of the comparative algorithms. Finally, regarding the performance metric MBE, the RF algorithm shows smaller values that are closer to zero in both training and testing sets, signifying that the RF algorithm predictions are largely consistent with the actual values.

Furthermore, the comparative analysis presented in Tables 2–4 demonstrates the significance of processing steps such as removing low-frequency words, keyword extraction, and feature extraction. This approach not only helps avoid the pitfalls associated with high-dimensional data but also allows for the selection of key participles that effectively characterize job construction scenarios, as evidenced by the comparison between Tables 2 and 3. The resulting low-dimensional vector representation of the field operation plan text is achieved while maintaining prediction accuracy. Additionally, FastICA effectively extracts the latent information from the keyword vector matrix, enabling further feature extraction and dimensionality reduction. A comparison of Tables 3 and 4 indicates that this dimensionality reduction operation produces a vector representation that significantly enhances the probability of violation risk prediction.

In summary, a comprehensive comparison of various algorithms reveals that the RF algorithm developed in this study achieves an R2 value of 92.37% on the testing set. It indicates that the model effectively leverages information from current operational scenarios to predict violation risks in real-time. Furthermore, it provides relevant early warning information to operational staff when issuing commands, specifically regarding common violations observed in similar historical cases. This proactive approach significantly reduces the likelihood of violations and ensures the safe operation of the power grid. In comparison to other algorithms, the proposed method demonstrates superior performance in terms of both accuracy and robustness.

4. Conclusions

This paper analyzes the multimodal grid historical violation operation data from January to July 2022 of a power company and constructs a historical case database. Based on the Jieba partition and bag-of-words model, the different modal data have been processed to construct the initial word vector space, which realizes the numerical structural description of the historical grid violation operation cases. In order to remove the redundancy of the initial lexicon, a novel key feature extraction strategy has been designed to extract the keywords based on mutual information, construct the keyword vector space of historical cases, and introduce the ICA algorithm to realize the dimensionality reduction characterization of the word vector space. After that, a RF algorithm is utilized to construct a grid operation violation risk prediction model. The proposed method can realize real-time assessment of violation risk, so as to effectively reduce the occurrence of violations and provide a safety guarantee for power grid operation.

Author Contributions: Conceptualization, L.M. and J.Z.; methodology, S.Z.; software, X.Z.; validation, Y.W., L.M. and S.L.; formal analysis, Y.W.; investigation, S.L.; resources, J.Z.; data curation, X.Z.; writing—original draft preparation, L.M. and J.Z.; writing—review and editing, S.Z. and S.L.; supervision, L.M.; project administration, X.Z. funding acquisition, L.M. All authors have read and agreed to the published version of the manuscript.

Funding: This research was funded by Guizhou Power Grid Limited Liability Company, grant number GZKJXM20222320.

Data Availability Statement: Data is available upon request from the corresponding authors or works cited in the article.

Conflicts of Interest: Author Lingwen Meng were employed by the Electric Power Scientific Research Institute of Guizhou Power Grid Guizhou. The remaining authors declare that the research was conducted in the absence of any commercial or financial relationships that could be construed as a potential conflict of interest.

References

1. Marot, A.; Kelly, A.; Naglic, M.; Barbesant, V.; Cremer, J.; Stefanov, A.; Viebahn, J. Perspectives on Future Power System Control Centers for Energy Transition. *J. Mod. Power Syst. Clean Energy* **2022**, *10*, 328–344. [CrossRef]
2. Sun, Y.; Zhu, J.; Yu, X.; Ye, C.; Fan, G. Research on early warning of power grid construction safety based on PSO-SVM model. *J. Phys. Conf. Ser.* **2020**, *1449*, 012037. [CrossRef]
3. AlHaddad, U.; Basuhail, A.; Khemakhem, M.; Eassa, F.E.; Jambi, K. Towards sustainable energy grids: A machine learning-based ensemble methods approach for outages estimation in extreme weather events. *Sustainability* **2023**, *15*, 12622. [CrossRef]
4. Zhang, Z.; Lin, S.; Ye, Y.; Xu, Z.; Zhao, Y.; Zhao, H.; Sun, J. A hybrid MCDM model for evaluating the market-oriented business regulatory risk of power grid enterprises based on the bayesian best-worst method and MARCOS approach. *Energies* **2022**, *15*, 2978. [CrossRef]
5. Reason, J. *The Human Contribution: Unsafe Acts, Accidents and Heroic Recoveries*; CRC Press: Boca Raton, FL, USA, 2017.
6. Memarian, B.; Mitropoulos, P. Safety incidents and high-risk activities of masonry construction. In Proceedings of the Construction Research Congress 2012: Construction Challenges in a Flat World, West Lafayette, Indiana, 21–23 May 2012; pp. 2510–2519.
7. Peng, G.; Lei, Y.; Li, H.; Wu, D.; Wang, J.; Liu, F. CORY-Net: Contrastive res-YOLOv5 network for intelligent safety monitoring on power grid construction sites. *IEEE Access* **2021**, *9*, 160461–160470. [CrossRef]
8. Chen, B.; Zhang, H.; Li, Y.; Fang, X.; Chen, H. A lightweight detection scheme for the safety of human behaviors in power construction sites. *J. Circuits Syst. Comput.* **2023**, *32*, 2350230. [CrossRef]
9. Peng, B.; Zhang, W.; Lin, X.; Qiu, H. A novel risk assessment method and application for the on-site power operations. *Vibroengineering Procedia* **2021**, *37*, 47–53. [CrossRef]
10. Qaiser, S.; Ali, R. Text mining: Use of TF-IDF to examine the relevance of words to documents. *Int. J. Comput. Appl.* **2018**, *181*, 25–29. [CrossRef]

11. Midya, M.; Ganguly, P.; Datta, T.; Chattopadhyay, S. ICA-Feature-Extraction-Based Fault Identification of Vehicular Starter Motor. *IEEE Sens. Lett.* **2023**, *7*, 1–4. [CrossRef]
12. Zhou, F.; Lu, H.; Jiang, C. Violation with concerns of safety: A study on non-compliant behavior and the antecedent and consequent effects in power grid construction. *Saf. Sci.* **2024**, *170*, 106353. [CrossRef]
13. Awasthi, S.; Singh, G.; Ahamad, N. Classifying electrical faults in a distribution system using k-nearest neighbor (knn) model in presence of multiple distributed generators. *J. Inst. Eng. Ser. B* **2024**, *105*, 621–634. [CrossRef]
14. Tan, M.; Hu, C.; Chen, J.; Wang, L.; Li, Z. Multi-node load forecasting based on multi-task learning with modal feature extraction. *Eng. Appl. Artif. Intell.* **2022**, *112*, 104856. [CrossRef]
15. Chang, Z.; Deng, Y.; Wu, J.; Xiong, X.; Chen, M.; Wang, H.; Xie, X. Safety risk assessment of electric power operation site based on variable precision rough set. *J. Circuits Syst. Comput.* **2022**, *31*, 2250254. [CrossRef]
16. Skydt, M.R.; Bang, M.; Shaker, H.R. A probabilistic sequence classification approach for early fault prediction in distribution grids using long short-term memory neural networks. *Measurement* **2021**, *170*, 108691. [CrossRef]
17. Rai, P.; Londhe, N.D.; Raj, R. Fault classification in power system distribution network integrated with distributed generators using CNN. *Electr. Power Syst. Res.* **2021**, *192*, 106914. [CrossRef]
18. Ding, Y.; Teng, F.; Zhang, P.; Huo, X.; Sun, Q.; Qi, Y. Research on text information mining technology of substation inspection based on improved Jieba. In Proceedings of the 2021 International Conference on Wireless Communications and Smart Grid (ICWCSG), Aachen, Germany, 25–28 October 2021; IEEE: Piscataway, NJ, USA; pp. 561–564.
19. Hyvärinen, A.; Oja, E. Independent component analysis: Algorithms and applications. *Neural Netw.* **2000**, *13*, 411–430. [CrossRef]
20. Hyvarinen, A. Fast and robust fixed-point algorithms for independent component analysis. *IEEE Trans. Neural Netw.* **1999**, *10*, 626–634. [CrossRef] [PubMed]
21. Zhang, S. Challenges in KNN classification. *IEEE Trans. Knowl. Data Eng.* **2021**, *34*, 4663–4675. [CrossRef]
22. Bian, Z.; Vong, C.M.; Wong, P.K.; Wang, S. Fuzzy KNN method with adaptive nearest neighbors. *IEEE Trans. Cybern.* **2020**, *52*, 5380–5393. [CrossRef]
23. Breiman, L. Random forests. *Mach. Learn.* **2001**, *45*, 5–32. [CrossRef]
24. Yu, M.; Niu, D.; Gao, T.; Wang, K.; Sun, L.; Li, M.; Xu, X. A novel framework for ultra-short-term interval wind power prediction based on RF-WOA-VMD and BiGRU optimized by the attention mechanism. *Energy* **2023**, *269*, 126738. [CrossRef]

energies

MDPI

Article

Dynamic Prediction of Proton-Exchange Membrane Fuel Cell Degradation Based on Gated Recurrent Unit and Grey Wolf Optimization

Xiangdong Wang [1], Zerong Huang [1], Daxing Zhang [1], Haoyu Yuan [1], Bingzi Cai [1], Hanlin Liu [1], Chunsheng Wang [2], Yuan Cao [2], Xinyao Zhou [2] and Yaolin Dong [2,*]

[1] Huizhou Power Supply Bureau, Guangdong Power Grid Corporation, Huizhou 516000, China
[2] School of Automation, Central South University, Changsha 410083, China
* Correspondence: dyl0812@csu.edu.cn

Abstract: This paper addresses the challenge of degradation prediction in proton-exchange membrane fuel cells (PEMFCs). Traditional methods often struggle to balance accuracy and complexity, particularly under dynamic operational conditions. To overcome these limitations, this study proposes a data-driven approach based on the gated recurrent unit (GRU) neural network, optimized by the grey wolf optimizer (GWO). The integration of the GWO automates the hyperparameter tuning process, enhancing the predictive performance of the GRU network. The proposed GWO-GRU method was validated utilizing actual PEMFC data under dynamic load conditions. The results demonstrate that the GWO-GRU method achieves superior accuracy compared to other standard methods. The method offers a practical solution for online PEMFC degradation prediction, providing stable and accurate forecasting for PEMFC systems in dynamic environments.

Keywords: proton-exchange membrane fuel cells; degradation prediction; durability test; gated recurrent unit; grey wolf optimizer; accuracy; complexity

Citation: Wang, X.; Huang, Z.; Zhang, D.; Yuan, H.; Cai, B.; Liu, H.; Wang, C.; Cao, Y.; Zhou, X.; Dong, Y. Dynamic Prediction of Proton-Exchange Membrane Fuel Cell Degradation Based on Gated Recurrent Unit and Grey Wolf Optimization. *Energies* **2024**, *17*, 5855. https://doi.org/10.3390/en17235855

Academic Editor: Abdul-Ghani Olabi

Received: 16 October 2024
Revised: 12 November 2024
Accepted: 14 November 2024
Published: 22 November 2024

1. Introduction

As a leading technology in the field of clean energy, proton-exchange membrane fuel cells (PEMFCs) offer a promising solution for sustainable power generation due to their high efficiency and low emissions [1–3]. The advantages of PEMFCs include high power density, compact size, being lightweight, and having a low operating temperature [4]. Therefore, PEMFCs have been widely applied in various fields, from transportation to stationary power generation and portable electronics [5,6].

Nevertheless, widespread adoption of PEMFCs is hindered by issues related to degradation over time, which impacts their performance and reliability [7]. The degradation of a PEMFC stack is mainly caused by the degradation of membranes, including physical defects and chemical degradation. Physical defects, such as cracks and pinholes, mainly caused by repeated swelling and shrinking of the membrane during operation, can exacerbate fuel permeation and reduce proton selectivity. Chemical degradation can be attributed to the attack of side reaction products such as hydrogen peroxide on the weak parts of the membrane, reducing the proton conductivity and overall durability of membranes [8]. Factors such as membrane degradation, catalyst poisoning, and water and thermal management challenges contribute to the gradual decline in efficiency and output of PEMFC systems [9,10]. The degradation not only reduces the operational lifespan of the fuel cells but also increases maintenance costs and the risk of unexpected failures. Therefore, accurate prediction of PEMFC degradation is crucial for enhancing the longevity and efficiency of PEMFCs, as well as for optimizing maintenance schedules and ensuring operational safety.

In the realm of PEMFC degradation prediction, a number of methodologies have been explored. These methods can be broadly categorized into three main approaches: model-driven, data-driven, and hybrid approaches [11].

Model-driven approaches predict the degradation trend of PEMFCs by creating degradation models based on physical and chemical principles. These models are obtained through analyzing the ageing mechanism [12] or fitting empirical formulas [13]. Pei et al. [14] revealed the important effect of hydrogen crossover on fuel cell ageing, then developed a series of linear and nonlinear formulae to evaluate the lifetime of fuel cells. Wang et al. [15] proposed a new degradation model based on polarization resistance which was combined with a particle filter (PF) to forecast the future degradation trend of PEMFC. Ao et al. [16] utilized a frequency-domain Kalman filter to predict PEMFC voltage degradation under constant and dynamic conditions and concluded that the FDKF method offers superior accuracy and reduced computation time compared to the extended Kalman filter (EKF) method. Model-driven approaches offer detailed and explicable predictions of PEMFC degradation, but existing models have difficulties in explaining the nonlinear features in a degradation profile.

Data-driven approaches utilize machine learning and artificial intelligence techniques to forecast degradation trends. These approaches analyze and predict the ageing process of PEMFCs based on historical data without establishing complex physical or mathematical models. Many sorts of neural network approaches are extensively utilized in PEMFC prognostics [17]. Convolutional Neural Networks (CNNs) can effectively extract local degradation features from measurement data while maintaining high computational efficiency, and they were first utilized for PEMFC prediction in 2020 [18]. Benaggoune et al. [19] proposed a multi-step-ahead prediction of PEMFC degradation trends based on a stacked dilated CNN with an attention block, then developed a conditional CNN to track long-term ageing trends. Recurrent neural networks (RNNs) have attracted widespread interest in regard to fuel cell performance prediction due to their ability to handle sequential data and capture temporal dependencies. However, traditional RNN structures face significant challenges like gradient explosion and gradient vanishing [20]. To overcome this, various variants of RNNs have been utilized, such as long short-term memory (LSTM) [21,22], gated recurrent unit (GRU) [23], and echo state network (ESN) [24,25]. Zuo et al. [26] applied an attention mechanism to RNNs, comparing the predictive performance of two RNN architectures, LSTM and GRU, in long-term durability tests of fuel cells. It was demonstrated that the attention-based GRU showed the highest precision. Recently, the transformer model has also been gradually applied to PEMFC prediction. Lv et al. [27] utilized the transformer model for PEMFC long-term degradation prediction for the first time. Meng et al. [28] employed the segmented filtering and masking mechanisms to tackle the issues of the recovery of reversible voltage, which significantly improved the prediction accuracy. In addition, a variety of other deep learning methods have been applicated in PEMFC degradation prediction. For instance, Wu et al. [29] proposed an advanced self-adaptive relevance vector machine (RVM) for PEMFC performance prediction. While data-driven approaches are model-free, their reliance on data quality and availability can be a limitation.

Hybrid approaches integrate the model-driven approaches and data-driven approaches to improve prediction accuracy and robustness. Ma et al. [30] utilized a hybrid method based on extended Kalman filter (EKF) and LSTM to forecast the PEMFC degradation trend under multiple operational conditions. To overcome the drawback of LSTM in parallel computing, Hu et al. [31] utilized a transformer model to forecast the trend and local fluctuation information of the ageing factor, while the ageing factor was described by a Wiener process model. Peng et al. [32] emphasized the impact of health recovery phenomena on PEMFC prognostics. The entire prediction process was divided into voltage prediction with recovery identification, trend prediction, and fluctuation prediction, and the results were then obtained through a hybrid approach based on particle filter (PF) and random forest

(RF). Hybrid approaches offer enhanced accuracy compared to model-driven or data-driven approaches but face problems in terms of high complexity and computational cost.

Traditional methods for predicting PEMFC degradation often face limitations in terms of striking a balance between accuracy and complexity, particularly when dealing with complex, dynamic operating conditions. This paper aims at presenting a data-driven method with high precision and simplicity to perform PEMFC degradation prediction. Among the variants of RNNs, LSTM is the most commonly applied in PEMFC degradation prediction. However, a GRU exhibits comparable performance with a simpler structure and fewer parameters. Considering these advantages, this paper establishes a prediction framework based on GRU.

The precision of the degradation prediction is not only related to sample selection but also to the hyperparameter of the GRU network. To further improve the predictive performance, this paper introduced a grey wolf optimizer (GWO) to optimize the hyperparameters of the GRU network. Thus, this article presents a gated recurrent unit model integrated with a GWO algorithm (GWO-GRU) to predict the PEMFC degradation. The main contributions of this paper are given as follows:

(a) This paper establishes a data-driven framework based on a gated recurrent unit (GRU) network to model the PEMFC degradation process. Compared with other RNNs, GRU has a more concise architecture and can provide a similar nonlinear prediction performance.
(b) A grey wolf optimizer (GWO) is integrated with the GRU framework to automatically adjust the parameters and structure of the network, which effectively improve prediction performance and balance the precision and complexity.
(c) The performance of the proposed method was validated based on durability test data of PEMFCs under actual operating conditions, exhibiting high precision and stable generalization performance.

The remainder of this paper is organized as follows. Section 2 provides a detailed description of the PEMFC degradation data and data-processing procedure. Section 3 explains the prediction method in this article and its detailed implementation steps. Section 4 validates the performance of the presented method and provides an analysis of the prediction results. Finally, in Section 5, conclusions are drawn.

2. Data Acquisition and Preprocessing

The data utilized in this paper were acquired from the IEEE PHM 2014 Data Challenge, provided by the FCLAB Research Federation [33]. FCLAB conducted two experiments on the PEMFC test bench shown in Figure 1a, and critical data were recorded for developing a new technique in PEMFC prognostics and health management. The first experiment is conducted under constant load conditions. A 1 kW PEMFC stack consisting of five cells operated under a constant current of 70 A for more than 1100 h. The second experiment is a durability test under dynamic load conditions, and the experimental parameters are illustrated in Table 1. In the second experiment, a 70 current with 7 A high frequency ripples was applied to another identical PEMFC stack for more than 1000 h, and the data collected were utilized in this paper for model training and validation. The current curves in the two experiments are illustrated in Figure 1b.

Table 1. Experimental parameters during the second experiment.

Parameter	Value
Number of cells	5
Active area	100 cm^2
Stack rated current	70 A with 7 A oscillations
Temperature	54 °C
Hydrogen pressure	1.3 bar
Relative humidity	50%

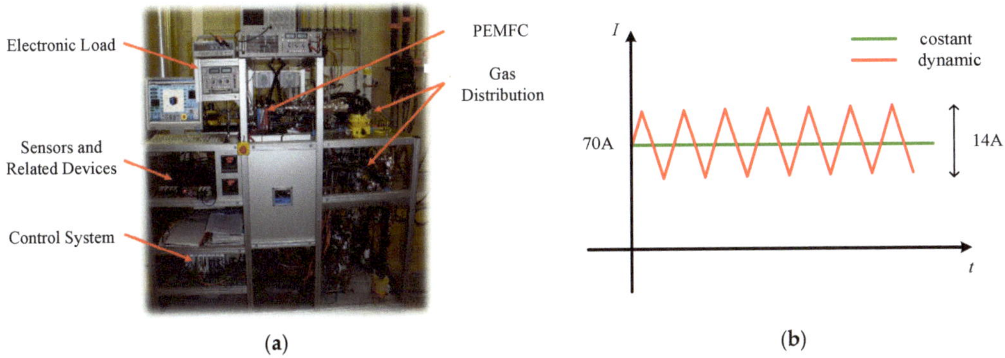

Figure 1. PEMFC durability test. (**a**) Test bench in FCLAB. (**b**) Constant and dynamic currents in the two experiments.

Some key physical parameters were recorded through the test bench, including voltage, temperature, pressure, etc. The voltage curves of the two experiments are shown in Figure 2. The initial voltage under dynamic load is slightly higher than that under constant load, which may be attributed to the different initial responses of the active load under two different conditions. Nevertheless, it does not affect the voltage degradation trend. It is obvious that the stack voltage of both PEMFC stacks decrease with increasing experimental time and that dynamic load conditions accelerate the ageing process of the fuel cell stack. Therefore, the voltage degradation can indicate the degradation process of the PEMFCs, to some extent. During the test, characterization experiments were conducted approximately every week, including the polarization curve and electrochemical impedance spectroscope (EIS) measurement. After the characterizations, the PEMFC system recovered from reversible failures, leading to noticeable voltage recovery phenomena. These recoveries significantly enhanced the nonlinear features of the voltage profiles, especially under dynamic operational conditions.

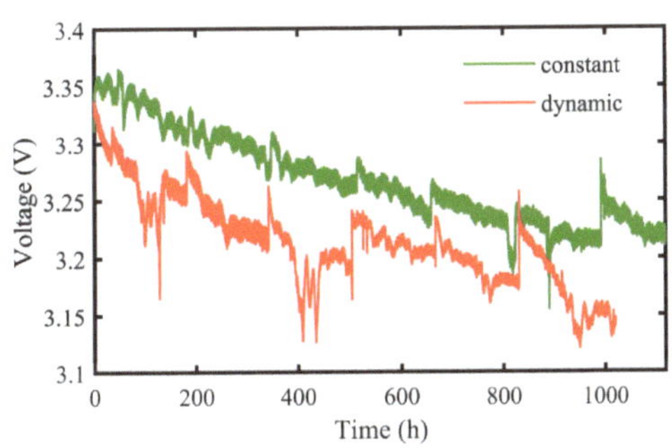

Figure 2. Degradation voltage profiles under constant and dynamic load conditions.

In the PEMFC durability test under dynamic load conditions, more than 120,000 data points are obtained, and the voltage signal shows many noises and spikes, presenting challenges in terms of accurate prediction. Therefore, the original data have to be preprocessed before being utilized in model training. The raw data are reconstructed by calculating the average values per half hour; then, the data volume is reduced to 2042.

Subsequently, a Savitzky–Golay filter is applied for data smoothing. The S-G filter utilized a $k - 1$ order polynomial to fit the voltage data:

$$v_i = \sum_{j=0}^{k-1} c_j t_j \tag{1}$$

where v_i and t_i are the voltage and time samples, respectively, and c_j is the fitting coefficient.

Extend the filter window to $n = 2m + 1$, then the above fitting process can be expressed in matrix form as:

$$V_{(2m+1)\times 1} = T_{(2m+1)\times k} \cdot C_{k\times 1} + E_{(2m+1)\times 1} \tag{2}$$

where V and T represent voltage and time, respectively. C represents the fitting coefficients and E is the fitting error.

The filtered output voltage value $\hat{V}$ can be obtained by:

$$\hat{V} = T \cdot C = T \cdot \left(T^{\mathrm{T}} \cdot T\right)^{-1} \cdot T^{\mathrm{T}} \cdot V \tag{3}$$

The raw, reconstructed, and smoothed voltage data under dynamic load conditions are demonstrated in Figure 3. After preprocessing, abnormal signals and high-frequency noise in the raw data are eliminated, thus avoiding their negative impacts on the prediction process. The processed data can still represent the original trend of the raw data, which will be utilized for model training in Section 3.

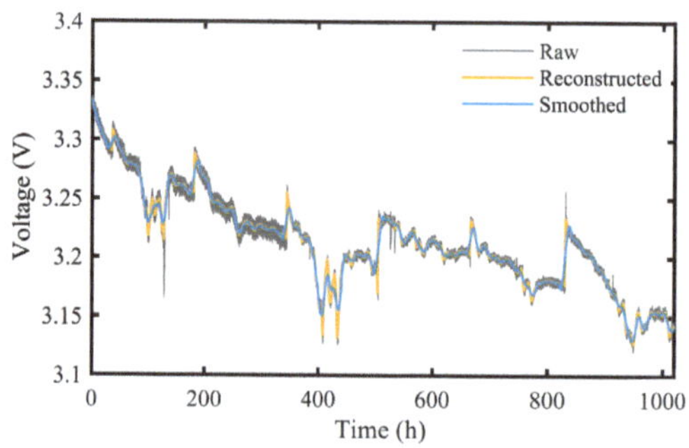

Figure 3. Raw voltage data and processed voltage data under dynamic load conditions.

3. Methodology

In this section, a data-driven framework based on gated recurrent unit (GRU) and a grey wolf optimizer (GWO) is presented to forecast the PEMFC degradation process. The proposed data-driven framework establishes a mapping relationship between time, stack temperature, historical voltage, and PEMFC degradation voltage. The voltage and temperature data utilized were obtained through the data preprocessing method described in Section 2.

By incorporating simplified gate mechanisms, the GRU method is advantageous in terms of reaching a balance between precision and simplicity. Additionally, a GWO algorithm is utilized to adjust and optimize the parameters of GRU to further enhance the predictive precision. The following contents provide a detailed description of the presented GWO-GRU approach.

Based on long short-term memory (LSTM), the GRU method further improves the structure of recurrent neural networks (RNNs). The GRU method combines the cell state and hidden state and replaces the forget gate and input gate in LSTM with a single update gate. Therefore, the training cost of the GRU model is significantly reduced compared to the LSTM model [34].

The structure of GRU cells is illustrated in Figure 4. In the structure, the update gate z_t determined whether the data should be retained or updated and reset gate r_t determined how much of the past information should be ignored. The outputs of the update gate and reset gate are formulated as follows, respectively:

$$z_t = \sigma(W_{xz}x_t + W_{hz}h_{t-1} + b_z) \tag{4}$$

$$r_t = \sigma(W_{xr}x_t + W_{hr}h_{t-1} + b_r) \tag{5}$$

where x_t represents the input at the current time step and h_{t-1} represents the hidden state from the previous time step.

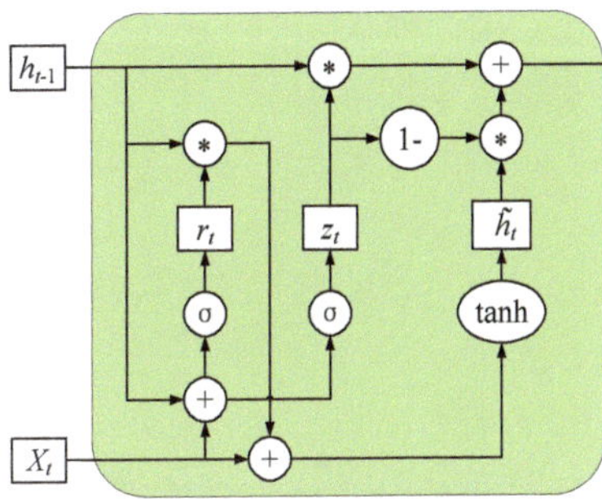

Figure 4. Architecture of GRU cells.

With the output of reset gate, the candidate state $\widetilde{h}_t$ state can be calculated by

$$\widetilde{h}_t = tanh(W_{xh}x_t + W_{hh}(r_t * h_{t-1}) + b_h) \tag{6}$$

where the $*$ the represents the multiplication of the vector elements.

Finally, the hidden state h_t can be obtained by

$$h_t = z_t * h_{t-1} + (1 - z_t) * \widetilde{h}_t \tag{7}$$

where W_{xz}, W_{hz}, W_{xr}, W_{hr}, W_{xh}, W_{hh} are the weight matrices and b_z, b_r, b_h are the corresponding bias.

The grey wolf optimizer is a metaheuristic algorithm inspired by the social hierarchy and hunting behaviour of grey wolves. In the wild, grey wolves usually act in small groups that consist of 5–12 individuals with different statuses. The wolves can be separated into four types, ranked by their leadership roles: alpha (α), beta (β), delta (Δ), and omega (ω) [35]. The alpha wolves are the leaders, guiding the pack and making decisions. The beta wolves are subordinate to the alphas and assist in leadership decisions. The delta wolves rank third in the wolf pack, and they can control omega wolves. The omega wolves are the lowest-ranking members.

The hunting process of grey wolves can be described by the following equations

$$\vec{D} = \left| \vec{C} \cdot \vec{X}_p(t) - \vec{X}(t) \right| \tag{8}$$

$$\vec{X}(t+1) = \vec{X}_p(t) - \vec{A} \cdot \vec{D} \tag{9}$$

where $\vec{X}$ indicates the wolf position and $\vec{X}_p$ indicates the target position. $\vec{D}$ represents the distance between the wolf and its prey. $\vec{A}$ and $\vec{C}$ are the coefficient vectors that control the convergence of the solution, calculated by

$$\vec{A} = 2\vec{a} \cdot \vec{r}_1 - \vec{a} \tag{10}$$

$$\vec{C} = 2 \cdot \vec{r}_2 \tag{11}$$

where $\vec{r}_1$ and $\vec{r}_2$ are random vectors in [0, 1]. $\vec{a}$ decreases from 2 to 0 linearly during iterations.

In a GWO, the wolves are considered as search individuals and the prey are considered as an optimal solution. However, unlike prey in the real world, the position of the optimal solution in abstract space is usually unknown. Therefore, the top three best solutions are considered as alpha (α), beta (β), and delta (Δ), respectively, and the other search individuals (ω) update their positions based on the positions of α, β and Δ. These processes are described by the following equations:

$$\vec{D}_\alpha = \left|\vec{C}_1 \cdot \vec{X}_\alpha - \vec{X}\right|, \vec{D}_\beta = \left|\vec{C}_2 \cdot \vec{X}_\beta - \vec{X}\right|, \vec{D}_\delta = \left|\vec{C}_3 \cdot \vec{X}_\delta - \vec{X}\right| \tag{12}$$

$$\vec{X}_1 = \vec{X}_\alpha - \vec{A}_1 \cdot (\vec{D}_\alpha), \vec{X}_2 = \vec{X}_\beta - \vec{A}_2 \cdot (\vec{D}_\beta), \vec{X}_3 = \vec{X}_\delta - \vec{A}_3 \cdot (\vec{D}_\delta) \tag{13}$$

$$\vec{X}(t+1) = \frac{\vec{X}_1 + \vec{X}_2 + \vec{X}_3}{3} \tag{14}$$

where $\vec{X}$ represents the position of the ω.

As the iteration progresses, the fluctuation range of the coefficient vector A decreases and the fluctuation range of the coefficient vector $\vec{A}$ decreases. As shown in Figure 5, when $|\vec{A}| > 1$, the search individuals tend to depart from the current optimal solution and keep searching. The attacking phase begins as the wolves close in on the prey, which means $|\vec{A}| < 1$ forces the individuals to approximate the optimal solution. In addition, the coefficient vector $\vec{C}$ effectively enhances the randomness of the algorithm and avoids local solutions.

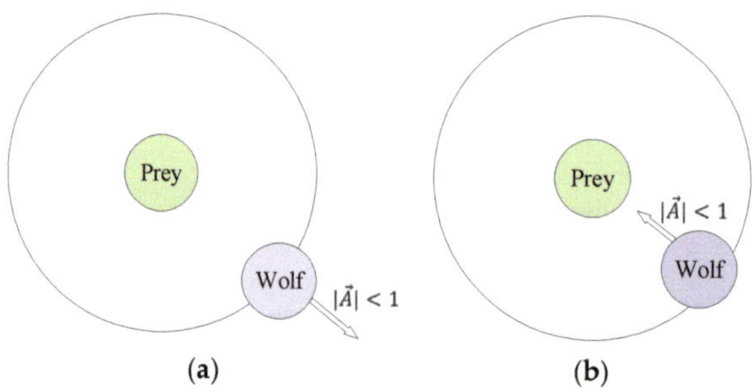

Figure 5. Searching for prey versus attacking prey, (**a**) searching, (**b**) attacking.

The weights and biases of the GRU network are random initialized and automatically updated during model iterations, which can be considered as trainable parameters. In contrast, hyperparameters are parameters that are set to fixed values before the model training process, including learning rate, batch size, number of neurons, epochs, etc. These hyperparameters largely determine the structure, learning process, and final performance of the network, especially for deep neural networks with multiple hidden layers. It is evident that adjusting the hyperparameters of GRU networks through optimization algorithms is far more efficient than through experience. In this article, the GWO is introduced into the GRU network to optimize the learning rate, the number of neurons in hidden layers and the max epochs. Based on the GWO's strong global convergence, the performance of the GRU in dynamic PEMFC prediction is significantly improved. The schematic diagram of the presented GWO-GRU method is shown in Figure 6, and a detailed optimization process is described as follows.

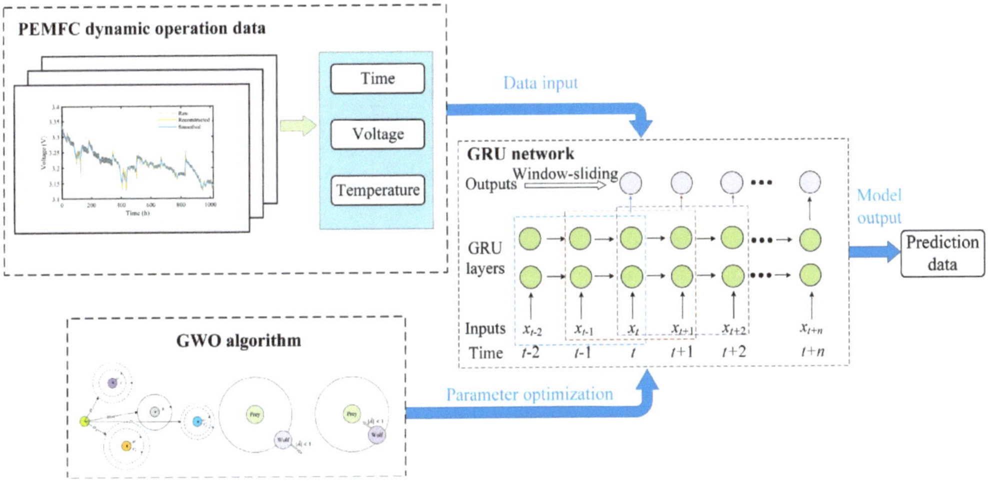

Figure 6. GWO-GRU schematic diagram.

(1) **Model initialization:**

Build the GRU network and initialize the critical parameters based on the characteristics of the PEMFC ageing data. Construct multidimensional vectors utilizing hyperparameters that require optimization.

(2) **Population initialization:**

Initialize the parameters of the GWO algorithm, including the number of wolf packs, initial positions, and maximum number of iterations.

(3) **Fitness calculation:**

Calculate the fitness of each individual by training the GRU network with the specified hyperparameters determined by current positions. The fitness is evaluated by the mean squared error (MSE).

$$fitness = \frac{1}{M}\sum_{i=1}^{M}(v_i - \hat{v}_i)^2 \tag{15}$$

where M is the number of samples. v_i and $\hat{v}_i$ represent the actual values and predicted values, respectively.

(4) **Update the population:**

Update the positions of alpha, beta and delta wolves, and then update the positions of all individuals in the wolf pack. Iterate repeatedly until the termination condition is reached.

(5) **Model establishment:**

Obtain the optimal hyperparameter vector from the GWO algorithm and apply it to the GRU framework to build the PEMFC degradation model.

4. Results

This section presents the predictive results of the PEMFC degradation process under dynamic load conditions. The prediction is multi-step, and the input and output window of the model are set to 40 h and 10 h, respectively. The training lengths are set to 50%, 60%, 70% and 80% of the entire dataset, respectively; thus, four groups of predictive results are obtained. To prove the effectiveness of the presented GWO-GRU method, the predictive results based on two standard methods are presented for comparison.

4.1. Performance Evaluation Metric

To evaluate the predictive performance, three statistical indicators, including the mean absolute percentage error (*MAPE*), the root mean square error (*RMSE*) and the coefficient of determination (R^2) are applied as evaluation metrics:

$$RMSE = \sqrt{\frac{1}{M}\sum_{t=1}^{M}(v_i - \hat{v}_i)^2} \tag{16}$$

$$MAPE = \frac{1}{M}\sum_{t=1}^{M}\left|\frac{v_i - \hat{v}_i}{v_i}\right| \tag{17}$$

$$R^2 = 1 - \frac{\sum_{i=1}^{M}(v_i - \hat{v}_i)^2}{\sum_{i=1}^{M}(v_i - \overline{v})^2} \tag{18}$$

in which M represents the amount of data, v_i and $\hat{v}_i$ are the actual values and predicted values, respectively, and $\overline{v}$ represents the average of the actual values.

Furthermore, the absolute percentage error (APE) is utilized to evaluate the variation in prediction error over time, formulated as follows

$$APE = \left|\frac{v_i - \hat{v}_i}{v_i}\right| \times 100\% \tag{19}$$

4.2. Experimental Results and Discussion

In order to validate the accuracy of the proposed GWO-GRU method, the prediction results of two additional standard methods, i.e., GRU and PSO-GRU, are provided in this section. The hyperparameters of the GRU method are manually adjusted before the training process, while the hyperparameters of the other two methods are given by optimization algorithms. Figure 7 shows the predicted PEMFC voltage degradation trend under different methods and training lengths, and Figure 8 shows the APE results. The five percent line in Figure 7 is a critical line which means a five percent drop from the initial voltage. When the voltage of PEMFC stack is lower than the critical line, the ageing fault can be declared. The detailed evaluation indicators are illustrated in Table 2.

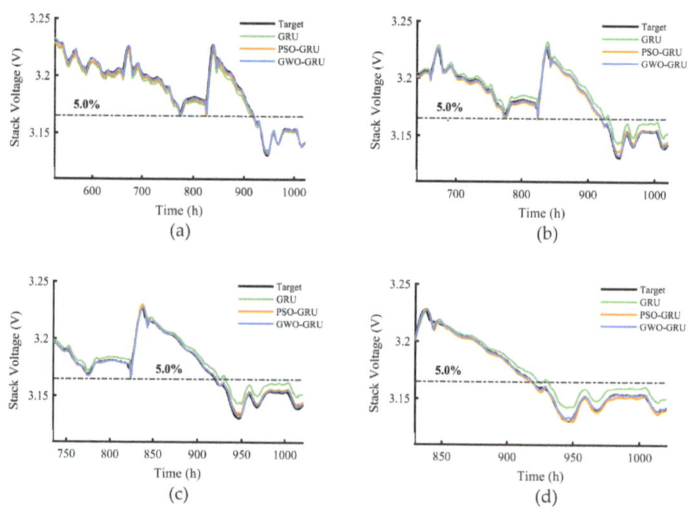

Figure 7. Prediction results under different methods and training lengths. (**a**) Fifty percent training length. (**b**) Sixty percent training length. (**c**) Seventy percent training length. (**d**) Eighty percent training length.

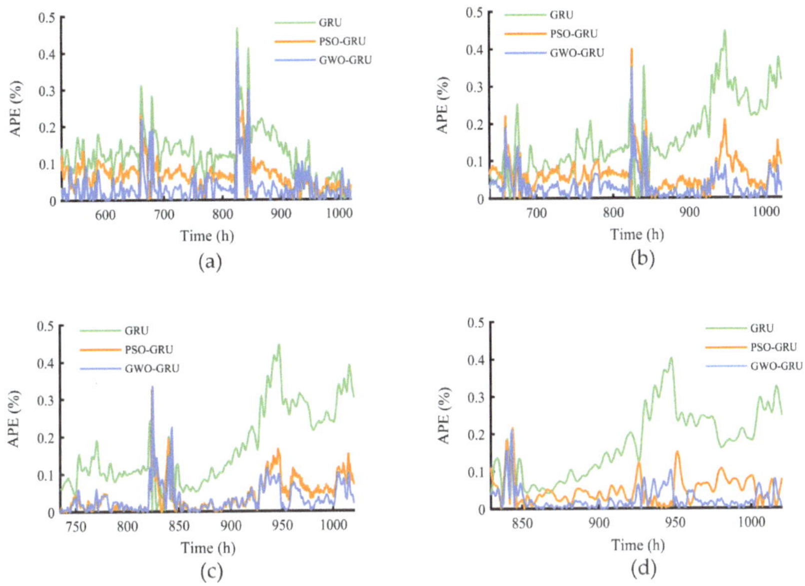

Figure 8. Absolute percentage error results under different methods and training lengths. (**a**) Fifty percent training length. (**b**) Sixty percent training length. (**c**) Seventy percent training length. (**d**) Eighty percent training length.

Table 2. Comparison of prediction accuracy under different methods and training lengths.

Training Length	Method	*RMSE*	*MAPE*	R^2
50%	GRU	0.0044654	0.12139%	0.97088
	PSO-GRU	0.0032765	0.071766%	0.98432
	GWO-GRU	0.0020784	0.040252%	0.99369
60%	GRU	0.0057774	0.15616%	0.94509
	PSO-GRU	0.002468	0.06466%	0.98998
	GWO-GRU	0.0016032	0.034031%	0.99577
70%	GRU	0.0061923	0.16631%	0.93256
	PSO-GRU	0.0023485	0.046628%	0.9903
	GWO-GRU	0.0017118	0.035812%	0.99485
80%	GRU	0.0064539	0.177%	0.9455
	PSO-GRU	0.0028158	0.058697%	0.98962
	GWO-GRU	0.0020843	0.034444%	0.99432

The comparison between the standard GRU, PSO-GRU, and GWO-GRU methods demonstrates a clear improvement in prediction accuracy when optimization algorithms are applied. Additionally, across all training lengths, the predicted values of the presented GWO-GRU method are generally closer to the true values. For example, with 50% of the data used for training, GWO-GRU achieves an *RMSE* of 0.0020784, which is significantly lower than the 0.0044654 of the standard GRU model and even surpasses the performance of PSO-GRU (*RMSE* of 0.0032765). The same pattern can also be observed from the other two evaluation indicators, *MAPE* and R^2.

It can be found in Figure 8 that the presented method not only has the minimal relative error but also the minimal error fluctuations, exhibiting high stability and accuracy. The data show that the GWO algorithm leads to a better configuration of the GRU network compared to the PSO algorithm and shows a stronger effectiveness in PEMFC degradation prediction.

The superior performance of the presented method can be attributed to the GWO's stronger global search capabilities, allowing the model to better capture the complex, nonlinear degradation patterns in PEMFC systems. In Table 2, the evaluation indicators shows that the presented method has reduced the prediction error by about 60–70% compared to the standard GRU model and by about 12–35% compared to the PSO-GRU model.

Upon analyzing the provided data in Table 2, it becomes evident that increasing the training length does not consistently lead to improvements in prediction accuracy, regardless of the method used. In fact, the results suggest that adding more training data can, in some cases, slightly degrade performance. For example, in the GRU model, the RMSE increases from 0.0044654 at 50% training length to 0.0064539 at 80%, while GWO-GRU shows a similar trend, with RMSE increasing from 0.0016032 at 60% training length to 0.0020843 at 80%.

This counterintuitive effect is likely due to the inherent characteristics of PEMFC degradation data. In the early stages of operation, PEMFC degradation follows a more predictable pattern, making it easier for models to learn from relatively smaller datasets. However, as degradation progresses and becomes more complex—due to factors like membrane thinning, catalyst degradation, and water management issues—the prediction task becomes more challenging. As training length increases, the GRU networks become more prone to overfitting, especially if the data in the later stages include more nonlinear features. Nevertheless, adjusting the network configuration through the GWO algorithm can significantly reduce the possibility of overfitting and improve the accuracy of degradation prediction.

In summary, the presented GWO-GRU method exhibits better performance than other standard methods in PEMFC degradation prediction and presents stable prediction results under different sizes of training samples, achieving a balance between accuracy and complexity.

4.3. Comparison with Other Methods in the Literature

To further validate the effectiveness of the presented GWO-GRU method, the error results in Table 2 are compared with some existing methods in recent studies. The comparison results are listed in Table 3. It is evident that the proposed method shows the best performance in Table 3, which further proves its superiority.

Table 3. Comparison with other methods in the literature.

Method	Train Length	*RMSE*	*MAPE*
FDKF [16]	50%	0.0325	0.8545%
	60%	0.0090	0.3664%
	70%	0.0195	0.4595%
LSTM RNN [36]	60%	0.0058	0.17%
	70%	0.0054	0.14%
	80%	0.0062	0.15%
VAE-DGP [37]	450 h (43.9%)	0.02549	0.160%
	600 h (58.8%)	0.02780	0.186%
Hybrid [32]	515 h (50.5%)	0.0275	0.657%
	615 h (60.3%)	0.0248	0.618%
	715 h (70.1%)	0.0266	0.694%
Presented method	50%	0.00209	0.040%
	60%	0.00160	0.034%
	70%	0.00171	0.036%

5. Conclusions

In this paper, a novel approach integrating the GRU neural network with a grey wolf optimizer (GWO) was proposed to predict PEMFC degradation process. The GWO-GRU method successfully addresses key challenges in accurately predicting PEMFC degradation, particularly under dynamic operational conditions. By adjusting hyperparameters of the network automatically through GWO, the method effectively balances prediction precision and complexity, demonstrating a significant improvement over standard methods.

The results obtained from validation utilizing a PEMFC durability test data show the superior performance of the GWO-GRU approach. The model reduced prediction error by approximately 60–70% compared to the standard GRU method and by 12–35% compared to the PSO-GRU method, showing its robustness and accuracy across various training data sizes. Additionally, the analysis revealed that increasing the training length did not consistently improve prediction accuracy, suggesting the importance of optimizing the model structure to avoid overfitting, especially in later degradation stages.

The presented GWO-GRU method shows strong potential for practical applications in PEMFC prognostics, where accurate and real-time predictions are essential for optimizing maintenance schedules and ensuring fuel cell system reliability. Future work will explore extending the method to more complex operational scenarios and further improving its scalability for larger datasets.

Author Contributions: Conceptualization, X.W. and H.Y.; methodology, B.C.; software, Y.D.; validation, Z.H., H.L. and X.Z.; formal analysis, D.Z.; investigation, X.W.; resources, D.Z.; data curation, B.C.; writing—original draft preparation, X.W. and Z.H.; writing—review and editing, C.W. and Y.C.; visualization, Y.D.; supervision, Y.D.; project administration, D.Z.; funding acquisition, Y.C. All authors have read and agreed to the published version of the manuscript.

Funding: This research was supported by the Science and Technology Project of the Huizhou Power Supply Bureau, Guangdong Power Grid Corporation. (project number: 031300KC23030008, contract number: 0313002023030103B200031).

Data Availability Statement: The original contributions presented in the study are included in the article; further inquiries can be directed to the corresponding author.

Conflicts of Interest: Authors Xiangdong Wang, Zerong Huang, Daxing Zhang, Haoyu Yuan, Bingzi Cai, Hanlin Liu were employed by the Guangdong Power Grid Corporation. The remaining authors declare that the research was conducted in the absence of any commercial or financial relationships that could be construed as potential conflicts of interest.

References

1. Yue, M.; Jemei, S.; Gouriveau, R.; Zerhouni, N. Review on Health-Conscious Energy Management Strategies for Fuel Cell Hybrid Electric Vehicles: Degradation Models and Strategies. *Int. J. Hydrogen Energy* **2019**, *44*, 6844–6861. [CrossRef]
2. Olabi, A.G.; Wilberforce, T.; Abdelkareem, M.A. Fuel Cell Application in the Automotive Industry and Future Perspective. *Energy* **2021**, *214*, 118955. [CrossRef]
3. Lin, R.-H.; Xi, X.-N.; Wang, P.-N.; Wu, B.-D.; Tian, S.-M. Review on Hydrogen Fuel Cell Condition Monitoring and Prediction Methods. *Int. J. Hydrogen Energy* **2019**, *44*, 5488–5498. [CrossRef]
4. Mei, J.; Meng, X.; Tang, X.; Li, H.; Hasanien, H.; Alharbi, M.; Dong, Z.; Shen, J.; Sun, C.; Fan, F.; et al. An Accurate Parameter Estimation Method of the Voltage Model for Proton Exchange Membrane Fuel Cells. *Energies* **2024**, *17*, 2917. [CrossRef]
5. Wang, T.; Li, Q.; Qiu, Y.; Chen, W.; Breaz, E.; Ravey, A.; Gao, F. Power Optimization Distribution Method for Fuel Cell System Cluster Comprehensively Considering System Economy. *IEEE Trans. Ind. Electron.* **2022**, *69*, 12898–12911. [CrossRef]
6. Sun, X.; Xie, M.; Fu, J.; Zhou, F.; Liu, J. An Improved Neural Network Model for Predicting the Remaining Useful Life of Proton Exchange Membrane Fuel Cells. *Int. J. Hydrogen Energy* **2023**, *48*, 25499–25511. [CrossRef]
7. Liu, J.; Li, Q.; Han, Y.; Zhang, G.; Meng, X.; Yu, J.; Chen, W. PEMFC Residual Life Prediction Using Sparse Autoencoder-Based Deep Neural Network. *IEEE Trans. Transp. Electrif.* **2019**, *5*, 1279–1293. [CrossRef]
8. Li, X.; Ye, T.; Meng, X.; He, D.; Li, L.; Song, K.; Jiang, J.; Sun, C. Advances in the Application of Sulfonated Poly(Ether Ether Ketone) (SPEEK) and Its Organic Composite Membranes for Proton Exchange Membrane Fuel Cells (PEMFCs). *Polymers* **2024**, *16*, 2840. [CrossRef]
9. He, K.; Zhang, C.; He, Q.; Wu, Q.; Jackson, L.; Mao, L. Effectiveness of PEMFC Historical State and Operating Mode in PEMFC Prognosis. *Int. J. Hydrogen Energy* **2020**, *45*, 32355–32366. [CrossRef]
10. Chen, J.; Zhou, D.; Lyu, C.; Lu, C. A Novel Health Indicator for PEMFC State of Health Estimation and Remaining Useful Life Prediction. *Int. J. Hydrogen Energy* **2017**, *42*, 20230–20238. [CrossRef]
11. He, W.; Liu, T.; Ming, W.; Li, Z.; Du, J.; Li, X.; Guo, X.; Sun, P. Progress in Prediction of Remaining Useful Life of Hydrogen Fuel Cells Based on Deep Learning. *Renew. Sustain. Energy Rev.* **2024**, *192*, 114193. [CrossRef]

12. Futter, G.A.; Latz, A.; Jahnke, T. Physical Modeling of Chemical Membrane Degradation in Polymer Electrolyte Membrane Fuel Cells: Influence of Pressure, Relative Humidity and Cell Voltage. *J. Power Sources* **2019**, *410–411*, 78–90. [CrossRef]
13. Zhang, X.; Yang, D.; Luo, M.; Dong, Z. Load Profile Based Empirical Model for the Lifetime Prediction of an Automotive PEM Fuel Cell. *Int. J. Hydrogen Energy* **2017**, *42*, 11868–11878. [CrossRef]
14. Pei, P.; Chen, D.; Wu, Z.; Ren, P. Nonlinear Methods for Evaluating and Online Predicting the Lifetime of Fuel Cells. *Appl. Energy* **2019**, *254*, 113730. [CrossRef]
15. Wang, P.; Liu, H.; Chen, J.; Qin, X.; Lehnert, W.; Shao, Z.; Li, R. A Novel Degradation Model of Proton Exchange Membrane Fuel Cells for State of Health Estimation and Prognostics. *Int. J. Hydrogen Energy* **2021**, *46*, 31353–31361. [CrossRef]
16. Ao, Y.; Laghrouche, S.; Depernet, D.; Chen, K. Proton Exchange Membrane Fuel Cell Prognosis Based on Frequency-Domain Kalman Filter. *IEEE Trans. Transp. Electrif.* **2021**, *7*, 2332–2343. [CrossRef]
17. Wang, D.; Min, H.; Zhao, H.; Sun, W.; Zeng, B.; Ma, Q. A Data-Driven Prediction Method for Proton Exchange Membrane Fuel Cell Degradation. *Energies* **2024**, *17*, 968. [CrossRef]
18. Li, H.; Zhao, W.; Zhang, Y.; Zio, E. Remaining Useful Life Prediction Using Multi-Scale Deep Convolutional Neural Network. *Appl. Soft Comput.* **2020**, *89*, 106113. [CrossRef]
19. Benaggoune, K.; Yue, M.; Jemei, S.; Zerhouni, N. A Data-Driven Method for Multi-Step-Ahead Prediction and Long-Term Prognostics of Proton Exchange Membrane Fuel Cell. *Appl. Energy* **2022**, *313*, 118835. [CrossRef]
20. Tao, Z.; Zhang, C.; Xiong, J.; Hu, H.; Ji, J.; Peng, T.; Nazir, M.S. Evolutionary Gate Recurrent Unit Coupling Convolutional Neural Network and Improved Manta Ray Foraging Optimization Algorithm for Performance Degradation Prediction of PEMFC. *Appl. Energy* **2023**, *336*, 120821. [CrossRef]
21. Wang, F.-K.; Cheng, X.-B.; Hsiao, K.-C. Stacked Long Short-Term Memory Model for Proton Exchange Membrane Fuel Cell Systems Degradation. *J. Power Sources* **2020**, *448*, 227591. [CrossRef]
22. Yu, Y.; Yu, Q.; Luo, R.; Chen, S.; Yang, J.; Yan, F. Deep Learning with Dual-Stage Attention Mechanism for Interpretable Prediction of Proton Exchange Membrane Fuel Cell Performance Degradation. *Int. J. Hydrogen Energy* **2024**, *58*, 902–911. [CrossRef]
23. Li, C.; Lin, W.; Wu, H.; Li, Y.; Zhu, W.; Xie, C.; Gooi, H.B.; Zhao, B.; Zhang, L. Performance Degradation Decomposition-Ensemble Prediction of PEMFC Using CEEMDAN and Dual Data-Driven Model. *Renew. Energy* **2023**, *215*, 118913. [CrossRef]
24. Hua, Z.; Zheng, Z.; Péra, M.-C.; Gao, F. Remaining Useful Life Prediction of PEMFC Systems Based on the Multi-Input Echo State Network. *Appl. Energy* **2020**, *265*, 114791. [CrossRef]
25. Deng, Z.; Chen, Q.; Zhang, L.; Zhou, K.; Zong, Y.; Liu, H.; Li, J.; Ma, L. Degradation Prediction of PEMFCs Using Stacked Echo State Network Based on Genetic Algorithm Optimization. *IEEE Trans. Transp. Electrif.* **2022**, *8*, 1454–1466. [CrossRef]
26. Zuo, J.; Lv, H.; Zhou, D.; Xue, Q.; Jin, L.; Zhou, W.; Yang, D.; Zhang, C. Deep Learning Based Prognostic Framework towards Proton Exchange Membrane Fuel Cell for Automotive Application. *Appl. Energy* **2021**, *281*, 115937. [CrossRef]
27. Lv, J.; Yu, Z.; Zhang, H.; Sun, G.; Muhl, P.; Liu, J. Transformer Based Long-Term Prognostics for Dynamic Operating PEM Fuel Cells. *IEEE Trans. Transp. Electrif.* **2024**, *10*, 1747–1757. [CrossRef]
28. Meng, X.; Mei, J.; Tang, X.; Jiang, J.; Sun, C.; Song, K. The Degradation Prediction of Proton Exchange Membrane Fuel Cell Performance Based on a Transformer Model. *Energies* **2024**, *17*, 3050. [CrossRef]
29. Wu, Y.; Breaz, E.; Gao, F.; Paire, D.; Miraoui, A. Nonlinear Performance Degradation Prediction of Proton Exchange Membrane Fuel Cells Using Relevance Vector Machine. *IEEE Trans. Energy Convers.* **2016**, *31*, 1570–1582. [CrossRef]
30. Ma, R.; Xie, R.; Huangfu, Y. A Hybrid Prognostic Method for PEMFC With Aging Parameter Prediction. *IEEE Trans. Transp. Electrif.* **2021**, *7*, 2318–2331. [CrossRef]
31. Hu, Y.; Zhang, L.; Jiang, Y.; Peng, K.; Jin, Z. A Hybrid Method for Performance Degradation Probability Prediction of Proton Exchange Membrane Fuel Cell. *Membranes* **2023**, *13*, 426. [CrossRef] [PubMed]
32. Peng, W.; Wei, Z.; Huang, C.-G.; Feng, G.; Li, J. A Hybrid Health Prognostics Method for Proton Exchange Membrane Fuel Cells With Internal Health Recovery. *IEEE Trans. Transp. Electrif.* **2023**, *9*, 4406–4417. [CrossRef]
33. Gouriveau, R.; Hilairet, M.; Hissel, D.; Jemei, S.; Jouin, M.; Lechartier, E.; Morando, S.; Pahon, E.; Pera, M.C.; Zerhouni, N. IEEE PHM 2014 Data Challenge: Outline, Experiments, Scoring of Results, Winners. In Proceedings of the IEEE Conference on Prognostic and Health Management, Cheney, WA, USA, 22–25 June 2014; pp. 1–6.
34. Liu, L.; Sun, W.; Yue, C.; Zhu, Y.; Xia, W. Remaining Useful Life Estimation of Lithium-Ion Batteries Based on Small Sample Models. *Energies* **2024**, *17*, 4932. [CrossRef]
35. Mirjalili, S.; Mirjalili, S.M.; Lewis, A. Grey Wolf Optimizer. *Adv. Eng. Softw.* **2014**, *69*, 46–61. [CrossRef]
36. Ma, R.; Li, Z.; Breaz, E.; Liu, C.; Bai, H.; Gao, F. Data-Fusion Prognostics of Proton Exchange Membrane Fuel Cell Degradation. *IEEE Trans. Ind. Appl.* **2019**, *55*, 4321–4331. [CrossRef]
37. Deng, H.; Hu, W.; Cao, D.; Chen, W.; Huang, Q.; Chen, Z.; Blaabjerg, F. Degradation Trajectories Prognosis for PEM Fuel Cell Systems Based on Gaussian Process Regression. *Energy* **2022**, *244*, 122569. [CrossRef]

energies

Article

Data-Driven-Based Full Recovery Technology and System for Transformer Insulating Oil

Feng Chen [1], Li Wang [1], Zhiyao Zheng [2], Bin Pan [1], Yujia Hu [2,*] and Kexin Zhang [3,4,*]

1. Materials Branch, State Grid Zhejiang Electric Power Co., Ltd., Hangzhou 310015, China
2. Zhejiang Huadian Equipment Testing and Research Insitute Co., Ltd., Hangzhou 310015, China
3. Huzhou Institute of Zhejiang University, Huzhou 313000, China
4. Institute of Cyber-Systems and Control, Zhejiang University, Hangzhou 310027, China

* Correspondence: xjtuhyj@stu.xjtu.edu.cn (Y.H.); zhangkexin@zju.edu.cn (K.Z.)

Abstract: This study aims to develop an efficient recovery solution for waste transformer insulating oil, addressing the challenge of incomplete separation of residual oil in existing recovery technologies. A multi-module integrated system is constructed, comprising a waste oil extraction module, a residual oil vaporization module, an exhaust gas treatment module, and an online monitoring module. By combining steps such as oil extraction, residual oil absorption, hot air circulation heating, and negative-pressure low-frequency induction heating, the complete recovery of waste oil is achieved. The recovery process incorporates oil–gas saturation monitoring and an oil–gas precipitation assessment algorithm based on neural networks to enable intelligent control, ensuring thorough recovery of residual oil from transformers. The proposed system and methods demonstrate excellent recovery efficiency and environmental protection effects during the pre-treatment of waste transformer oil. Experiments conducted on 50 discarded transformers showed an average recovery efficiency exceeding 99%, with 49 transformers exhibiting no damage to core components after the recovery process. From a theoretical perspective, this research introduces monitoring and control methods for transformer insulating oil recovery, providing significant support for the green processing and reutilization of discarded transformer insulating oil. From an application value perspective, the recovery process helps reduce environmental pollution and facilitates the disassembly of transformers. This enables better analysis of transformer operating characteristics, thereby enhancing the reliability and safety of power systems.

Keywords: transformer insulating oil recovery; data-driven monitoring; thermal cycle of insulating oil; recovery rate

Citation: Chen, F.; Wang, L.; Zheng, Z.; Pan, B.; Hu, Y.; Zhang, K. Data-Driven-Based Full Recovery Technology and System for Transformer Insulating Oil. *Energies* **2024**, *17*, 6345. https://doi.org/10.3390/en17246345

Academic Editor: Xiongbo Wan

Received: 21 November 2024
Revised: 12 December 2024
Accepted: 13 December 2024
Published: 17 December 2024

1. Introduction

The purification of waste oil is one of the necessary stages in the post-treatment of many devices, such as transformer and hydraulic [1–3]. Furthermore, thermal oil recovery is also used in the drilling and fracturing process of oil fields [4,5]. In the power industry, a transformer is a critical node connecting the transmission network with end-users, ensuring efficient transmission and safe use of electrical energy through voltage level conversion. Mineral insulating oil is the primary insulating and cooling medium for oil-immersed transformers, composed of hydrocarbons, naphthenes, aromatics, and paraffin compounds [6]. Throughout the transformer's lifecycle, insulating oil is subjected to mechanical, electrical, and thermal stresses, leading to decomposition, oxidation, and other complex chemical reactions that degrade key indicators such as breakdown voltage and dielectric loss factor [7]. When transformer oil no longer meets the operational requirements of the power system, it needs to be recovered and reused. However, this process is challenging because the prolonged degradation of transformer oil produces many toxic and harmful substances, such as Polychlorinated Biphenyls (PCBs) [8], Polycyclic Aromatic Hydrocarbons (PAHs), hazardous gases, and heavy metals, which can have adverse environmental impacts if disposed of improperly [9–11]. With the expansion of power systems and the increase in transformer installations, the recycling and reuse of waste transformer oil (WTO) has become a crucial issue in the field of environmental protection [12].

Currently, there are two main methods for transformer oil recovery. One is the drainage method, where the oil is pumped out of the oil tank and various types of oil are collected in the same pool. The other is the drying method, in which the transformer body is lifted out and placed in a drying chamber to be dried or air-dried by natural wind. The first method recovers no more than 80% of the oil, and it is particularly difficult to extract oil that has been in contact with the inner walls of the tank, windings, core, and insulation materials. Moreover, during subsequent processes such as core lifting, residual oil can remain on the factory floor, on the workers, etc., causing pollution, further impacting the workers' health, and contributing to land pollution. The second method is energy-intensive because there are many transformer models, and the size of the drying chamber is often much larger than the transformer itself, leading to increased energy consumption. Additionally, the oil tank requires an extra drying process, making it more complex. Furthermore, there is also a significant amount of residual oil inside the oil tank that cannot be recovered, approximately 3–4 L.

The recovery process of transformer oil is mainly divided into two key steps: pre-treatment and post-treatment. Pre-treatment involves extracting degraded insulating oil from waste transformers, while post-treatment refers to reusing or harmlessly processing the degraded insulating oil. Currently, researchers have conducted extensive work on the post-treatment process, such as removing impurities from the oil using semi-permeable membranes [13], activated material [14–16], and surfactants [17–19], enhancing the dielectric strength of insulating oil with microwave-assisted methods [20], or converting waste transformer oil into an alternative fuel for energy production, suitable for low-calorific power generators, diesel engines, and more [21,22]. These post-treatment technologies significantly improve the environmental and economic value of waste transformer oil, preventing major environmental pollution issues.

However, there is currently little research on pre-treatment technologies for waste transformer oil, making it challenging to eliminate its negative environmental impacts. Due to the complex internal structure of transformers, which includes an iron core stacked with silicon steel sheets, copper wire windings, insulation materials, and more, the high-viscosity insulating oil adheres to these components' surfaces and is difficult to fully separate through conventional oil extraction methods. The residual oil poses significant risks to human health and the environment. For instance, waste transformer oil contains toxic substances like PAHs and PCBs. During transformer disassembly, workers are exposed to these residual oils, which can lead to the accumulation of toxins in their bodies and related health issues, such as skin rashes, kidney damage, or even cancer [23,24]. Additionally, as transformers are dismantled, unrecovered residual oil may enter natural environments, including the air, soil, and water sources, leading to increased levels of toxic gases like methane and carbon monoxide in the atmosphere [25]; alterations in soil pH and acidity, which raises the concentration of heavy metals absorbed by plants [26]; and contamination of millions of liters of urban groundwater, rendering it undrinkable [27]. Therefore, achieving full recovery of waste transformer oil at the pre-treatment stage, completely separating and extracting the residual oil from the internal structures of transformers, is also critically important.

In summary, this study developed a full recovery system for insulating oil in waste transformers, achieving complete recovery of waste transformer oil through key steps, including oil extraction and residual oil absorption, hot air vaporization and circulation heating, negative pressure extraction of oil–gas mixtures, and low-frequency induction heating under negative pressure. Additionally, the XGBoost model was used for real-time monitoring of oil–gas saturation, and in combination with a Transformer model for output power prediction of the low-frequency induction heating device, intelligent monitoring and management of the recovery process were achieved. In experiments involving the recovery of insulating oil from 50 waste transformers, the overall residual oil rate was less than 1%, and no damage to the core components was observed in 49 of the transformers after recovery. Compared with existing methods, the method proposed in this article leverages the sealing properties of the transformer body itself, allowing the entire recovery process to take place in a sealed environment to prevent any oil leakage. Our system also enables the collection of insulation oil from the same model of transformer in separate boxes and barrels. The transformer body is closely integrated with the internal components, which reduces the energy consumption for drying and allows the internal components and oil tank to be dried simultaneously.

The major contributions are as follows:

- This paper first proposes a system for achieving complete recovery of waste transformer oil. The system includes a waste oil extraction module, a residual oil vaporization module, an exhaust gas treatment module, and an online monitoring module.
- This paper proposes a data-driven oil and gas saturation inference model and a low-frequency induction heating device output power prediction model, enabling precise monitoring of the transformer oil recovery process.
- This paper conducted extensive experimental validation on numerous real waste transformers, demonstrating the superiority of the proposed system and methods in recovery performance.

The rest of the article is organized as follows. Section 2 describes the insulating oil recovery system and methods. Section 3 gives the details of the data-driven models. Comprehensive experiments are implemented in Section 4. The conclusions are presented in Section 5.

2. Insulating Oil Recovery System and Method

2.1. *Overview of the Transformer Insulating Oil Recovery Method*

The full recovery process of insulating oil from decommissioned transformers is shown in Figure 1. First, oil extraction and residual oil absorption operations are performed on the decommissioned transformer. Real-time monitoring of the oil flow rate is conducted, and this step ends when the flow rate drops to zero. The insulating oil obtained from this step undergoes degassing, regeneration treatment, and performance testing, and, if qualified, can be reused. Due to the complex internal structure of transformers, such as windings and iron cores, a large amount of residual oil cannot be recovered through extraction alone. Therefore, hot air vaporization and circulation heating are further applied to fully vaporize residual oil on the surfaces of the inner tank wall, windings, and other structures.

After maintaining a stable pressure and temperature within the tank for a period, the heated oil–gas mixture is extracted. This heating and extraction cycle is repeated multiple times until the oil–gas saturation detection results remain unchanged from the previous cycle. Finally, a low-frequency induction heating method, combined with negative pressure inside the oil tank, is used to thoroughly dry residual oil from the windings and insulation materials, thus achieving complete recovery of insulating oil from decommissioned transformers.

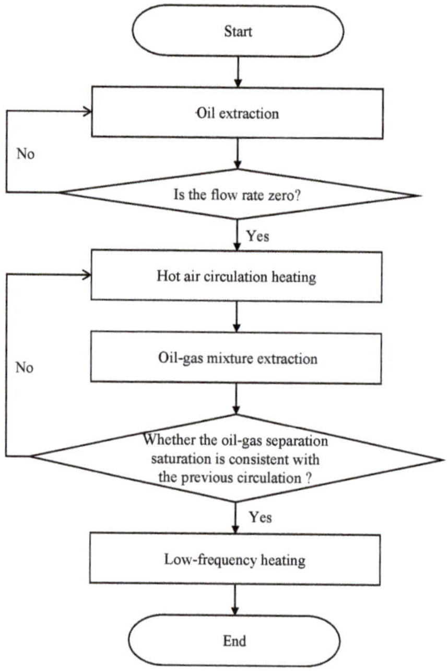

Figure 1. Flowchart of the full recovery process for insulating oil.

2.2. Transformer Insulating Oil Recovery System

This paper developed a comprehensive recovery system for waste transformer insulating oil based on the aforementioned methods. The structural schematic is shown in Figure 2, comprising the waste oil extraction module, residual oil vaporization module, exhaust gas treatment module, and online monitoring module. The integrated recovery system for waste transformer insulating oil, along with the low-frequency induction heating device, is depicted in Figure 3, with the specifications of each component listed in Table 1.

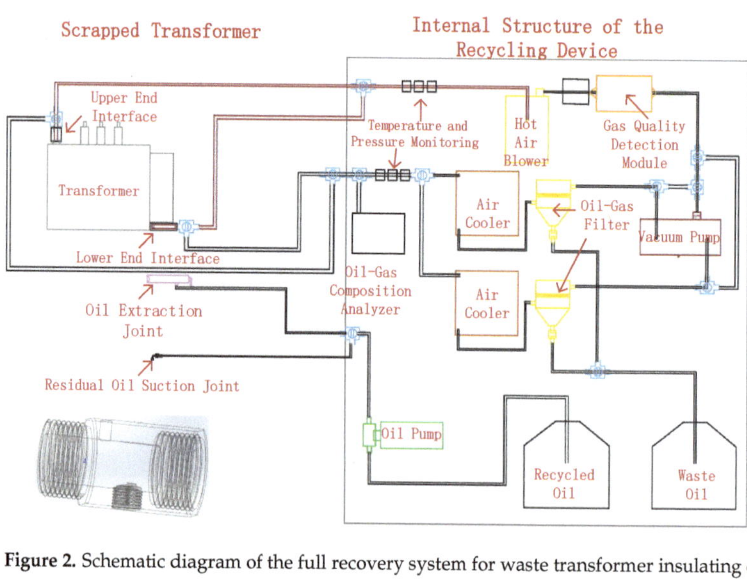

Figure 2. Schematic diagram of the full recovery system for waste transformer insulating oil.

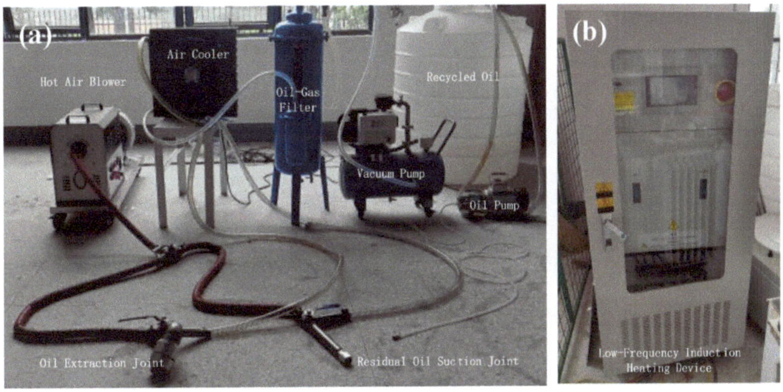

Figure 3. Full recovery system for waste transformer insulating oil and low-frequency induction heating device. (**a**) Full recovery system. (**b**) Low-frequency induction heating device.

Table 1. Parameters of the recovery system for waste transformer insulating oil.

Device	Model	Key Parameters	Manufacturer	City
Hot Air Blower	KMS-DRGF-3380-10	Heating Power: 10 kW Fan Power: 1.5 kW Maximum Air Pressure: 17 kPa Maximum Airflow: 210 m^3/h	Shanghai Meng Shang Industrial Hot Air Blower Co., Ltd.	Shanghai, China
Air Cooling Unit	3HP Air-Cooled	Power: 3 kW Flow Rate: 3 m^3/min Pressure Resistance: 1.25 MPa	Suzhou Aoubo Machinery Co., Ltd.	Suzhou, China
Oil-Gas Filter	FS-A30	Flow Rate: 3 m^3/min Pressure Resistance: 1.25 MPa	FuSheng Industrial (Shanghai) Co., Ltd.	Shanghai, China
Vacuum Pump	FKMVP-25	Power: 0.75 kW Flow Rate: 25 m^3/h Vacuum Level: −101 kPa	Zhejiang KaiMeng Mechanical Technology Co., Ltd.	Hangzhou, China
Oil Pump	WCB-75	Power: 750 W Flow Rate: 40 L/min Pressure: 0.15 MPa	Zhejiang HuanYao Machinery Co., Ltd.	Yiwu, China
Temperature Sensor	WRNT-02	Temperature Range: 0–600 °C	Jiangsu Mingyang Electrical Appliance Instrument Co., Ltd.	Xinghua, China
Pressure Sensor	PT129	Pressure Range: 0.1–1 MPa	Shanghai Yamen Automation Instrument Co., Ltd.	Shanghai, China

The waste oil extraction module includes an oil extraction connector, a residual oil suction connector, an oil pump, and an oil pump flow sensor, operating during the oil extraction and residual oil suction phases. The oil extraction connector is used for high-flow extraction of residual oil from the upper interface of the transformer. The residual oil suction connector, designed with a honeycomb column structure at the front end, ensures maximum suction efficiency for the remaining oil at the bottom. Additionally, to enhance the residual oil suction effect, the transformer oil tank can be tilted by 10–20 degrees during this phase, allowing the remaining oil at the bottom to converge toward the lower interface.

The oil pump flow sensor monitors the residual oil extraction process and signals the transition to the next phase when the flow rate reaches zero.

The residual oil vaporization module includes components such as a hot air blower, a vacuum pump, and a low-frequency induction heating device. It operates during the hot air vaporization circulation heating and negative-pressure low-frequency induction heating phases. The hot air blower and vacuum pump are connected to the upper and lower interfaces of the waste transformer via high-temperature steel-wire flexible hoses, with electromagnetic valves controlling the on–off states of each pipeline. The hot air blower initially delivers hot air into the transformer oil tank, vaporizing the residual oil adhered to the surfaces of internal structural components and the inner walls of the tank. Pressure and temperature sensors in the online monitoring module monitor the internal pressure and temperature of the oil tank, controlling the hot air blower to maintain values close to but not exceeding the rated limits (temperature not exceeding the transformer oil flash point of 135 °C, and pressure not exceeding 200 kPa). Once the pressure and temperature stabilize for 1 min, the hot air blower is turned off, and the vacuum pump is activated to extract the oil–gas mixture. The vacuum pump operates until the pressure drops below 10 kPa, after which it is turned off.

The above describes a single hot air vaporization and circulation heating cycle. The insulating oil recovery process involves performing at least two such cycles until the saturation level of the oil–gas mixture extracted in the current cycle shows no further change compared to the previous cycle. The process then transitions to the vacuum low-frequency induction heating stage, where the low-frequency induction heating device, as shown in Figure 3b, is used to heat the metal components inside the transformer. This device facilitates the extraction of residual oil from the complex structures of components such as silicon steel sheets and windings. The low-frequency induction heating device operates on the principle of electromagnetic induction and consists of components such as induction coils and a power control system. When alternating current passes through the induction coil, an alternating magnetic field is generated around it. Metal components placed within this magnetic field produce eddy currents due to electromagnetic induction. As these eddy currents flow through the metal, they generate Joule heat, which rapidly raises the temperature of the metal components. Coupled with a vacuum pump that creates a negative pressure environment inside the oil tank, the boiling point of oil vaporization decreases, enhancing the drying efficiency of the transformer's internal structures. Additionally, the vacuum environment reduces oxidation on the surfaces of the metal components, improving the quality of the recovered transformer. During this phase, the vacuum pump continuously extracts the oil–gas mixture until the saturation level of the extracted oil–gas mixture reaches zero, indicating that all residual oil has been completely recovered. At this point, the recovery process is complete.

The exhaust gas treatment module includes an air cooler, an oil–gas filter, and a gas quality detection module. After the oil–gas mixture is extracted by the vacuum pump, it contains a significant amount of pyrolysis products such as alkanes, alkenes, alkynes, sulfides, and a small amount of solid particulates. Direct discharge into the air would cause environmental pollution. Therefore, the high-temperature oil–gas mixture first passes through the air cooler, where it is cooled to condense some of the large molecular clusters. Next, the oil–gas filter removes solid particulates from the exhaust gas using fibrous filter materials. Activated carbon adsorbents are then used to absorb gaseous pollutants. Finally, the gas quality detection module ensures that the exhaust emissions meet regulatory standards before releasing them into the atmosphere. In the exhaust gas treatment module, we mainly monitor the following substances: acetylene (C_2H_2), ethylene (C_2H_4), hydrogen (H_2), carbon monoxide (CO), and hydrocarbon gases. The emission standards applied in our system are as follows: C_2H_2 content generally does not exceed 0.5 ppm, C_2H_4 content usually does not exceed 100 ppm, H_2 content typically does not exceed 1000 ppm, the CO content generally does not exceed 10 ppm, and the hydrocarbon gas content generally does not exceed 150 ppm.

The online monitoring module includes pressure sensors, temperature sensors, and an oil–gas composition analyzer, which monitor the internal pressure, temperature, and oil–gas saturation of the transformer oil tank throughout the recovery process. This module controls the output power of devices such as the low-frequency induction heating system, hot air blower, and vacuum pump. The goal is to ensure the efficient and high-quality recovery of residual oil without damaging the oil tank or internal components. The installation locations of the pressure and temperature sensors are shown in Figure 4a. Both sensors are fixed via threads onto specially designed metal connectors. These connectors, as part of the oil–gas mixture circulation loop, are threaded onto the upper and lower interfaces of the waste transformer, enabling real-time monitoring of the internal pressure and temperature of the transformer oil tank.

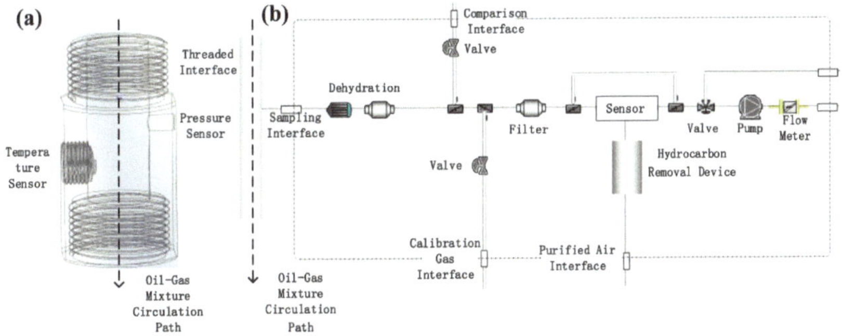

Figure 4. (**a**) Diagram of the pressure and temperature sensor installation and (**b**) structural diagram of the oil–gas analyzer.

The schematic of the internal structure of the oil–gas composition analyzer is shown in Figure 4b. Gas samples are extracted from the oil–gas mixture circulation path via a sampling interface, then processed through a dehumidifier and filters. The analyzer detects oil–gas macromolecules as well as the concentrations of gases such as methane, ethane, acetylene, ethylene, and carbon monoxide using sensors. Based on a neural network-driven oil–gas saturation analysis, the system monitors the recovery status of the residual oil within the transformer and adjusts the output power of the low-frequency induction heating system accordingly. This prevents overheating, which could cause damage to components such as windings and silicon steel sheets, and avoids adverse impacts on the dismantling efficiency, as well as the economic and environmental value of the recovery process. Exhaust gas analyzed by the sensors is purified through a hydrocarbon removal device before being discharged into the atmosphere.

3. Analysis of Oil and Gas Saturation and Control of the Low-Frequency Induction Heating Device

3.1. Oil and Gas Saturation Evaluation Algorithm Based on the XGBoost Model

The oil and gas composition detector is a key component of the waste transformer insulation oil recycling device, determining the time and power settings for the hot air circulation and low-frequency induction heating during system operation. Its role is to effectively control the accuracy of the heating process, avoiding two common problems: on the one hand, preventing residual transformer oil due to insufficient heating, and on the other hand, avoiding thermal damage to the internal structure of the transformer (such as windings, core, etc.) due to overheating. Therefore, the oil and gas composition detector not only directly affects the performance of the recycling device but also relates to the stability and safety of the entire recycling process.

The oil and gas composition detector monitors the various oil and gas components produced during the heating and cracking process of transformer oil, including the content of large molecular groups of oil and gas, methane, ethane, acetylene, ethylene, carbon monoxide, etc., in real time. Combined with the oil and gas saturation evaluation module based on convolutional neural networks, the oil and gas composition detector can instantly assess the state of oil and gas release, thereby accurately controlling the heating process.

The oil and gas saturation evaluation algorithm based on the XGBoost model captures the complex nonlinear relationship between different gas concentrations (such as methane, ethane, acetylene, ethylene, carbon monoxide, etc.) and oil and gas saturation to achieve high-precision saturation prediction. The model's input is the concentration values of various gases monitored, and the output is the predicted value of oil and gas saturation. To ensure the accuracy and robustness of the model, a comprehensive dataset must first be prepared, ensuring that the dataset includes the concentration values of various gases as features, while the oil and gas saturation is used as the target label. To ensure the model's generalization ability, the dataset is divided into training and testing sets to evaluate the model's performance. In addition, in data preprocessing, cleaning of missing or abnormal values is performed, such as filling missing values with mean or median values, and removing or smoothing extreme or abnormal values to ensure data quality and stability of model input.

Specifically, in this paper, we collect oil–gas saturation data in the following way: the used transformer is placed in a sealed state for heating. First, it is kept sealed for 30 min, and after 30 min, the air is drawn out using an exhaust fan, and data are collected. The total gas volume inside the oil tank is 100 L, and the exhaust fan's air flow rate is 10 L per second. The data at 0 s are considered as 100% oil–gas saturation, and the data at 10 s are considered as 0% oil–gas saturation. The intermediate data are obtained through linear fitting, with a fitting dataset size of 1000. We repeat the experiment 50 times, and the final size of the training dataset is 50,000. In our method, when encountering outliers, we first remove the outliers and then fill the missing data using the average of the preceding and following data points.

3.2. Output Power Prediction Model for Low-Frequency Induction Heating Devices Based on the Transformer Model

In the output power prediction task of low-frequency induction heating devices, the transformer model can effectively process multi-dimensional time-series data and capture the complex relationships between input features. The model's inputs include the evolution trends of oil and gas saturation precipitation, multi-point temperature evolution trends of the transformer, and the evolution trends of the transformer's equivalent resistance (i.e., input voltage divided by current). These features collectively influence the heating process's efficiency and output power.

The transformer model consists of multiple encoders and decoders, making it particularly suitable for processing sequence data. Its self-attention mechanism can capture long-range dependencies. First, the input features are standardized and embedded, forming vector representations of the same dimension for easier processing by the model. Through the self-attention layers, the model can establish associations between different time steps, thereby recognizing temporal relationships between features. For example, it can understand how temperature changes at one moment affect subsequent power output. The self-attention mechanism can be expressed as:

$$\text{Attention}(Q, K, V) = \text{softmax}(\frac{QK^T}{\sqrt{d_k}}V) \tag{1}$$

where Q, K, and V represent the query, key, and value matrices, respectively, and d_k is the dimension of the key, which is used for scaling to prevent excessively large values.

Subsequently, the output of the self-attention layer undergoes a nonlinear transformation through a feedforward neural network to further extract features and enhance the model's expressive capability. The computation of the feedforward network is typically represented as:

$$FFN(x) = \max(0, xW_1 + b_1)W_2 + b_2. \tag{2}$$

At the same time, the transformer model introduces residual connections and layer normalization to address the vanishing gradient problem in deep network training, ensuring smoother information flow through the network. After passing through multiple encoder layers, the model outputs the predicted power value for the low-frequency induction heating device. This process can be implemented through a linear regression layer, mapping the high-dimensional features to the output power value. During the training phase, the model fits using historical data and evaluates the difference between the predicted values and the true power values using loss functions such as Mean Squared Error (MSE). The loss function is represented as:

$$\text{MSE} = \frac{1}{n}\sum_{i=1}^{n}(y_i - \hat{y}_i)^2. \tag{3}$$

The model parameters are adjusted through optimization algorithms (such as Adam) to improve prediction accuracy. During the evaluation phase, metrics like MSE and Root Mean Squared Error (RMSE) can be used to assess the model's performance on the test set, ensuring its generalization ability.

Using the Transformer model for output power prediction of low-frequency induction heating devices can effectively capture the complex temporal relationships between oil and gas saturation, transformer temperature, and equivalent resistance. Its powerful self-attention mechanism and flexible architecture enable the model to demonstrate superior performance when handling multi-dimensional time-series data, providing strong support for the optimization of the low-frequency induction heating process.

4. Experiments

4.1. Parameters of the Waste Transformer

This study conducts an insulating oil full recovery test on a 10/0.4 kV, 400 kVA waste transformer, with its physical appearance shown in Figure 5a. The exterior of the waste transformer is intact, with no signs of destructive damage or oil leakage, meeting the requirements for the full oil recovery test. The test site is shown in Figure 5b. The detailed factory parameters of the waste transformer are listed in Table 2.

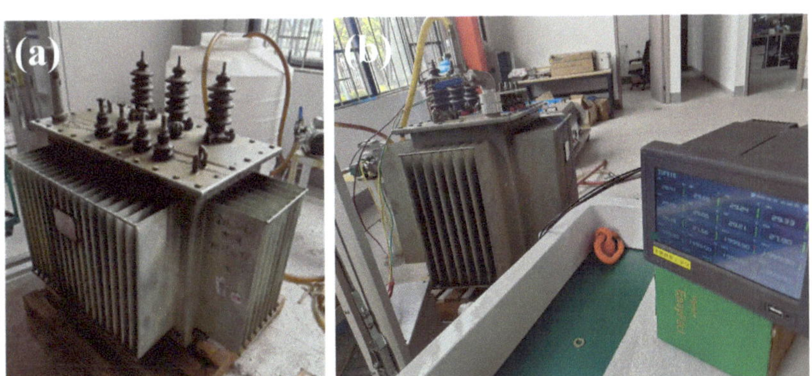

Figure 5. The waste transformer and the full oil recovery test. (**a**) The waste transformation. (**b**) The full oil recovery test.

Table 2. Parameters of the waste transformer.

Parameter	Value
Model	10/0.4–400
Rated Capacity (kVA)	400
High Voltage (kV)	10
Low Voltage (kV)	0.4
No-load Loss (kW)	0.8
Load Loss (kW)	4.3
No-load Current (%)	1
Impedance voltage (%)	4
Oil weight (kg)	275
Core and winding weight (kg)	905
Total weight (kg)	1465
Length (mm)	1545
Width (mm)	1120
Height (mm)	1558

4.2. Electrical Performance and Sealing Test of the Waste Transformer

Before conducting the full oil recovery experiment based on vacuum low-frequency induction heating, it is necessary to test the electrical performance and sealing integrity of the transformer to ensure that it meets the operational requirements of the recovery system. The DC resistance of the windings and the ground insulation of the waste transformer were measured according to the standard (GB 1094.1-2013) [28] and the results are shown in Table 3.

Table 3. Parameter estimation of the waste transformer.

Parameter	Value
Winding resistance	R_{AB} = 7.826 Ω R_{BC} = 8.035 Ω R_{CA} = 8.046 Ω R_{ab} = 6.572 mΩ R_{bc} = 6.573 mΩ R_{ca} = 6.618 mΩ R_{an} = 3.403 mΩ R_{bn} = 3.399 mΩ R_{cn} = 3.445 mΩ
Ground resistance (low voltage to high voltage and ground)	28.2 GΩ
Ground resistance (high voltage to low voltage and ground)	32.2 GΩ
Ground resistance (low voltage and high voltage to ground)	30.0 GΩ

In Table 3, R_{AB} is the winding resistance between phases A and B on the high-voltage side, R_{ab} is the winding resistance between phases a and b on the low-voltage side, and R_{an} is the winding resistance between phase a and ground on the low-voltage side. The notation for other winding resistances is the same. From Table 3, it can be seen that the winding resistance and insulation condition of each phase of the waste transformer are good. Furthermore, a sealing test was conducted on the waste transformer. A vacuum pump was connected to the upper interface, and the internal air was extracted until the pressure reached 10 kPa and maintained for 10 min. No significant fluctuations were observed in the pressure data, proving that the sealing performance is good. In conclusion, the electrical performance and sealing integrity of the waste transformer meet the experimental requirements for the transformer full oil recovery system.

4.3. Full Oil Recovery Experiment

Firstly, the degraded insulating oil inside the waste transformer's oil tank was extracted. The factory parameters indicate that the total amount of insulating oil is approximately 275 kg. The flow rate of the recovery system's oil pump is 40 L/min. The oil was extracted using both the upper interface for oil extraction and the lower interface for residual oil suction simultaneously, until the oil pump flow sensor indicated a flow rate of 0. The total time taken was 12 min.

Next, a hot air vaporization circulation heating method was used to fully vaporize the residual oil on the inner wall of the oil tank, windings, and other structures. Since the hot air blower used in the recovery system was not powerful enough to raise the internal temperature and pressure of the oil tank to the maximum values (temperature not exceeding the transformer oil flashpoint of 135 °C, pressure not exceeding 200 kPa), the hot air blower was operated at maximum power during the heating process, with a heating power of 10 kW and a fan power of 1.5 kW. After each one-minute heating cycle, the vacuum pump was started to extract the oil–gas mixture until the pressure dropped below 10 kPa, after which the vacuum pump was turned off. This cycle was repeated 13 times, with a total time of 25 min.

Finally, a low-frequency induction heating method under vacuum was used to heat the metal components inside the transformer, allowing the residual oil inside the complex structures of the silicon steel sheets, windings, and other components to be extracted, achieving full recovery of the degraded insulating oil. The voltage and current waveforms input to the low-frequency induction heating device are shown in Figure 6. The rapid decrease in current, voltage, and power after 60 min is due to the winding temperature approaching the maximum temperature set in our experiment. If high power is maintained further, there is a risk of damaging the transformer windings. At the same time, after the power drops to a lower level, the heat generation and dissipation will balance out, keeping the temperature of the transformer's internal components close to the critical value, which helps improve efficiency. After 100 min, the delay in the data decrease is to ensure that the temperature of the internal components of the transformer does not drop too quickly, allowing all the residual oil to fully evaporate.

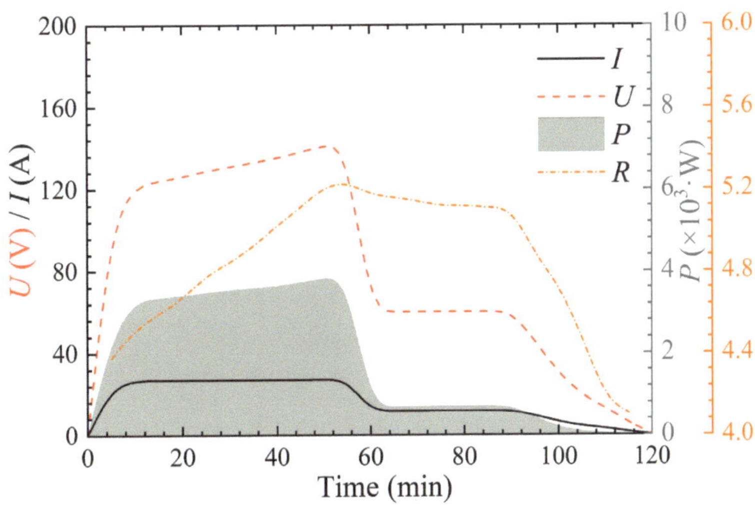

Figure 6. Current, voltage, power, and equivalent resistance waveforms of the low-frequency induction heating device during the recovery process.

The entire recovery process can be divided into four stages, each corresponding to different heating modes and oil–gas precipitation characteristics:

- Stage I: Voltage Rise Stage. After the low-frequency induction heating device is started, the voltage and current applied to the transformer windings rise rapidly, causing the transformer windings to heat up quickly. At 10 min, the input power reaches approximately 3.1 kW. Although the maximum heating power of the low-frequency induction heating device has not yet been reached, the equivalent resistance of the transformer windings rises rapidly, reaching approximately 1 Ω/h. Based on the copper resistivity–temperature curve, at 100 °C, the resistance is about 1.4 times that at room temperature (20 °C). Therefore, at this power level, the winding is expected to reach over 100 °C in about 1.5 h, which satisfies the requirement for the precipitation of residual oil in the transformer's internal structure. Typically, low-frequency induction heating of transformers should not exceed 80 °C to ensure that the equipment stays within safe operating limits. This heating rate also helps prevent localized overheating inside the transformer that could cause damage.
- Stage II: Constant Current Stage. During this stage, the current from the low-frequency induction heating device is kept constant at around 24 A. However, as the winding temperature gradually rises, the input power also increases, reaching a peak of approximately 3.6 kW at around 53 min. The oil–gas analyzer detects a rapid increase in the oil–gas saturation of the extracted mixed gas, indicating that the coil temperature is sufficient to meet the conditions for rapid vaporization of the residual oil in the transformer's internal structure. Therefore, the input voltage is reduced to enter the next stage.
- Stage III: Constant Voltage Stage. The input voltage is gradually reduced to approximately 61 V, and the equivalent resistance of the transformer windings begins to decrease slowly, at a rate of around 0.2 Ω/h. During this stage, the input voltage of the low-frequency induction heating device is kept constant. As the equivalent resistance decreases, the input power increases, allowing the transformer windings to remain at a high temperature for an extended period. This ensures the rapid precipitation of residual oil while minimizing energy loss.
- Stage IV: Voltage Reduction Stage. By around 90 min, the oil–gas analyzer detects a decrease in the oil–gas saturation of the extracted mixed gas, indicating that the main oil–gas precipitation process is nearly complete. The low-frequency induction heating device then begins to gradually reduce the input power until the oil–gas saturation reaches zero. At approximately 90 min, both the voltage and current drop to zero, marking the end of the entire recovery process.

Finally, a core removal treatment is performed on the waste transformer to evaluate the final effect of full oil recovery. As shown in Figure 7, it can be observed that both the left and right sides of the transformer windings are completely dry, with no oil stains remaining. No oil traces are found on the inner side of the core and cover plate. In addition, using absorbent paper to wipe the winding insulation paper and the inner walls of the oil tank shows no residual oil marks. These observations demonstrate that the recovery system effectively achieves full oil recovery, ensuring that no oil stains remain on any surface of the waste transformer components, thereby verifying the high efficiency and reliability of the recovery system.

In the subsequent statistical analysis of full oil recovery from 50 waste transformers, the following evaluation indicators were used to comprehensively assess the reliability of the full oil recovery system:

- Recovery Time: The total time from the start to the completion of the recovery process;
- Residual Oil Rate: The completion rate of full oil recovery;
- Safety Assessment: The impact on the core components of the transformer during the recovery process.

Figure 7. Full oil recovery effect of the waste transformer.

Table 4 presents the comparison results with traditional methods. It should be noted that direct comparison with traditional methods is quite difficult because the field we are studying is highly engineering-oriented. A fair experimental comparison requires tests on transformers with exactly the same parameters. The transformers we used are 10 KV, 400 KVA oil-immersed transformers. However, based on our research, we have not found any papers or work that use the same transformer for recovery. Therefore, we apologize for being unable to provide highly accurate comparative experimental data. However, based on our research and past engineering experience, comparisons can be made from several aspects. Traditional methods are mainly divided into direct oil draining and drying methods. First, in terms of recovery rate, the direct oil draining method generally has a recovery rate of no more than 80%, because the insulating oil inside the inner walls of the tank, the windings, the core, and the insulating materials cannot be directly drained by gravity. The recovery rate of the drying method is about 90%, as traditional methods use a separate drying process, involving handling and hoisting. During this process, residual oil drips onto the factory floor and onto the workers, leading to about a 10% loss.

Regarding recovery time, the average time for full oil recovery from the 50 waste transformers was approximately 132.1 min. As for the residual oil rate, oil stains were found on internal structural parts of over 30 transformers. Small amounts of insulating oil droplets were found in the core layers of 11 transformers, and partial residual oil was found on the inner side of the cover plate of 13 transformers. The overall residual oil rate was less than 1%.

Table 4. Comparison results with traditional draining and drying methods.

Method	Recovery Rate (%)	Recovery Time (min)
Draining	<80	<20
Drying	<90	>120
Ours	>99	>120

The experimental results show that in the full oil recovery experiment of 49 used transformers, the system did not cause any damage to core components. The oil–gas saturation precipitation algorithm and the low-frequency induction heating output prediction model effectively prevented damage to critical components. However, there was one failure case in the recovery process, and the recovery result is shown in Figure 8.

In the failure case, excessive heating occurred during the recovery process of the used transformer, leading to unsatisfactory recovery results. Specifically, the left and right sides of the transformer windings showed obvious burn marks, the surface of internal structural components underwent oxidation, and there was paint peeling. Additionally, although the inner wall of the oil tank was severely damaged, a large amount of insulating oil remained at the bottom, which had not been fully recovered. The test results indicated that the transformer's oil tank shell had undergone severe aging and damage due to long-term use, which caused the failure to maintain the internal vacuum during the recovery process. This poor sealing condition allowed cold air to seep into the oil tank while extracting the mixed gas. As a result, the vaporized insulating oil condensed when it came into contact with the cold air, reattaching to the surfaces of the oil tank's internal structural components or accumulating at the bottom of the oil tank, rather than being extracted by the vacuum pump. Despite the overheating inside the oil tank, the residual oil could not be effectively expelled due to condensation, which significantly impacted the overall recovery effect.

Figure 8. Full oil recovery failure case.

This case demonstrates that the aging of the oil tank shell led to a decline in sealing performance, making it impossible to maintain a vacuum environment during the insulating oil recovery process, thereby exacerbating the condensation and redeposition of oil vapors. It is evident that ensuring the integrity and sealing of the oil tank structure is crucial in the insulating oil recovery process to prevent the ingress of cold air, thereby improving recovery efficiency and avoiding unnecessary condensation and deposition issues. Generally, the issue of aging and poor sealing is divided into two situations. The first is poor sealing at the interface, which can be addressed by creating universal interface components for sealing. The second is when cracks appear on the transformer's body, leading to poor sealing. In this case, pouring thermal conductive epoxy resin for sealing may be considered.

4.4. Scalability Explanation

Since our system uses the transformer itself as the sealed container, the site requirements are relatively low. For larger transformers, a higher power heating device, hot air blower, and related connection equipment may be needed. To improve the scalability of the system for transformers of different sizes or configurations, we will further integrate our equipment to form a system similar to an engineering vehicle or assembly line, making it more adaptable to different scenarios and transformer capacities. In addition, the device we designed for interfacing with the transformer oil port is independent, allowing it to be compatible with various transformer models.

4.5. Discussion of Key Factors

Here, we discuss several key factors, including ambient temperature, transformer transformation efficiency, and derived voltage, and analyze their impact on the recovery process.

In fact, the ambient temperature, transformer transformation efficiency, and derived voltage have little effect on the recovery efficiency of transformer insulating oil, but they do affect the heating time and the quality of the insulating oil. Ambient temperature does not directly affect the recovery efficiency, but it can influence the heating duration, thereby indirectly affecting the recovery time, although the impact is minimal. Similarly, transformer transformation efficiency and derived voltage have little impact on recovery efficiency. However, low transformer transformation efficiency means that more energy is lost as heat during operation. During the insulating oil recovery process, this heat can affect the physical and chemical properties of the insulating oil, ultimately impacting its quality. For example, if the transformer generates excessive heat due to low transformation efficiency, the temperature of the insulating oil will rise. High temperatures may accelerate the aging of the insulating oil, producing more oxidation products and impurities. The derived voltage could also lead to local discharge phenomena in the insulating oil. When the voltage is too high, impurities and small particles in the insulating oil may become the starting point for discharge. Local discharges will cause chemical changes in the insulating oil, generating new decomposition products such as hydrogen, acetylene, and other gases. The generation of these gases will alter the composition of the insulating oil, reducing its insulating properties and ultimately affecting its quality.

In terms of energy consumption, transformers with low transformation efficiency consume more energy, increasing operational costs. In the insulating oil recovery process, sustained high energy consumption will increase the overall cost of recovery. Furthermore, additional cooling measures may be needed to deal with the excess heat generated by the transformer, which will also add to equipment costs and energy consumption. For example, extra cooling equipment may be required to lower the temperature of the insulating oil to ensure it is within an appropriate temperature range for recovery.

5. Conclusions

This paper proposes a transformer insulating oil full recovery technology based on collaborative intelligence. In response to the challenges of oil recovery, especially the difficulty of fully extracting residual oil from complex structures, a recovery system incorporating methods such as hot air vaporization, circulation heating, and low-frequency induction heating under vacuum conditions was designed. Experimental results show that the system can achieve over 95% oil vaporization through 2–3 hot air cycles, while keeping the oil tank temperature below 135 °C and the pressure below 200 kPa. The remaining residual oil is then further extracted under vacuum conditions via low-frequency induction heating, achieving full recovery of the residual oil. After recovery, the oil–gas saturation approaches zero, ensuring that no oil remains inside the discarded transformer. Additionally, by integrating the XGBoost model-based oil–gas saturation intelligent monitoring module, the system can adjust heating power and duration in real-time to avoid overheating damage to the transformer's internal structure. Overall, this study achieves efficient, safe, and environmentally friendly recovery of transformer insulating oil, providing significant technical support for the recycling of used transformer oil, with notable environmental and economic benefits.

Author Contributions: Conceptualization, F.C. and L.W.; methodology, F.C.; software, B.P.; validation, L.W.; formal analysis, L.W.; investigation, Z.Z.; resources, Y.H.; data curation, B.P.; writing—original draft preparation, F.C.; writing—review and editing, F.C. and K.Z.; visualization, K.Z.; supervision, Y.H.; project administration, Y.H.; funding acquisition, Z.Z. All authors have read and agreed to the published version of the manuscript.

Funding: This paper is funded by the State Grid Zhejiang Electric Power Science and Technology Project (5211WF240001 Research on Key Technology of Fine Dismantling and Green Full-volume Recovery of Distribution Transformer Based on Artificial Intelligence).

Data Availability Statement: The raw data supporting the conclusions of this article will be made available by the authors on request.

Conflicts of Interest: Authors Feng Chen, Li Wang and Bin Pan were employed by the company State Grid Zhejiang Electric Power Co., Ltd. Authors Zhiyao Zheng and Yujia Hu were employed by the company Zhejiang Huadian Equipment Testing and Research Insitute Co., Ltd. The remaining authors declare that the research was conducted in the absence of any commercial or financial relationships that could be construed as a potential conflict of interest.

References

1. Slobodina, Y.N.; Rakhmankulov, S.F.; Vorkunov, O.V.; Garifullin, M.S.; Chernyshov, V.; Korenkov, D. Efficiency of Adsorbent Cleaning of Transformator Oils Exposed to Distribution Processes. In Proceedings of the 2024 6th International Youth Conference on Radio Electronics, Electrical and Power Engineering (REEPE), Moscow, Russia, 29 February–2 March 2024; pp. 1–5.
2. Brazhenko, V.; Qiu, Y.; Mochalin, I.; Zhu, G.; Cai, J.C.; Wang, D. Study of hydraulic oil filtration process from solid admixtures using rotating perforated cylinder. *J. Taiwan Inst. Chem. Eng.* **2022**, *141*, 104578. [CrossRef]
3. Vu, T.T.; La, T.V.; Pham, V.T.; Vu, T.G. The highly efficient purification capacity of rGO-zeolite composites for aged oil in transformer machines. *Arab. J. Chem.* **2022**, *15*, 103683. [CrossRef]
4. Li, Q.; Li, Q.; Wang, F.; Wu, J.; Wang, Y. The Carrying Behavior of Water-Based Fracturing Fluid in Shale Reservoir Fractures and Molecular Dynamics of Sand-Carrying Mechanism. *Processes* **2024**, *12*, 2051. [CrossRef]
5. Li, Q.; Li, Q.; Han, Y. A Numerical Investigation on Kick Control with the Displacement Kill Method during a Well Test in a Deep-Water Gas Reservoir: A Case Study. *Processes* **2024**, *12*, 2090. [CrossRef]
6. Karthik, R.; Sree Renga Raja, T. Investigations of transformer oil characteristics. *IEEJ Trans. Electr. Electron. Eng.* **2012**, *7*, 369–374. [CrossRef]
7. Othman, N.A.; Zainuddin, H.; Yahaya, M.S.; Azis, N.; Ibrahim, Z. Prerequisites for Accelerated Aging of Transformer Oil: A Review. *IEEE Trans. Dielectr. Electr. Insul.* **2024**, 1. [CrossRef]
8. Gherdaoui, C.E.; Oumeddour, H.; Aldoori, H.; Alaoui, C.; Delbarre, J.P.; Bouberka, Z.; Supiot, P.; Lerari, D.; Maschke, U. Improving transformer oil decontamination: A synergistic approach integrating adsorption and radiative treatment for polychlorinated biphenyls. *Sep. Purif. Technol.* **2024**, *334*, 126028. [CrossRef]
9. Sarathkumar, D.; Srinivasan, M.; Raj, R.A. An extensive critique on transformer oil reclamation techniques. In Proceedings of the 2022 IEEE 7th International Conference on Recent Advances and Innovations in Engineering (ICRAIE), Mangalore, India, 1–3 December 2022; Volume 7, pp. 438–443.
10. Tiwari, R.; Agrawal, P.; Bawa, S.; Karadbhajne, V.; Agrawal, A.J. Soil contamination by waste transformer oil: A review. *Mater. Today Proc.* **2023**, *72*, 306–310. [CrossRef]
11. Tiwari, R.; Agrawal, P.S.; Belkhode, P.N.; Ruatpuia, J.V.; Rokhum, S.L. Hazardous effects of waste transformer oil and its prevention: A review. *Next Sustain.* **2024**, *3*, 100026. [CrossRef]
12. Lyutikova, M.N.; Korobeynikov, S.M.; Ridel, A.V. Method for Reducing Sediment Formation in Transformer Oil. *IEEE Trans. Dielectr. Electr. Insul.* **2024**, *31*, 2208–2215. [CrossRef]
13. Ingebrigtsen, S.; Dahlund, M.; Hansen, W.; Linhjell, D.; Lundgaard, L. Solubility of carboxylic acids in paper (Kraft)-oil insulation systems. In Proceedings of the The 17th Annual Meeting of the IEEE Lasers and Electro-Optics Society, 2004, LEOS 2004, Boulder, CO, USA, 20–20 October 2004; pp. 253–257.
14. Zhang, H.; Tong, Z.; Wei, T.; Tang, Y. Removal characteristics of Zn(II) from aqueous solution by alkaline Ca-bentonite. *Desalination* **2011**, *276*, 103–108. [CrossRef]
15. Behmadi, R.; Mokhtarian, M.; Ghadrian, K.; Davoodi, A.; Hosseinpour, S. Development of a low-cost activated mesoporous bauxite for the reclamation of used transformer oil. *Sep. Purif. Technol.* **2022**, *280*, 119826. [CrossRef]

16. Fu, P.; Wu, J.; Qian, Y.; Chen, J.; Huang, J.; Wu, J.; Zhao, Y.; Xia, Q. Adsorption performance of activated bentonite, zeolite 13X and activated carbon for aminophenazone in transformer oil. *Sep. Purif. Technol.* **2022**, *280*, 119913. [CrossRef]
17. Kouassi, K.D.; Fofana, I.; Cissé, L.; Hadjadj, Y.; Yapi, K.M.L.; Diby, K.A. Impact of low molecular weight acids on oil impregnated paper insulation degradation. *Energies* **2018**, *11*, 1465. [CrossRef]
18. Duraisamy, S.; Murugesan, S.; Murugan, K.; Raj, R.A. Reclamation of Aged Transformer Oil Employing Combined Adsorbents Techniques Using Response Surface for Transformer Applications. *IEEE Trans. Dielectr. Electr. Insul.* **2023**, *30*, 769–776. [CrossRef]
19. Qian, Y.; Luo, H.; Zhao, Y.; Liu, Y.; Huang, J.; Yu, L.; Xia, Q. Efficient removal of dibutyl phthalate from transformer oils by iron/activated carbon adsorbent. *J. Clean. Prod.* **2024**, *447*, 141476. [CrossRef]
20. Rodiah, Y.; Haryono, T.; Danang Wijaya, F. Transformer Oil Dielectric Characteristics in Microwave Assisted Reconditioning Processes. *J. Electr. Eng. Technol.* **2020**, *15*, 1261–1267. [CrossRef]
21. Hassanpour, M. Transformer Oil Generation and Regeneration Techniques Based on Recent Developments (A Review). *Int. J. Pet. Technol.* **2021**, *8*, 15–33. [CrossRef]
22. Belkhode, P.N.; Ganvir, V.N.; Shende, A.C.; Shelare, S.D. Utilization of waste transformer oil as a fuel in diesel engine. *Mater. Today Proc.* **2022**, *49*, 262–268. [CrossRef]
23. Esser, A.; Ziegler, P.; Kaifie, A.; Kraus, T.; Schettgen, T. Estimating plasma half-lives of dioxin like and non-dioxin like polychlorinated biphenyls after occupational exposure in the German HELPcB cohort. *Int. J. Hygen. Environ. Health* **2021**, *232*, 113667. [CrossRef]
24. Lankatilake, K.; Samaranayake, D.; Piyathunga, K. Exposure to polychlorinated biphenyls (PCBs) among welders in Sri Lanka. *Int. J. Occup. Environ. Health* **2012**, *18*, 110–115. [CrossRef] [PubMed]
25. Salama, M.M.; Mansour, D.E.A.; Daghrah, M.; Abdelkasoud, S.M.; Abbas, A.A. Thermal performance of transformers filled with environmentally friendly oils under various loading conditions. *Int. J. Electr. Power Energy Syst.* **2020**, *118*, 105743. [CrossRef]
26. Ogunlana, R.; Korode, A.; Ajibade, Z. Assessing the level of heavy metals concentration in soil around transformer at Akoko community of OndoState, Nigeria. *J. Appl. Sci. Environ. Manag.* **2020**, *24*, 2183–2189. [CrossRef]
27. Fagbote, E.; Olanipekun, E. Levels of polycyclic aromatic hydrocarbons and polychlorinated biphenyls in sediment of bitumen deposit impacted area. *Int. J. Environ. Sci. Technol.* **2010**, *7*, 561–570. [CrossRef]
28. *GB/T 1094.1-2013*; Power Transformers—Part 1: General Principles. China Electrical Equipment Industry Association: Hefei, China, 2013.

energies

Article

Model Predictive Hybrid PID Control and Energy-Saving Performance Analysis of Supercritical Unit

Qingfeng Yang [1], Gang Chen [2,3], Mengmeng Guo [4], Tingting Chen [1], Lei Luo [1] and Li Sun [4,*]

1 Dongfang Electric Qineng (Shenzhen) Technology Co., Ltd., Shenzhen 518000, China; yangqf@yeah.net (Q.Y.); chentt9196@dongfang.com (T.C.); dreisteine262@163.com (L.L.)
2 State Key Laboratory of Low-Carbon Smart Coal-Fired Power Generation and Ultra-Clean Emission, Nanjing 210023, China; 15905166361@163.com
3 Guoneng Nanjing Electric Power Test & Research Limited, Nanjing 210000, China
4 National Engineering Research Center of Power Generation Control and Safety, School of Energy and Environment, Southeast University, Nanjing 210096, China; guomengmeng@seu.edu.cn
* Correspondence: sunli12@seu.edu.cn; Tel.: +86-18001206210

Abstract: In response to the escalating challenges of rapid load fluctuations and intricate operating environments, supercritical power units demand enhanced control efficiency and adaptability. To this end, this study introduces a novel model predictive hybrid PID control strategy that integrates PID with model predictive control (MPC), leveraging the operational characteristics of multi-loop systems. The proposed strategy adeptly marries the swift response of PID controllers with the foresight and optimization capabilities of MPC. A dynamic model of a supercritical unit is constructed using the subspace identification method. The model's high precision is confirmed by its alignment with field data. Load change simulations demonstrate that the PID–MPC hybrid controller shows faster response times and more precise tracking capabilities compared to the feedforward-PID strategy. It achieves substantial improvements in the IAE index for three loops, with increases of 29.2%, 54.1%, and 57.3% over the feedforward-PID controller. An energy-saving performance analysis indicates that the proactive control actions of both the PID–MPC and MPC strategies lead to dynamic exergy efficiency and coal consumption rates with a broader range of dynamic process changes. The disturbance scenario simulation regarding the proposed controller achieves faster settling times and minimizes control deviation compared to the traditional controller.

Keywords: supercritical unit; coordination control; model predictive control; energy-saving analysis; rapid load change

Citation: Yang, Q.; Chen, G.; Guo, M.; Chen, T.; Luo, L.; Sun, L. Model Predictive Hybrid PID Control and Energy-Saving Performance Analysis of Supercritical Unit. *Energies* **2024**, *17*, 6356. https://doi.org/10.3390/en17246356

Academic Editor: Francesco Nocera

Received: 30 October 2024
Revised: 19 November 2024
Accepted: 26 November 2024
Published: 17 December 2024

1. Introduction

In the age of swift renewable energy expansion, while celebrated for their cleanliness and low-carbon profiles [1], the inherent unpredictability and intermittency of these sources present substantial challenges to the stability and dependability of electrical grids. Consequently, coal-fired power plants are evolving from their traditional single-energy supply paradigm towards a more agile and efficient energy supply model [2]. Supercritical coal-fired power plants (SCFPPs), recognized for their high efficiency in power generation, are assuming an increasingly stable role within the power production sector [3,4]. Nevertheless, with escalating and fluctuating power demands, particularly the prevalence of rapid load shifts and complex operating scenarios, the control efficiency and flexibility of supercritical units are subjected to more stringent demands.

The performance of a power unit's control system is directly linked to the safety, economy, and stability of the unit [5]. Among the various control challenges in SCFPPs, the coordinated control of boilers and turbines stands out as one of the core issues that urgently requires innovative solutions. For a long time, simple and easy-to-implement PID controllers have dominated the coordinated control systems of boiler–turbine units [6,7].

However, when faced with complex, nonlinear, and multivariable coupled control systems, the limitations of PID controllers become increasingly apparent, making it difficult to achieve the desired control effects. Consequently, academia and industry have been continuously exploring new control strategies, such as gain-scheduling [8], sliding mode [9], and adaptive control [10]. Among the many advanced control strategies, model predictive control (MPC) stands out due to its unique predictive mechanism and ability to handle multivariable coupling and complex constraints [11–13]. It has demonstrated significant advantages in improving dynamic response and operational efficiency over a wide range of load conditions for power units.

Many scholars have explored the application of MPC in the control systems of power plants. Wu et al. [14] presented a distributed MPC approach for coal-fired power plants that incorporates carbon capture technology, which led to a significant improvement of 33.1% in the load tracking performance. In another study, He and Lima [15] designed an enhanced MPC algorithm based on modified sequential quadratic programming (SQP) and successfully implemented it in an SCFPP. Zhu et al. [16] developed a three-term MPC strategy and successfully applied it to a 1030 MW supercritical coal power plant. Compared with the existing PID controller, the new strategy significantly improves the performance of load and steam pressure control.

However, while MPC theoretically demonstrates significant potential for managing complex control systems and optimizing performance, it encounters substantial challenges in practical implementation. These include high computational complexity and the difficulty in meeting real-time requirements. Particularly in the operation of supercritical units, which are subject to rapid load changes and complex working conditions, the question of how to effectively reduce computational costs and enhance system response speed without compromising control precision has become a critical issue that needs to be addressed.

To tackle the aforementioned challenges, an innovative model predictive hybrid PID control strategy (PID–MPC) was developed. Some scholars have studied the feasibility of PID–MPC hybrid control. Sanama, Xia, and Nguepnang [17] implemented PID–MPC to enhance the performance of a chiller-fan coil unit. Sen, Singh, and Ramachndran [18] designed a hybrid MPC–PID controller. Compared with the pure PID controller, the results show that the performance of the control loop is improved under the hybrid control system. Existing studies have demonstrated the application potential of a PID–MPC hybrid strategy.

In this study, the application of PID–MPC hybrid control in power plant units is explored. This approach aims to seamlessly integrate the immediate feedback adjustments of PID control with the predictive optimization capabilities of MPC. The strategy leverages the complementary strengths of these control techniques through multi-loop decoupling, particularly under complex operating conditions. The paper commences by constructing a high-precision dynamic simulation model of a supercritical unit using the subspace identification method. Subsequently, the PID–MPC hybrid control strategy is meticulously designed based on this model. To assess the control performance, a set of control and energy-saving performance indicators are proposed. Ultimately, the effectiveness and robustness of the proposed strategy are comprehensively evaluated through extensive multi-scenario simulation tests.

2. Supercritical Unit Coordination System

2.1. System Description

The schematic diagram of the supercritical unit in this study is shown in Figure 1. Coal is burned in the furnace as fuel to release heat energy. The feed water enters the water wall after being preheated by an economizer and becomes a two-phase flow after being heated at a high temperature in the furnace. The steam is output by the separator and then heated into high-temperature and high-pressure steam through the multi-stage heating surface. Steam turbines and motors convert thermal energy from product steam into electrical energy for power generation. The entire energy conversion process within a power plant is a complex orchestration involving the synchronized operation of numerous components.

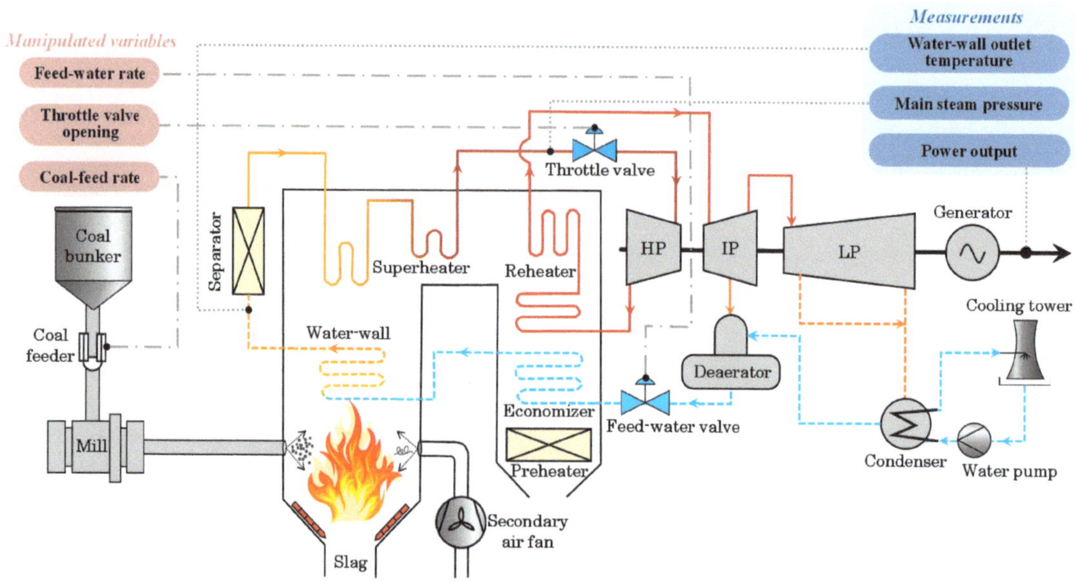

Figure 1. The structural diagram of the subcritical power plant.

2.2. Coordinated Control System

To maintain the stability and efficiency of supercritical units, effective boiler–turbine coordination is crucial. However, the multivariable coupling nature of these systems demands sophisticated control strategies. Traditionally, a combination of feedforward functions and PID control has been used, as depicted in Figure 2a. While this approach is straightforward and adaptable, it falls short as grid flexibility demands and unit capacities increase. Frequent load changes exacerbate internal variable coupling, heightening sensitivity to disturbances and uncertainties. Moreover, the need for rapid power plant load adjustments necessitates a control system that is capable of swift, accurate responses and flexible adaptation to dynamic changes, presenting a significant challenge to the conventional control strategy.

This study presents a novel PID–MPC hybrid control strategy to tackle the challenges faced by supercritical units, as shown in Figure 2b. The implementation of a PID–MPC hybrid control strategy in industrial applications faces challenges such as system nonlinearity, external disturbances, and the technical limitations of sensors and actuators. Our strategy effectively addresses these challenges by integrating the rapid response of PID with the predictive and optimization capabilities of MPC. The predictive and optimization features of MPC enable the direct consideration of operational constraints, a capability not present in traditional PID control, which is particularly crucial when dealing with industrial systems that have hard constraints. The PID controller swiftly adjusts the main steam valve in response to load changes, while MPC precisely controls coal and water feed to optimize pressure and temperature.

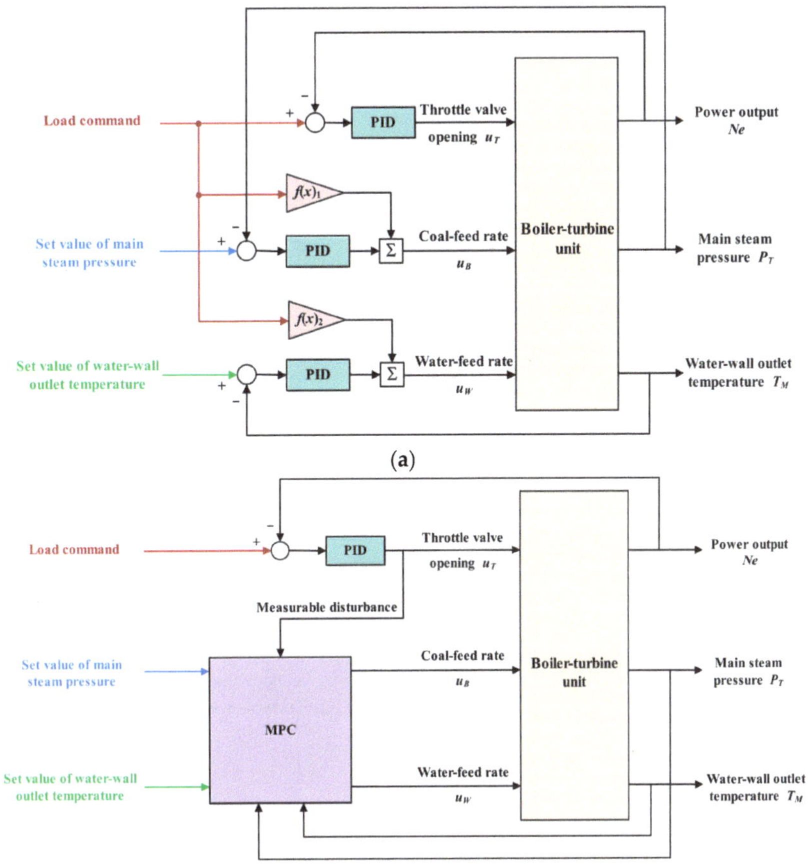

Figure 2. Coordinated control structure of supercritical unit. (**a**) Traditional feedforward-PiD control, (**b**) Novel PiD-MPC hybrid control.

3. Methods

3.1. Subspace Identification Modeling

Subspace identification (SID) has become extensively used in the realm of system identification [19]. Assuming that the n-th order continuous linear state-space equation of a CFPP can be expressed in the following form:

$$\begin{cases} \dot{x}(t) = A_c x(t) + B_c u(t) + K_c e(t) \\ y(t) = C_c x(t) + D_c u(t) + e_c(t) \end{cases} \quad (1)$$

where A_c, B_c, C_c, D_c, and K_c are the continuous state-space matrices. $u(t)$ are the inputs, and $y(t)$ are the outputs. $e(t)$ are the disturbances, and $x(t)$ are the vectors of n-th states. For dynamic systems, D is fixed to zero by default, meaning that the system has no feedthrough [19].

The parameters in system model (1) are determined by SID method, and the Hankel matrix composed of input and output data blocks is as follows:

$$U_{(0|2i-1)} \underset{=}{def} \left[\begin{array}{cccc} u(0) & u(1) & \cdots & u(j-1) \\ u(1) & u(2) & \cdots & u(j) \\ \vdots & \vdots & \ddots & \vdots \\ u(i-1) & u(i) & \cdots & u(i+j-2) \\ \hline u(i) & u(i+1) & \cdots & u(i+j-1) \\ u(i+1) & u(i+2) & \cdots & u(i+j) \\ \vdots & \vdots & \ddots & \vdots \\ u(2i-1) & u(2i) & \cdots & u(2i+j-2) \end{array}\right] = \left[\frac{U(0|i-1)}{U(i|2i-1)}\right] = \left[\frac{U_p}{U_f}\right] \tag{2}$$

$$Y_{(0|2i-1)} \underset{=}{def} \left[\begin{array}{cccc} y(0) & y(1) & \cdots & y(j-1) \\ y(1) & y(2) & \cdots & y(j) \\ \vdots & \vdots & \ddots & \vdots \\ y(i-1) & y(i) & \cdots & y(i+j-2) \\ \hline y(i) & y(i+1) & \cdots & y(i+j-1) \\ y(i+1) & y(i+2) & \cdots & y(i+j) \\ \vdots & \vdots & \ddots & \vdots \\ y(2i-1) & y(2i) & \cdots & y(2i+j-2) \end{array}\right] = \left[\frac{Y(0|i-1)}{Y(i|2i-1)}\right] = \left[\frac{Y_p}{Y_f}\right] \tag{3}$$

where i and j are user-defined parameters for adjusting identification accuracy. Y_p and Y_f represent past and future outputs, respectively. U_p and U_f represent past and future inputs, respectively.

Similarly, a corresponding definition of the state matrix is formulated to derive the generalized input–output matrix equation as follows:

$$X_f = A^i X_p + \Delta_i U_p \tag{4}$$

$$Y_p = \Gamma_i X_p + H_i U_p \tag{5}$$

$$Y_f = \Gamma_i X_f + H_i U_f \tag{6}$$

where X_p and X_f represent past and future state variables, respectively. Γ_i is the generalized observable matrix, and H_i is the low-dimensional partitioned triangular Toeplitz matrix.

By combining Formulars (4)–(6),

$$\begin{aligned} \hat{Y}_f &= L_w W_p + L_u U_f \\ L_w &= \left[\begin{array}{cc} A^i \Gamma_i^{\dagger} & \Delta_i - A^i \Gamma_i^{\dagger} H_i \end{array}\right] \\ W_p &= \left[\begin{array}{c} Y_p \\ U_p \end{array}\right] \\ L_u &= H_i \end{aligned} \tag{7}$$

The current state can be characterized by historical input and output data, with L_w representing the state subspace matrix and L_u denoting the deterministic input subspace matrix.

The subspace matrices L_w and L_u can be identified from the Hankel matrix employing the least squares method. Subsequently, leveraging a linear observer grounded in (7), the system matrix [A, B, C, D] is solved by least squares method as follows:

$$[A, B, C, D] = \arg \min_{A,B,C,D} \left\| \begin{pmatrix} \widetilde{X}_{i+1} \\ Y_i \end{pmatrix} - \begin{pmatrix} A & B \\ C & D \end{pmatrix} \begin{pmatrix} \widetilde{X}_{i+1} \\ U_{i|i} \end{pmatrix} \right\|_F^2 \tag{8}$$

To reflect the dynamic operation characteristics of the unit as the coordinated control object, this study is based on the historical data of a cooperative power plant in China as the identification data. In order to optimize the control design, the input variables were selected as coal feed, water feed, and throttle valve opening, and the output variables include power, main steam pressure and water wall outlet temperature. The SID identification method in MATLAB R2022 B system identification toolbox is used to identify the system state space model.

3.2. PID–MPC Hybrid Control Strategy

3.2.1. PID Controller for Output Power

In the scenario of power generation control, the PID controller accurately regulates the opening of the main steam valve through real-time power feedback signals collected so as to achieve efficient and stable control of the power generation of the supercritical unit. The expression is as follows:

$$u(t) = K_P \cdot e(t) + K_I \cdot \int e(t)\mathrm{d}t + K_D \frac{\mathrm{d}e(t)}{\mathrm{d}t} \tag{9}$$

where K_P is the proportional coefficient, K_I is the integral coefficient, and K_D is the integral coefficient. After controller parameters are adjusted, the value of K_P is 58, K_I is 0.1, and K_D is 0.3.

3.2.2. Design of MPC

In view of the nonlinear, time-varying, and strong coupling characteristics of supercritical unit coordination system, MPC control strategy is proposed to regulate the main steam pressure and the outlet temperature of water wall in the coordination system.

Utilizing the general form (1) of the identified state-space model, discretization and difference processing are implemented to derive an augmented model as follows:

$$\begin{cases} \underbrace{\begin{bmatrix} \Delta x(k+1) \\ y(k+1) \end{bmatrix}}_{x(k+1)} = \underbrace{\begin{bmatrix} A_m & 0_m^T \\ C_m A_m & 1 \end{bmatrix}}_{A} \underbrace{\begin{bmatrix} \Delta x(k) \\ y(k) \end{bmatrix}}_{x(k)} + \underbrace{\begin{bmatrix} B_m \\ 0 \end{bmatrix}}_{B} \Delta u(k) \\ y(k) = \underbrace{\begin{bmatrix} 0_m & 1 \end{bmatrix}}_{C} \begin{bmatrix} \Delta x(k) \\ y(k) \end{bmatrix} \end{cases} \tag{10}$$

where $0_m \in \mathbb{R}^{n_x}$ represents the 0 matrix with dimension $\mathrm{n_x}$. $\Delta u(k)$ is the increment of control input at time k.

Define the vectors of the predicted output Y and the control input ΔU as

$$\begin{aligned} Y &= \left[y(k+1|k)^T y(k+2|k)^T \cdots y(k+N_P|k)^T\right]^T \\ \Delta U &= \left[\Delta u(k)^T \Delta u(k+1)^T \cdots \Delta u(k+N_C-1)^T\right] \end{aligned} \tag{11}$$

Based on model (10), recursively in the prediction horizon N_P, the set of prediction output Y is obtained as follows:

$$\begin{gathered} Y(k) = F\hat{x}(k) + \Phi \Delta U(k) \\ F = \begin{bmatrix} CA \\ CA^2 \\ \vdots \\ CA^{N_P} \end{bmatrix}, \Phi = \begin{bmatrix} CB & 0 & \cdots & 0 \\ CAB & CB & \cdots & 0 \\ \vdots & & \ddots & \vdots \\ CA^{N_P-1}B & CA^{N_P-2}B & \cdots & CA^{N_P-N_C}B \end{bmatrix} \end{gathered} \tag{12}$$

where F and Φ represent the constructed coefficient matrix, respectively.

For the coordinated control problem of supercritical power plant in this study, the goal of controller design is to minimize the absolute difference between the control output and the set point in the future prediction horizon N_P. Define the cost function of the quadratic form reflecting the control objective as follows:

$$J = (R-Y)^T S(R-Y) + \Delta U^T \overline{R} \Delta U$$
$$R^T = \underbrace{[11\cdots 1]}_{N_P} r_i(k) \tag{13}$$
$$\overline{R} = r_w I_{N_C \times N_C}$$

where J is the cost function of energy-aware MPC. $r_i(k)$ and $y_i(k)$ are the i-th reference and measure outputs at time k, respectively. R is the reference value matrix. S and $\overline{R}$ are error and control weight matrix, respectively. r_w is used as a tuning parameter for the desired response speed.

Necessitating the substitution of (12) with (13),

$$J = \left(R - F\hat{\bar{x}}(k)\right)^T S_q\left(R - F\hat{\bar{x}}(k)\right) - 2\Delta U^T \Phi^T S_q\left(R - F\hat{\bar{x}}(k)\right) + \Delta U^T\left(\Phi^T S_q \Phi + S_r\right)\Delta U \tag{14}$$

To facilitate optimal solution, (14) is transformed into quadratic programming (QP) form as follows:

$$\begin{cases} \min\limits_{\Delta U} J = \min\limits_{\Delta U}\left(\Delta U^T H \Delta U - 2f^T \Delta U\right) \\ H = \Phi^T S \Phi + \overline{R} \\ f = \left(R - F\hat{\bar{x}}(k)\right)^T S\Phi \end{cases} \tag{15}$$

where H is the symmetric matrix of the quadratic term and f is the matrix of the linear term.

In the realm of coal-fired power plant operations, safety regulations and equipment specifications necessitate the imposition of upper and lower bounds on control variables. For regulating variables, coal feed rate, water feed rate, and valve opening, the constraints are set as follows:

$$U^{\min}(k) \le U(k) \le U^{\max}(k) \tag{16}$$

$$\Delta U^{\min}(k) \le \Delta U(k) \le \Delta U^{\max}(k) \tag{17}$$

The MPC optimization problem is composed of objective function (15) and constraint conditions (16) and (17). The optimal control sequence is obtained by solving the optimization problem at k time. According to the MPC principle, the MPC rolling optimization control is realized when only the first item in the prediction sequence is applied to the system.

After many tests in the simulation environment, the key parameters of MPC applied to the control of the main steam pressure of the coordination stage and the outlet temperature of the water wall are set in Table 1.

Table 1. Parameter settings of MPC.

Parameter	Value
Sampling time	4
Predictive horizon	25
Control horizon	10
Error weight of main steam pressure	1
Error weight of water wall outlet temperature	0.9

3.3. *Performance Index*

3.3.1. Control Performance

Considering the long-term operational performance of the control system and the comprehensive impact of error amplitude and persistence, the Integrated Absolute Error (IAE) index is employed as follows:

$$IAE_i = \sum_{k=1}^{N} |r_i(k) - y_i(k)| \Delta t \quad i = 1, 2, 3 \tag{18}$$

The maximum error (ME) is a comprehensive measure of the maximum difference between the actual output and the expected output. The formula is as follows:

$$ME_i = \max_{1 \le k \le N} [r_i(k) - y_i(k)] \quad i = 1, 2, 3 \tag{19}$$

3.3.2. Energy-Saving Performance

The exergy analysis method based on the second law of thermodynamics is a crucial tool for studying and optimizing the thermodynamic performance of systems [20,21]. The coal supply serves as the energy input source for an SCFPP. The electrical power is the product energy of SCFPP. Exergy represents the maximum potential of a system to perform work. The equations are as follows:

$$\dot{E}x_{\text{coal}} = \dot{m}_{\text{coal}} \cdot ex_{\text{coal}} \tag{20}$$

$$\dot{E}x^{W} = \dot{P} \tag{21}$$

where $\dot{E}x_{\text{coal}}$ is the dynamic exergy rate of coal feed, kW; $\dot{m}_{\text{coal}}$ is the mass flow rate of coal feed, kg/s; ex_{coal} is the specific exergy of coal, kJ/kg; $\dot{P}$ is the electric power, kW.

The dynamic exergy efficiency of the overall cycle of an SCFPP is precisely defined as follows:

$$\eta_{ex} = \frac{\dot{P}}{\dot{E}x_{\text{coal}}} \tag{22}$$

where η_{ex} represents dynamic exergy efficiency, %.

Coal consumption is a vital consideration in the operation and optimization of CFPPs. As a key indicator of energy saving and carbon reduction, the coal consumption rate (CCR) accurately reflects the amount of coal consumed per kilowatt-hour of electricity generated

$$\text{CCR} = \frac{M_{\text{coal}}}{\dot{P} \times t_d} \tag{23}$$

where CCR is the coal consumption rate, kg/MW·h; M_{coal} is the total coal consumption during operation period, kg; t_d is the operation period, h.

4. Results and Discussion

4.1. Model Validation

In this study, a dataset comprising 4500 samples, each taken at 4 s intervals, is extracted from a 350 MW supercritical unit at a cooperative power plant in China for the purpose of system identification modeling. To assess the model's validity, 9000 additional unidentified data are employed for testing. Figure 3 shows the model identification and testing results. Figure 3 demonstrates that the model's output closely aligns with the actual field data, and the trend of the three output groups shown in Figure 3b is consistent with the power plant's data, thereby validating the model's high accuracy and reliability.

4.2. Wide-Range Load Change

To assess the variable load control performance of various control strategies, a comparative analysis of the PID–MPC hybrid controller, the classical MPC controller, and the feedforward-PID controller is conducted, as depicted in Figure 4. Figure 4 reveals that, while the feedforward-PID controller exhibits minimal overshoot during the response phase, it suffers from a significant delay in response time, underscoring its lack of responsiveness. Conversely, both the PID–MPC and MPC controllers display superior rapid response capabilities, effectively and swiftly addressing the load change signals. The performance index is detailed in Table 2. According to the table, the PID–MPC hybrid controller achieves a control performance nearly equivalent to that of MPC, with substantial improvements in the IAE index for the three loops, showing increases of 29.2%, 54.1%, and 57.3% over the feedforward-PID controller. Furthermore, the energy-saving performance depicted in Figure 4c indicates that the more proactive control actions of PID–MPC and MPC result in dynamic exergy efficiency and CCRs with a broader range of dynamic process changes.

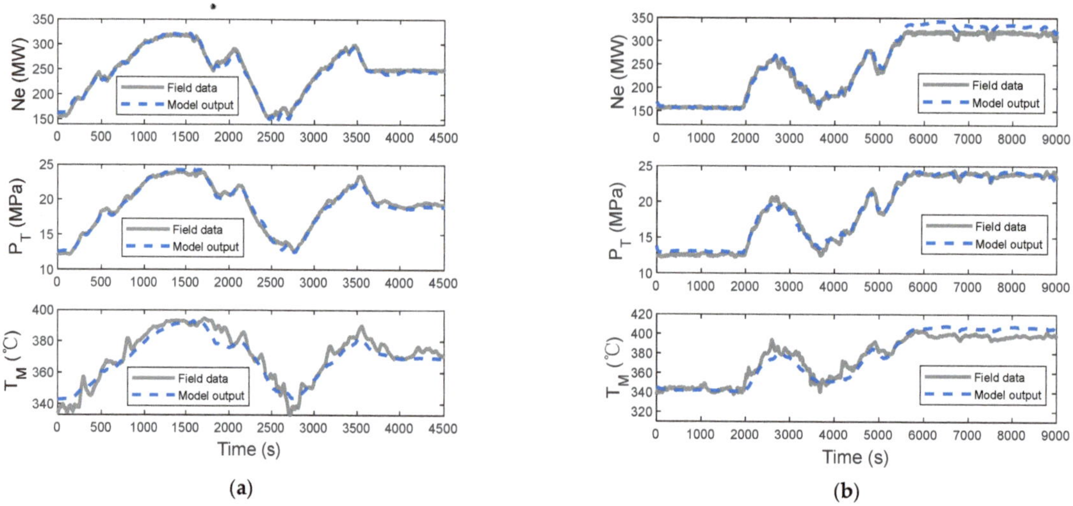

Figure 3. Model identification and test results. (**a**) Identification data; (**b**) test data.

Table 2. Performance indexes of different strategies.

Controller	IAE_1	IAE_2	IAE_3
PID–MPC	2320.4	452.7	1669.8
MPC	1998.2	430.1	2067.4
Feedforward-PID	3278.3	985.6	3909.9

4.3. Uncertainty Disturbances

To assess the performance of various control strategies in the presence of uncertain disturbances, step disturbances are introduced at 500 s, 4000 s, and 7500 s for three critical control variables: coal feed rate, water feed rate, and throttle valve opening. Figure 5 compares the responses of the PID–MPC hybrid controller, the MPC controller, and the feedforward-PID control to these disturbances. Figure 5a illustrates that, across the three disturbance scenarios, the feedforward-PID controller experiences the most significant overshoot. Its adjustment process is characterized by substantial fluctuations and a slow convergence to the set reference value. In contrast, the PID–MPC hybrid control strategy and the MPC controller demonstrate superior anti-interference capabilities. They can swiftly identify and respond to disturbances, markedly reducing settling time and effectively minimizing control deviation. The energy-saving performance following the disturbances, as influenced by the controllers, is depicted in Figure 5c. This figure indicates that the feedforward-PID exhibits more gradual changes, while PID–MPC and MPC show larger response peaks.

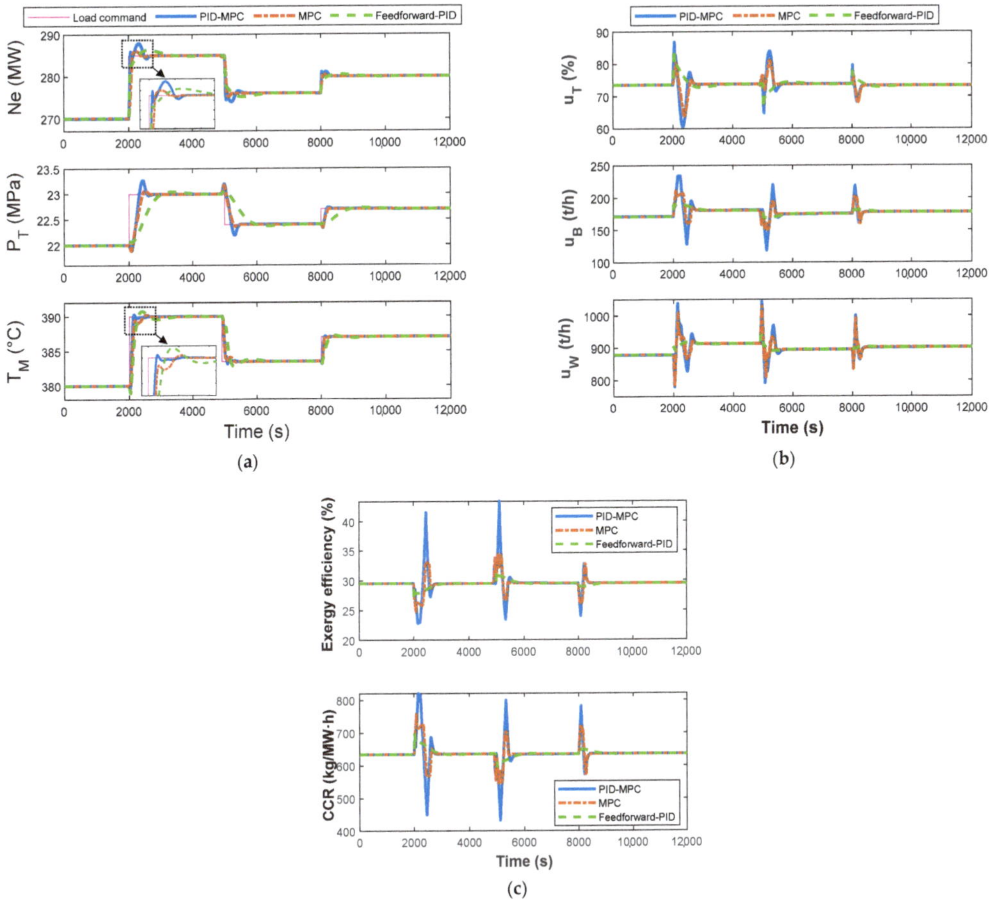

Figure 4. Simulation comparison of control strategies in the wide-range load tracking. (**a**) Process outputs; (**b**) control inputs; (**c**) energy-saving performance.

Table 3 provides a quantitative summary of the performance indicators, highlighting that, under three distinct interference conditions, the PID–MPC hybrid controller and the MPC controller significantly outperform the traditional feedforward-PID control method in terms of the ME and IAE indicators. Notably, as indicated in bold in Table 3, the MPC controller achieves the best performance in the indicators, while the PID–MPC hybrid control strategy closely matches MPC's performance but with a substantially reduced computational complexity. This reduction enhances the real-time performance and decreases resource consumption in practical applications, underscoring its significant potential and practical value in the optimization of control strategies.

Table 3. Performance index of different strategies.

Case	Controller	IAE_1	IAE_2	IAE_3	ME_1 (MW)	ME_2 (MPa)	ME_3 (°C)	$\eta_{ex,ave}$(%)	CCR_{ave} (kg/MW·h)
Case 1: Throttle valve opening disturbance (7500 s)	PID–MPC	34.2	20.5	114.7	0.10	0.02	−0.14	30.47	613.7
	MPC	**13.1**	**12.7**	**99.5**	**−0.03**	**0.01**	**−0.12**	30.47	613.7
	Feedforward-PID	327.4	103.7	201.9	−0.32	−0.09	−0.33	30.45	614.1

Table 3. *Cont.*

Case	Controller	IAE_1	IAE_2	IAE_3	ME_1 (MW)	ME_2 (MPa)	ME_3 (°C)	$\eta_{ex,ave}$(%)	CCR_{ave} (kg/MW·h)
Case 2: Coal feed rate disturbance (500 s)	PID–MPC	21.2	4.7	69.7	0.08	0.02	0.23	30.48	613.6
	MPC	**1.2**	**4.1**	**46.7**	**0.01**	**0.01**	**0.16**	30.48	613.6
	Feedforward-PID	73.5	20.8	550.9	0.09	−0.03	1.32	30.49	613.3
Case 3: Water feed rate disturbance (4000 s)	PID–MPC	56.2	3.8	25.4	**−0.10**	−0.01	−0.09	30.48	613.4
	MPC	**5.1**	**2.4**	**20.5**	−0.16	−0.01	**−0.06**	30.49	613.5
	Feedforward-PID	116.4	3.2	48.6	−0.20	0.01	−0.09	30.49	613.4

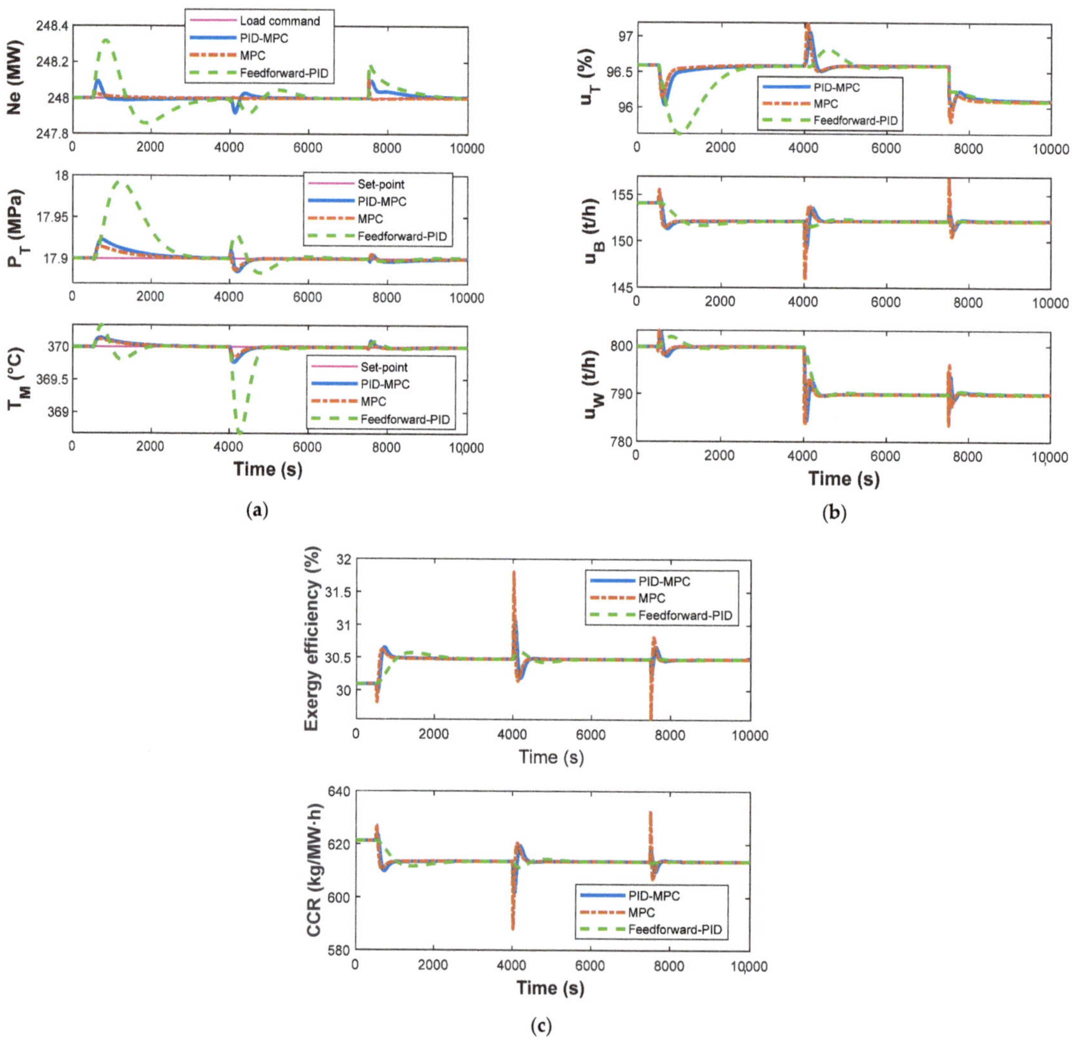

Figure 5. Simulation comparison of control strategies in the uncertainty disturbances case. (**a**) Process outputs; (**b**) control inputs; (**c**) energy-saving performance.

5. Conclusions

This paper introduces a novel PID–MPC hybrid control strategy that synergistically merges the immediate feedback adjustment capabilities of PID control with the predictive optimization features of MPC. The main conclusions are as follows:

- A dynamic model of a supercritical unit is constructed using the subspace identification method. The model's high precision is confirmed by its alignment with field data, providing a solid foundation for the design of subsequent control strategies.
- In variable load simulation tests, the PID–MPC hybrid control strategy shows faster response times and more precise tracking capabilities compared to the traditional feedforward-PID strategy. It achieves substantial improvements in the IAE index for three loops, with increases of 29.2%, 54.1%, and 57.3% over the feedforward-PID controller, closely matching the performance of a classic MPC controller and effectively enhancing the system's dynamic response performance.
- The energy-saving performance analysis indicates that the proactive control actions of both the PID–MPC and MPC strategies lead to dynamic exergy efficiency and CCRs with a broader range of dynamic process changes.
- Faced with a variety of control disturbance scenarios, the PID–MPC hybrid controller proves its robust anti-interference capabilities. This strategy consistently shows less overshoot, achieves faster settling times, and minimizes control deviation when compared to the conventional feedforward-PID controller.

Future research will place greater emphasis on directly considering the operational constraints in the energy efficiency and process optimization of MPC. Additionally, there will be an exploration of tighter integration between MPC and PID, along with more extensive application testing, particularly in industrial applications.

Author Contributions: Conceptualization, Q.Y., G.C. and L.S.; methodology, Q.Y., M.G. and L.S.; software, M.G. and L.L.; validation, M.G. and T.C.; investigation, G.C. and L.L.; resources, Q.Y., G.C. and L.S.; writing—original draft preparation, Q.Y. and M.G.; writing—review and editing, G.C., T.C. and L.S.; supervision, L.S.; funding acquisition, Q.Y. and L.S. All authors have read and agreed to the published version of the manuscript.

Funding: This research was funded by National Key Research and Development Program of China, grant number 2022YFB4100500, Jiangsu S&T Funding, grant numbers BK20240073 and BA2023005, and the Open Project Program of State Key Laboratory of Low-carbon Smart Coal-fired Power Generation and Ultra-clean Emission, grant number D2023FK111.

Data Availability Statement: The datasets used in this article are confidential.

Conflicts of Interest: Author Qingfeng Yang was employed by the Dongfang Electric Qineng (Shenzhen) Technology Co., Ltd. Author Gang Chen was employed by the Guoneng Nanjing Electric Power Test & Research Limited. The remaining authors declare that the research was conducted in the absence of any commercial or financial relationships that could be construed as a potential conflict of interest.

References

1. Forrester, N. Catching the Rays: My Part in Morocco's Renewable-Energy Revolution. *Nature* **2024**, *626*, 1156. [CrossRef] [PubMed]
2. Guo, M.; Liu, Y.; Chen, Y.; Gao, W.; Wang, D.; Tang, H.; Liu, H. Study on the Effect of Different Types and Distribution Ratio of Heat Users on Operation Performance of the Distributed Solar Centralized Heating System. *Energy Build.* **2022**, *270*, 112265. [CrossRef]
3. Tong, Y.; Duan, L.; Pang, L. Off-Design Performance Analysis of a New 300 MW Supercritical CO_2 Coal-Fired Boiler. *Energy* **2021**, *216*, 119306. [CrossRef]
4. Yang, M.; Zhou, Y.; Yang, J.; Bao, J.; Wang, D.; Yu, Q. Performance Analysis of an Efficient Waste Heat Utilization System in an Ultra-Supercritical Coal-Fired Power Plant. *Energy Rep.* **2022**, *8*, 5871–5882. [CrossRef]
5. Fan, H.; Su, Z.; Wang, P.; Lee, K.Y. A Dynamic Nonlinear Model for a Wide-Load Range Operation of Ultra-Supercritical Once-through Boiler-Turbine Units. *Energy* **2021**, *226*, 120425. [CrossRef]

6. Borase, R.P.; Maghade, D.K.; Sondkar, S.Y.; Pawar, S.N. A Review of PID Control, Tuning Methods and Applications. *Int. J. Dynam. Control* **2021**, *9*, 818–827. [CrossRef]
7. Çetin, G.; Özkaraca, O.; Keçebaş, A. Development of PID Based Control Strategy in Maximum Exergy Efficiency of a Geothermal Power Plant. *Renew. Sustain. Energy Rev.* **2021**, *137*, 110623. [CrossRef]
8. Sánchez, A.J.; Gallego, A.J.; Escaño, J.M.; Camacho, E.F. Event-Based MPC for Defocusing and Power Production of a Parabolic Trough Plant under Power Limitation. *Sol. Energy* **2018**, *174*, 570–581. [CrossRef]
9. Elsisi, M.; Bazmohammadi, N.; Guerrero, J.M.; Ebrahim, M.A. Energy Management of Controllable Loads in Multi-Area Power Systems with Wind Power Penetration Based on New Supervisor Fuzzy Nonlinear Sliding Mode Control. *Energy* **2021**, *221*, 119867. [CrossRef]
10. Tafti, H.D.; Sangwongwanich, A.; Yang, Y.; Pou, J.; Konstantinou, G.; Blaabjerg, F. An Adaptive Control Scheme for Flexible Power Point Tracking in Photovoltaic Systems. *IEEE Trans. Power Electron.* **2019**, *34*, 5451–5463. [CrossRef]
11. Stanger, L.; Bartik, A.; Hammerschmid, M.; Jankovic, S.; Benedikt, F.; Müller, S.; Schirrer, A.; Jakubek, S.; Kozek, M. Model Predictive Control of a Dual Fluidized Bed Gasification Plant. *Appl. Energy* **2024**, *361*, 122917. [CrossRef]
12. Dai, B.; Wang, F.; Chang, Y.; Chu, F.; Song, S. Process Model-Assisted Optimization Control Method Based on Model Predictive Control and Deep Reinforcement Learning for Coordinated Control System in Coal-Fired Power Unit. *IEEE Trans. Autom. Sci. Eng.* **2024**. [CrossRef]
13. Kim, D.; Wang, Z.; Brugger, J.; Blum, D.; Wetter, M.; Hong, T.; Piette, M.A. Site Demonstration and Performance Evaluation of MPC for a Large Chiller Plant with TES for Renewable Energy Integration and Grid Decarbonization. *Appl. Energy* **2022**, *321*, 119343. [CrossRef]
14. Wu, X.; Wang, M.; Shen, J.; Li, Y.; Lawal, A.; Lee, K.Y. Reinforced Coordinated Control of Coal-Fired Power Plant Retrofitted with Solvent Based CO2 Capture Using Model Predictive Controls. *Appl. Energy* **2019**, *238*, 495–515. [CrossRef]
15. He, X.; Lima, F.V. A Modified SQP-Based Model Predictive Control Algorithm: Application to Supercritical Coal-Fired Power Plant Cycling. *Ind. Eng. Chem. Res.* **2020**, *59*, 15671–15681. [CrossRef]
16. Zhu, Y.; Zhang, K.; Zhu, Y.; Jiang, P.; Zhou, J. Development of a Three-Term MPC and Its Application to an Ultra-Supercritical Coal Fired Power Plant. *Control Eng. Pract.* **2024**, *143*, 105787. [CrossRef]
17. Sanama, C.; Xia, X.; Nguepnang, M. PID-MPC Implementation on a Chiller-Fan Coil Unit. *J. Math.* **2022**, *2022*, 8405361. [CrossRef]
18. Sen, M.; Singh, R.; Ramachandran, R. A Hybrid MPC-PID Control System Design for the Continuous Purification and Processing of Active Pharmaceutical Ingredients. *Processes* **2014**, *2*, 392–418. [CrossRef]
19. Van Overschee, P.; De Moor, B. *Subspace Identification for Linear Systems: Theory—Implementation—Applications*; Springer Science & Business Media: Berlin, Germany, 2012.
20. Guo, M.; Hao, Y.; Nižetić, S.; Lee, K.Y.; Sun, L. Modelica-Based Heating Surface Modelling and Dynamic Exergy Analysis of 300 MW Power Plant Boiler. *Energy Convers. Manag.* **2024**, *312*, 118557. [CrossRef]
21. Chen, Y.; Guo, M.; Liu, Y.; Wang, D.; Zhuang, Z.; Quan, M. Energy, Exergy, and Economic Analysis of a Centralized Solar and Biogas Hybrid Heating System for Rural Areas. *Energy Convers. Manag.* **2023**, *276*, 116591. [CrossRef]

energies

Article

Intelligent Optimization Scheduling Strategy for Energy Consumption Reduction for Equipment in Open-Pit Mines Based on Enhanced Genetic Algorithm

Fudong Li [1,*], Zonghao Shi [1], Weiqiang Ding [2] and Yongjun Gan [1]

[1] School of Mechanical and Electrical Engineering, Beijing Information Science and Technology University, Beijing 100192, China; 2022020011@bistu.edu.cn (Z.S.); 2024020137@bistu.edu.cn (Y.G.)
[2] School of Energy and Power Engineering, Nanjing University of Science and Technology, Nanjing 210094, China; dingwq0425@njust.edu.cn
* Correspondence: 20182238@bistu.edu.cn

Abstract: To achieve a rational allocation of real-time operational equipment, such as excavators and dump trucks, in open-pit mines, and thereby enhance truck–shovel coordination, this paper addresses the challenges posed by unreasonable on-site scheduling, which includes excessive truck waiting times and prolonged excavator boom-and-dipper operations. Ultimately, the paper aims to attain optimal truck–shovel coordination efficiency. To this end, we construct a scheduling optimization model, with the production capacities of trucks and shovels serving as constraints. The objective functions of this model focus on minimizing transportation costs, reducing truck waiting times, and shortening excavator boom-and-dipper operation durations. To solve this model, we have developed an improved genetic algorithm that integrates roulette wheel selection and elite preservation strategies. The experimental results of our algorithm demonstrate that it can provide a more refined operational equipment scheduling scheme, effectively decreasing truck transportation costs and enhancing equipment utilization efficiency in open-pit mines.

Keywords: open-pit mine; transportation costs; truck waiting times; excavator boom-and-dipper operation durations; improved genetic algorithm

Academic Editor: David Borge-Diez

Received: 16 October 2024
Revised: 13 November 2024
Accepted: 15 November 2024
Published: 27 December 2024

Citation: Li, F.; Shi, Z.; Ding, W.; Gan, Y. Intelligent Optimization Scheduling Strategy for Energy Consumption Reduction for Equipment in Open-Pit Mines Based on Enhanced Genetic Algorithm. *Energies* **2025**, *18*, 60. https://doi.org/10.3390/en18010060

1. Introduction

Open-pit mining has been widely used in the development and utilization of modern mining practices due to the short construction period, high production capacity, low mining costs, and safe and reliable operating [1–4]. However, in opencast mining and large-scale earthwork projects, the scheduling management of most construction projects lags behind. Construction enterprises deploy multiple site managers who drive command vehicles to inspect the current construction status, identify unreasonable production scheduling, and make scheduling decisions to resolve the issues. However, inspections cannot identify problems in real-time, leading to many unreasonable production scheduling issues being concealed. Even when site managers discover production scheduling problems, they often only observe partial phenomena and lack a holistic perspective. Additionally, influenced by their work levels and experience, it is difficult for them to make optimal scheduling decisions [5–8]. This prevents construction machinery from operating at full productivity and increases the idling time of construction machinery [9], resulting in increased fuel consumption of engineering machinery and equipment [10].

Specifically, the problems in the scheduling of open-pit mining equipment include, but are not limited to, the lack of real-time and accuracy in scheduling decisions, resulting in low equipment utilization; scheduling strategies lacking global optimization, resulting in resource waste and increased energy consumption; and, a lack of experience or misjudgment of on-site management personnel leading to inefficient scheduling and increased production costs. These problems not only have a negative impact on the economic benefits of enterprises, but also exacerbate environmental pollution, such as the emission of large amounts of greenhouse gasses and dust, which pose a threat to the ecological environment.

Not only that, but open-pit mining equipment scheduling management also faces many other problems. Difficulties in fuel consumption control, delayed equipment maintenance, lack of security for goods, and low levels of intelligence not only lead to increased operating costs and low production efficiency for enterprises, but also exacerbate energy consumption and environmental pollution. Therefore, optimizing the scheduling of open-pit mining equipment through digital and intelligent means, improving scheduling efficiency, reducing operating costs, and reducing environmental impact have become urgent needs for achieving sustainable development of open-pit mines.

With the continuous improvement of industrialization and informatization levels in China, the trend towards intellectualization and digitization in the mining industry is inevitable [11–14]. To address the issues described above, digital and intelligent solutions can be employed to optimize the scheduling of trucks and excavators during the production process [15–18]. This aims to achieve optimal collaboration efficiency between trucks and excavators, thereby enhancing the production efficiency of enterprises and increasing their competitiveness in the mining and large-scale earthwork construction industries.

Currently, within the industry, truck dispatching systems primarily rely on manual scheduling. There is a lack of successful cases regarding automatic dispatching in the systematic operation of trucks and excavators. The intelligent scheduling algorithm developed through this project can effectively reduce the waiting time for excavators and trucks, improve the efficiency of earth stripping in opencast mines, decrease fuel consumption for excavators and dump trucks, and contribute to energy conservation, emission reduction, and the construction of green mines.

2. Related Work

The truck scheduling problem in opencast mines represents a highly intricate optimization challenge that has garnered significant attention over the past few decades. Researchers, including Wang Xu [19], have advanced a two-stage heuristic algorithm to tackle this issue, while Wei Ming [20] and others have introduced a discrete particle swarm optimization algorithm for the same purpose. To address the vehicle scheduling problem in opencast mines, certain scholars [21] have further refined the composition of transportation costs within a general model, incorporating factors such as truck wear and tear, tire damage, and driver salaries, thereby developing an optimization model aimed at minimizing total comprehensive costs. Additionally, other researchers [22] have established a time-constrained truck scheduling model to minimize total system costs and presented a simulated plant growth algorithm as a solution. Huo Xiaoyu [23] formulated a scheduling model focused on minimizing total truck operating costs and employed an adaptive mutation particle swarm optimization algorithm to solve it. Jian Jie [24] and colleagues constructed a scheduling model aimed at minimizing the combined costs of truck fuel consumption and fixed vehicle usage, developing a cloud-adaptive genetic algorithm tailored to this problem.

The aforementioned scholars primarily focused on single-truck-type scheduling problems. While the single-truck scheduling model excels in simplicity, ease of understanding, and low computational demand, particularly in small-scale or straightforward scheduling scenarios, its limitations in terms of scheduling efficiency and resource utilization become evident as the scale expands and requirements grow more complex. Conversely, the multi-truck scheduling model demonstrates superior performance in handling large-scale and complex scheduling demands, significantly boosting scheduling efficiency and resource utilization. Despite its higher computational load and model complexity, in practical settings such as opencast mines, optimizing scheduling strategies using this model can lead to substantial improvements in production efficiency. Therefore, some scholars have turned their attention to multi-truck-type scheduling. Notably, Ge Xianlong [25] designed a cloud-adaptive genetic algorithm to address dynamic demand issues in multi-truck-type scheduling. Subsequently, Ge Xianlong et al. [26] considered both fixed costs and fuel consumption costs of vehicle transportation, formulated a multi-truck-type scheduling model, and ultimately solved the problem using a quantum gene algorithm.

In the realm of genetic algorithms, the selection operation plays a crucial role in determining which individuals will serve as parents for the next generation. Currently, irrespective of whether the context involves multi-vehicle-type or single-vehicle-type scheduling, many algorithms utilize the roulette wheel selection method. However, this method is characterized by slower convergence and an increased risk of losing excellent individuals. In response to the challenge of truck scheduling in opencast mines, this paper introduces a genetic algorithm that integrates the elite preservation strategy with the roulette wheel selection method (see Table 1). To provide a comprehensive understanding of these two methods and their integration, we have included a table accompanied by detailed textual explanations.

Table 1. Introduction to Elite Preservation Strategy and Roulette Wheel Selection Method.

Method Name	Description	Advantages	Disadvantages
Elite Preservation Strategy	During the iteration process of the genetic algorithm, a portion of individuals with higher fitness are preserved to avoid the loss of excellent individuals.	Maintains elite genes in the population, accelerating the convergence process.	May lead to reduced population diversity.
Roulette Wheel Selection Method	Based on the fitness value of individuals, the probability of each individual being selected is calculated, and random selection is performed.	Maintains population diversity, avoiding premature convergence.	May result in the loss of excellent individuals.
Combination of Elite Preservation Strategy and Roulette Wheel Selection	Before roulette wheel selection, a portion of elite individuals are preserved, and the remaining individuals undergo roulette wheel selection.	Combines the advantages of both the elite preservation strategy and the roulette wheel selection method, maintaining population diversity while avoiding the loss of excellent individuals.	Higher implementation complexity.

The elite preservation strategy is a widely adopted enhancement technique for genetic algorithms. Its fundamental principle involves retaining a fraction of highly fit individuals during the genetic algorithm's iterative process to prevent the loss of superior solutions. This approach ensures the preservation of elite genes within the population, thereby accelerating the convergence rate. However, it may also compromise population diversity, increasing the likelihood of premature convergence.

The roulette wheel selection method, on the other hand, is a probabilistic selection technique grounded in individual fitness values. It assigns a selection probability to each individual based on its fitness and proceeds with a random selection process. This method preserves population diversity, mitigating the risk of premature convergence. Nevertheless, it may inadvertently lead to the elimination of excellent individuals, adversely impacting the algorithm's convergence speed and efficacy.

To address the limitations of these two methods, this paper introduces a genetic algorithm that integrates the elite preservation strategy with the roulette wheel selection method. In each iteration, a subset of elite individuals is preserved, while the remainder undergoes roulette wheel selection. This hybrid approach leverages the strengths of both methods, maintaining population diversity while safeguarding against the loss of superior individuals. Experimental results demonstrate that this method exhibits impressive convergence speed and effectiveness in tackling the truck scheduling problem in open-pit mines.

A comparative analysis of contemporary research methodologies for addressing the truck dispatching challenge in open-pit mines, as opposed to the methodology proposed in this paper, is presented in Table 2.

Table 2. Comparison of Research Methods on Open-Pit Mine Truck Dispatching.

Scholar	Research Method	Vehicle Type	Main Features/Contributions
Wang Xu et al. [19]	Two-stage heuristic algorithm	Single-truck-type	Proposed a heuristic approach for solving the problem
Wei Ming et al. [20]	Discrete particle swarm optimization (DPSO)	Single-truck-type	Introduced DPSO for solving the dispatching problem
Xu Maozeng et al. [21]	Comprehensive cost minimization model	Single-truck-type	Created a model considering various costs
Cao Qingkui et al. [22]	Simulated plant growth algorithm	Single-truck-type	Proposed an algorithm for solving the time-constrained problem
Huo Xiaoyu et al. [23]	Adaptive mutation particle swarm optimization (AMPSO)	Single-truck-type	Used AMPSO for cost minimization
Jian Jie et al. [24]	Cloud-adaptive genetic algorithm (CAGA)	Single-truck-type	Introduced CAGA for single-truck-type dispatching
Ge Xianlong et al. [25]	Cloud-adaptive genetic algorithm (for dynamic demands)	Multi-truck-type	Addressed dynamic demands in multi-truck-type dispatching
Ge Xianlong et al. [26]	Quantum gene algorithm (QGA)	Multi-truck-type	Included fixed and fuel costs in the model and solved it using QGA
This study	A Genetic Algorithm Incorporating Elite Preservation Strategy with Roulette Wheel Selection	Multi-truck-type	Proposes a method combining elite preservation strategy with roulette wheel selection, which avoids the loss of elite individuals and accelerates the convergence process.

3. Optimization Model for Truck Scheduling Objectives in Open-Pit Mines

3.1. Factors to Consider in Transportation Scheduling

The truck scheduling problem in open-pit mines aims to minimize transportation costs and equipment idleness by considering several critical factors. These include:

- Number and location of deployed excavators: Mines typically feature multiple excavation points, each with varying quantities of ore. Consequently, the determination of excavation locations and the number of excavators must be based on actual conditions. Typically, only one excavator is required at each excavation point (loading point).
- Number of required trucks and traffic frequency on each route: Due to factors such as equipment specifications and environmental conditions, the loading, unloading, and travel times of trucks are not fixed, posing challenges for precise operational planning. The optimized number of trucks and the delivery frequency on a given traffic route can only fulfill the basic scheduling requirements.
- Reducing truck waiting time: Larger trucks consume more fuel, and each restart consumes energy. Therefore, it is advisable to minimize truck restarts, ideally limiting them to shift changes and keeping trucks running as much as possible during other

times. Additionally, minimizing truck waiting time is crucial. When trucks are stopped waiting, their utilization rate decreases, leading to increased transportation costs.

By addressing these factors, effective truck scheduling in open-pit mines can be achieved, resulting in cost savings and improved operational efficiency.

3.2. Objective Functions

The truck scheduling process in open-pit mines is quite cumbersome, and its essence remains a vehicle sequencing problem. The mathematical model for vehicle scheduling in multiple depots can be expressed as follows:

At a mining site, a mixed fleet of trucks and excavators is used for construction operations. The mining site has n loading points D_i $(i = 1, 2, \ldots, n)$, with each loading point having a specific ore quantity q_i. The fleet departs from the depot, proceeds to each loading point to load ore, and then transports it to the unloading point for discharge.

During operation, trucks are not assigned to a specific excavator but are instead assigned to the required excavator within a specified time frame based on the actual situation of the operational objectives and plans. This approach enhances production efficiency and reduces the amount of required equipment and vehicle waiting times.

To achieve coordinated operation between trucks and excavators and reduce transportation costs, unified scheduling of the fleet is required, with a reasonable and efficient scheduling scheme design. Based on reducing transportation costs, truck waiting times, and excavator hook times, two objective functions are constructed: $F_1(S)$ represents the total transportation cost, and $F_2(S)$ represents the total number for truck waiting time and excavator hook time.

$$F_1(S) = \sum_k \sum_{(i,j)\in D} c_{ij}^k d_{ij} x_{ij}^k + \sum_g \sum_{(i,j)\in D} g_{ij}^g d_{ij} y_{ij}^g + \sum_k \sum_{(i,j)\in D} e_{ij}^k d_{ij} z_{ij}^k + \sum_g \sum_{(i,j)\in D} f_{ij}^g d_{ij} w_{ij}^g, \quad (1)$$

$$F_2(S) = \sum_k (T_k - \sum_{(i,j\in D)} t_{ij}^a x_{ij}^k - \sum_{(i,j\in D)} t_{ij}^b z_{ij}^k) + \sum_g (T_g - t_g). \quad (2)$$

In the formula,

c_{ij}^k represents the unit heavy-load transportation cost for trucks;
g_{ij}^g represents the unit heavy-load transportation cost for excavators;
e_{ij}^k represents the unit empty-load transportation cost for trucks;
f_{ij}^g represents the unit empty-load transportation cost for excavators;
d_{ij} represents the distance between nodes i and j.
x_{ij}^k and z_{ij}^k are indicators:
$x_{ij}^k = 1$ indicates that a truck is traveling from point i to point j with a heavy load; $x_{ij}^k = 0$ indicates that a truck does not travel from i to j;
$z_{ij}^k = 1$ indicates that a truck is traveling from point i to point j with an empty load; $z_{ij}^k = 0$ indicates that a truck does not travel from i to j empty;
y_{ij}^g and w_{ij}^g are indicators:
$y_{ij}^g = 1$ indicates that an excavator is traveling from point i to point j with a heavy load; $y_{ij}^g = 0$ indicates that an excavator does not travel from i to j;
$w_{ij}^g = 1$ indicates that an excavator is traveling from point i to point j with an empty load; $w_{ij}^g = 0$ indicates that an excavator does not travel from i to j empty.
i and j are node indices:
T_k represents the instance working duration of a truck;
t_{ij}^a represents the heavy-load operating time of a truck;
t_{ij}^b represents the empty-load operating time of a truck;
T_g represents the instance dwell time of an excavator;
t_g represents the instance working duration of an excavator.

3.3. Constraint Conditions

The constraint conditions of the model are as follows:

1. Path Constraint (Each demand point is served only once, and each path is traversed only once):

$$\sum_k \sum_{(i.j)\in D} x_{ij}^k = 1,\ i,\ j \in \{1,\ 2, \ldots, n\}, \tag{3}$$

$$\sum_g \sum_{(i.j)\in D} g_{ij}^g = 1,\ i,\ j \in \{1,\ 2, \ldots, n\}, \tag{4}$$

$$\sum_{(i,j)\in D} x_{ij} = \sum_{(i,j)\in D} x_{ji} \leq 1,\ i,\ j \in \{1,\ 2, \ldots, n\}. \tag{5}$$

2. The cargo quantity transported by each vehicle should not exceed its carrying capacity, and the loading quantity of the vehicle when leaving point i should not be less than the loading quantity when arriving at point i:

$$\sum_{(i,j)\in D} x_{ij}^k q_i^k \leq Q_{max}^k, \tag{6}$$

$$Q_i^k + q_i^k + M(1 - x_{ij}^k) \geq Q_j^k, \tag{7}$$

$$\sum_{(i,j)\in D} q_i^g y_{ij}^g \leq Q_{max}^g, \tag{8}$$

$$Q_i^g + q_i^g + M(1 - x_{ij}^k) \geq Q_j^g \tag{9}$$

In the formula, q_i represents the cargo-loading quantity of the vehicle at point i, Q_i represents the cargo-loading quantity of the vehicle when leaving point i, M represents an infinitely large number, Q_{max} represents the maximum carrying capacity of the vehicle, and Q_j represents the cargo-loading quantity of the vehicle when arriving at point j.

3. The total shipping volume of all vehicles is not less than the total demand at the demand points:

$$\sum_k \sum_{(i,j)\in D} q_i^k x_{ij}^k + \sum_g \sum_{(i,j)\in D} q_i^g y_{ij}^g \geq \sum_{i\in D} q_i. \tag{10}$$

4. The number of allocated vehicles does not exceed the total number of vehicles available:

$$\sum_{(i,j)\in D} x_{oi}^k + \sum_{(i.j)\in D} y_{oi}^g = \sum_{(i,j)\in D} Z_{j0}^k + \sum_{(i,j)\in D} w_{j0}^g \leq K + G. \tag{11}$$

In the formula, K and G represent the total number of trucks and excavators, respectively, with 0 denoting the departure point and unloading point.

5. Each excavator can only be assigned to one area within each time period:

$$\sum_{a\in A} \lambda_{g,a}^t = 1,\ \forall t \in T,\ g \in G. \tag{12}$$

In the formula,

$a \in A$ represents a specific mining area within the set of all mining areas;

$t \in T$ represents the time planning range;

$g \in G$ represents an excavator operating within the set of all excavators;

$\lambda_{g,a}^t$ indicates that if excavator g is located in area a during time period t, then $\lambda_{g,a}^t = 1$, otherwise $\lambda_{g,a}^t = 0$.

6. The constraint that an excavator is only allowed to operate within one region is as follows:

$$\sum_{g\in G} \lambda_{g,a}^t \leq G_a^{Max},\ \forall t \in T,\ a \in A. \tag{13}$$

In the formula, G_a^{Max} represents the maximum number of excavators allowed to operate within the same region.

7. Limitation on the number of trucks that need to be dispatched:

$$K^{Min} \leq k^t \leq K^{Max}, \forall t \in T. \quad (14)$$

In the formula, K^{Min} and K^{Max} represent the maximum and minimum number of trucks that need to be dispatched within any time period, respectively.

4. Design of Improved Genetic Algorithm Combining Roulette Wheel Selection with Elite Preservation Strategy

The Genetic Algorithm (GA) is an adaptive algorithm with global optimization capabilities, mimicking the genetic evolution process of organisms in nature. Owing to its robust global optimization ability, good resilience, and suitability for parallelization, GA has found widespread application in various domains, including vehicle sequencing, transportation, and supply chain network optimization. As a heuristic search algorithm, GA involves encoding the problem, initializing a population, calculating fitness values, selecting elite individuals, generating new individuals through crossover, introducing diversity through mutation, and checking termination conditions before outputting the optimal solution. The steps of applying GA to open-pit truck scheduling are illustrated in Figure 1. Initially, the problem solution is identified and formalized as a decoding process. An initial population is then created to generate a series of potential solutions. To evaluate the quality of each solution, the fitness of each decoder is calculated, quantifying the performance of each solution for subsequent selection. The algorithm terminates when the termination conditions are met, such as reaching a preset number of iterations or finding a satisfactory solution. If the conditions are not met, the algorithm continues to execute. In the selection step, based on the fitness values of the decoders, excellent individuals from the current population are chosen for replication, as they have higher fitness values and are more likely to produce better solutions in subsequent iterations. Finally, the decoders are enhanced through crossover and mutation operations, which combine the genetic information of different individuals and introduce random variations to generate new, potentially superior solutions. After multiple iterations and selections, the optimal solution is ultimately output, representing the best open-pit truck scheduling scheme.

In this paper, we introduce an enhanced genetic algorithm that incorporates an elite preservation strategy and a roulette selection method to address the challenges of premature convergence and sluggish convergence rates often encountered in traditional genetic algorithms during the search process. Traditional genetic algorithms generate new individuals through selection, crossover, and mutation operations, but this approach can occasionally result in the loss of high-quality genes during evolution, thereby compromising the algorithm's performance. To mitigate this issue, we introduce an elite preservation strategy. The core principle of this strategy is to retain the best individuals from each generation and directly propagate them to the next generation, ensuring that the algorithm does not lose high-quality solutions during the search process. This approach helps maintain population diversity, reduces the risk of premature convergence, and accelerates the convergence towards the global optimal solution.

Furthermore, we incorporate the roulette selection method to further refine our algorithm. Roulette selection allocates selection probabilities based on individuals' fitness ratios, with higher fitness individuals having a greater chance of being selected. This method not only ensures the inheritance of outstanding individuals but also maintains the competitiveness and diversity of the population. By effectively integrating the elite preservation strategy and the roulette selection method, our enhanced algorithm achieves a delicate balance between exploring new solution spaces and exploiting the current optimal solution, thereby significantly improving the global search efficiency and solution quality of the algorithm.

In conclusion, the improved genetic algorithm presented in this paper exhibits substantial advancements over traditional genetic algorithms through the incorporation of elite preservation strategies and roulette selection methods. These innovations not only enhance the algorithm's performance but also offer novel insights and methodologies for addressing complex optimization problems.

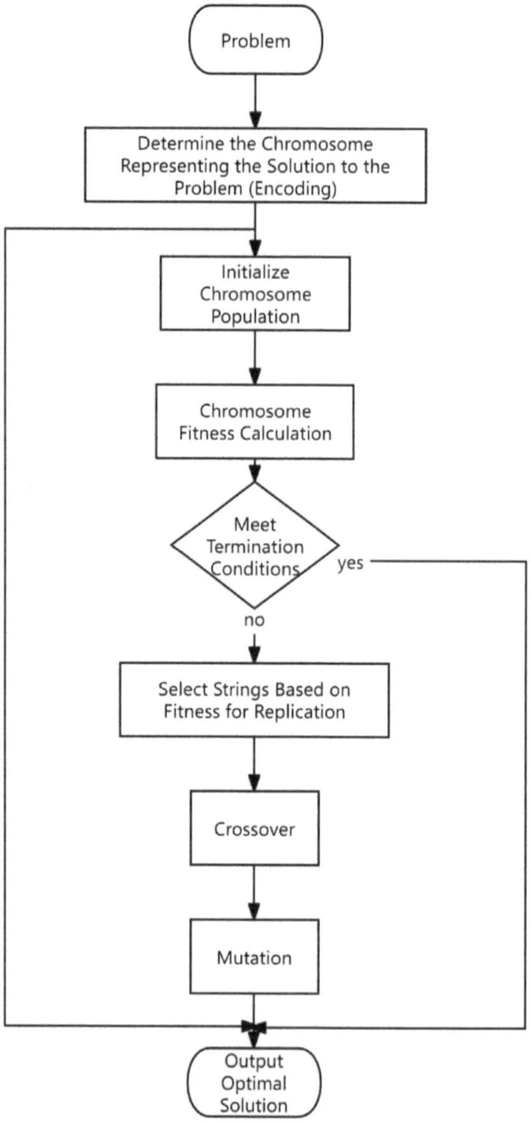

Figure 1. Flowchart of Genetic Algorithm for Open-Pit Truck Scheduling.

4.1. Encoding

When applying the genetic algorithm to solve the truck dispatching schedule problem in open-pit mines, selecting an appropriate chromosome encoding method is crucial as it directly influences the complexity and overall efficiency of system implementation. This paper chooses a symbolic encoding method that does not require decoding and is conducive to programming. This method refers to the use of symbols in the encoding process that, apart from $\{0,1\}$, do not carry any actual numerical significance. These symbols can be letters, numbers, or other characters, such as $\{a,\ b,\ c,\ d\ldots\}$, $\{1,\ 2,\ 3,\ 4\ldots\}$, or $\{a1,\ a2,\ a3,\ a4\ldots\}$.

From practical transportation scheduling, it can be observed that trucks generally travel back and forth between loading points and unloading points for transportation. Therefore, when encoding, it is necessary to first create three vector sets:

- Loading Point Set: *z_name*, represented by uppercase letters, for example, *z_name* = ['*ABC*'];
- Unloading Point Set: *x_name*, represented by lowercase letters, for example, *x_name* = ['*abc*'];
- Truck Set: *k_name*, represented by numbers, for example, *k_name* = [1, 2, 3, 4].

Then, within a certain period of time, a scheduling plan for all (for example, four) trucks can be represented as:

$$T = \begin{bmatrix} A & a & B & b & C & c \\ A & b & C & a & B & c \\ B & c & A & a & C & b \\ C & b & A & a & & \end{bmatrix}$$

That is, a vehicle scheduling plan for Truck 1 can be represented as 'AaBbCc', Truck 2 as 'AbCaBc', Truck 3 as 'BcAaCb', and Truck 4 as 'CbAa'.

In traditional genetic algorithms, the length of the chromosome (or individual) is typically fixed. However, for certain specific problems, a variable-length chromosome symbolic encoding may be more appropriate. A variable-length chromosome implies that the number of genes in different individuals may vary, depending on the specific requirements of the problem. To obtain such chromosomes, the following conditions must be met:

- Based on the unique circumstances of truck scheduling in open-pit mines, it is essential to first define the optimal working time period before developing the genetic algorithm for truck scheduling. This period can be one hour, one shift, or any other relevant time unit. During this specified time period, each program iteration will generate a vehicle scheduling plan. Once the set working period has elapsed, the program will reinitiate, repeating the process cyclically.
- When a truck completes loading at a loading point (excavator), its subsequent destination will be randomly selected from the unloading points that are currently unoccupied by other trucks (i.e., those that are free). Conversely, after a truck unloads at an unloading location, its next destination can only be randomly chosen from the loading points (excavators) that are not currently in use.

These conditions ensure that the variable-length chromosome encoding is tailored to the complexities of truck scheduling in open-pit mines, facilitating the development of an effective genetic algorithm for this specific application.

4.2. Initialization of Population

Firstly, a transportation scheduling plan is created for an arbitrary truck within a specified time period. This plan is then replicated and applied to other trucks to construct a chromosome. Subsequently, this chromosome is repeated N times to generate an initial population comprising N individuals.

This approach ensures that each individual in the initial population represents a potential solution to the truck scheduling problem in open-pit mines, laying the foundation for the application of genetic algorithms to optimize truck scheduling.

4.3. Fitness Function

In this study, the objective function is chosen as the fitness function, based on several key considerations. Firstly, as the cornerstone of the problem-solving process, the objective function directly embodies the optimization target we seek to achieve. By adopting it as the fitness function, we ensure that the genetic algorithm remains focused on this target throughout the search, thereby boosting the efficiency of identifying the optimal solution. Secondly, the objective function typically possesses a straightforward mathematical formulation, enabling swift calculation and comparison of individuals' fitness values. This simplifies the selection and reproduction operations within the genetic algorithm. Lastly, this choice streamlines the design and implementation of the algorithm, eliminating the need to devise a complex fitness function tailored to specific problems. In summary, utilizing the objective function as the fitness function offers numerous benefits, including clear goal orientation, computational efficiency, and simplicity in design, rendering it a prudent choice.

4.4. Selection

In the selection operation of the genetic algorithm for truck transportation scheduling, a combination of roulette wheel selection and elitist preservation is employed. The roulette wheel selection mechanism allocates selection opportunities based on the fitness proportion of individuals, where those with higher fitness have a greater chance of being selected. This mechanism embodies the natural principle of "survival of the fittest" and directs the algorithm towards evolving individuals with higher fitness.

The elitist preservation mechanism is a strategy designed to prevent the loss of optimal solutions. During the iteration process of the genetic algorithm, this mechanism identifies the top individuals (or several individuals with the highest fitness) in the current population and directly transfers them to the next generation, bypassing the crossover and mutation processes. This ensures continuous progress towards a superior solution domain during the evolution process and mitigates the risk of accidentally losing the current optimal solution due to crossover and mutation operations.

The combination of roulette wheel selection and elitist preservation maintains population diversity while ensuring that the algorithm converges to the global optimal solution. The specific implementation of this combined strategy is as follows:

1. Calculate the fitness values and directly retain the individuals with the highest fitness values into the next generation of the population:

Calculate the fitness value $f(x)$ for all individuals based on the fitness function $f(x_i)$, where $i = 1, 2, \ldots, n$, and n represents the population size.

This is usually achieved by evaluating the objective function value of each individual. The higher (or lower, depending on the nature of the problem) the objective function value, the higher the fitness of the individual. Then, calculate the sum of the fitness values for all individuals:

$$F = \sum_{i=1}^{n} f(x_i). \tag{15}$$

Based on the fitness values, select the optimal individuals (or several individuals with the highest fitness) in the current population as elite individuals. These elite individuals will be directly retained for the next generation of the population.

2. Calculate Selection Probabilities:

For individuals other than the elite individuals, the roulette wheel selection method is used. Based on the fitness value of each individual, calculate the probability of each individual being selected to pass its genes to the next generation of the population:

$$p_i = \frac{f(x_i)}{F}. \tag{16}$$

3. Calculate Cumulative Probabilities:

Calculate the cumulative probability for each chromosome:

$$q_i = \sum_{j=1}^{m} q_j. \tag{17}$$

where $j = 1, 2, \ldots, m$, m represents the number of iterations, and the cumulative probability q_i denotes the sum of the selection probabilities from the first individual to the i-th individual (inclusive of the i-th individual).

4. Roulette Wheel Selection:

Generate a uniformly distributed random number r within the interval [0, 1]. Then, starting from the first individual, compare the random number r with the cumulative probability q_i one by one. If r is less than or equal to q_i, select the i-th individual; otherwise, continue comparing r with the next cumulative probability q_{i+1} until the appropriate individual is found.

Through the roulette wheel method, individuals with higher fitness will have a greater probability of being selected to pass their genes to the next generation. Combine the elite individuals and the individuals selected through the roulette wheel method to form a new population for the next generation. During the merging process, ensure that the population size remains unchanged.

4.5. Crossover

The crossover operator in genetic algorithms mimics the process of genetic recombination in nature by combining the genetic information of different parent individuals to generate new offspring. This operation not only bolsters the genetic diversity of the population, leading to variations in the genetic composition of offspring, thereby aiding the algorithm in exploring a broader search space, but also fosters the dissemination of advantageous gene segments within the population. Through genetic recombination, new individuals with superior adaptability may emerge, accelerating the convergence of the algorithm towards the global optimal solution. Consequently, the crossover operator stands as a pivotal mechanism for achieving population evolution and optimized search in genetic algorithms.

Given the potential for task conflicts between different trucks, it is determined that crossover should be permitted solely between chromosomes of the same type, as opposed to traditional direct crossover, to mitigate the occurrence of such conflicts. Following the chromosome crossover operation, it is imperative to verify whether the individual satisfies the constraint conditions. If the conditions are unmet, the positions of trucks involved in task conflicts are iteratively adjusted and re-verified. If the conditions remain unmet, the individual must be regenerated. This process persists until the entire iteration is completed. Ultimately, the crossover results are evaluated, and any individual where the starting trip time of the last truck exceeds the predetermined scheduled time is removed from the schedule, as illustrated in Figure 2, where Truck #10, post-crossover, needs deletion until the starting time of the last truck's trip falls within the scheduled timeframe. In Figure 2, the symbol "#" denotes the vehicle number, with "#1" representing the first vehicle, for instance. The gray color signifies Individual 1 of the current generation, while the green color signifies Individual 2 of the current generation. The portion to the right of the dashed line indicates the section that exceeds the predetermined time.

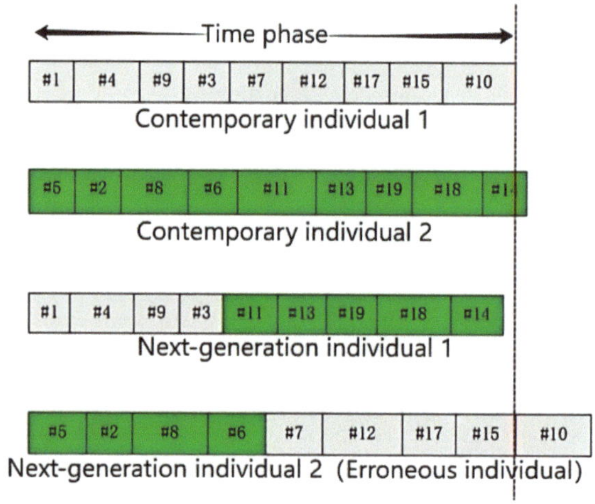

Figure 2. Schematic Diagram of Crossover Operation.

4.6. Mutation

In genetic algorithms, the mutation operation introduces random changes to the genes of an individual to generate new individuals, with the primary objective of enhancing the population's diversity. The core purpose of this operation is to maintain genetic diversity within the population, thereby mitigating the risk of premature convergence. As an auxiliary means of generating new individuals, mutation plays a crucial role. It can recover and supplement genetic information that may be lost during the crossover operation, preventing the algorithm from becoming trapped in local optimal solutions. This ensures that the genetic algorithm possesses robust local search capabilities.

In the application of genetic algorithms to truck transportation scheduling, the mutation operation is carried out as follows:

Set the population size of the genetic algorithm to N, and generate N random numbers a within the range defined by $(0, 1)$. On this basis, compare each of these N a values with the mutation probability P_m. If a b value is less than P_m, then for one randomly selected truck in its scheduling route, mutate either a loading point or an unloading point.

4.7. Generating a New Population

In genetic algorithms, new individuals are generated through the three mechanisms of selection, crossover, and mutation, collectively forming a new population of size N'. If condition $N' > N$ is met, the top N individuals with the highest fitness are selected to form the new population. If condition $N' < N$ is met, an additional $(N - N')$ individuals are randomly generated and added to the new population to ensure that its size remains consistent with the previous generation.

4.8. Termination Condition: Maximum Iteration Reached

Set a maximum number of iterations for the algorithm, and stop the algorithm when it has reached this specified number of iterations.

5. Model Application

5.1. Basic Data

Ultimately, the model is resolved using the specific context of an opencast coal mine in China. The mining area comprises a total of five loading points and three unloading points, with each loading point staffed by one excavator. Each unloading point necessitates a total of 3500 tons per shift. Figure 3 (The colors in the figure are only for aesthetic purposes and have no actual meaning) illustrates the mine's geographical map depicting the shovel points, while Table 3 enumerates the distances between each loading and unloading point. The garage houses 60 trucks (allocated in unequal numbers based on the daily work schedule), each with a maximum payload capacity of 50 tons. The operational costs for empty and loaded trucks are 40 CNY/km and 60 CNY/km, respectively, with average speeds of 30 km/h and 20 km/h. Similarly, the costs for empty and loaded excavators are 50 CNY/km and 70 CNY/km, respectively (considering that the excavators' positions are largely stationary, only the cost of one round trip is typically factored in). Based on the collected data, the excavator's loading time is 6 min, the truck's unloading time is 3 min, the average waiting time for trucks per shift is 1.25 h, and the average hook time for excavators per shift is 0.5 h.

Table 3. Distances Between Shovel Points and Unloading Points (km).

Site Location	Loading Point 1	Loading Point 2	Loading Point 3	Loading Point 4	Loading Point 5
Unloading point 1	1.6	3.4	3.3	2.2	2.5
Unloading point 2	2.0	3.6	2.9	1.6	1.4
Unloading point 3	1.7	3.2	3.1	1.7	1.6

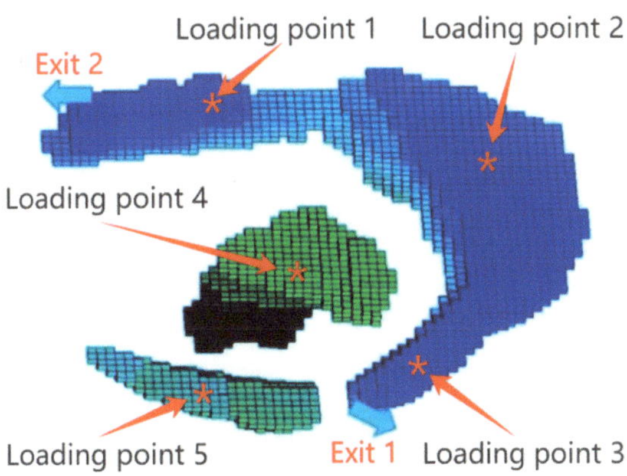

Figure 3. Geographic Map of the Mining Area with Shovel Points.

5.2. Result Analysis

By setting the population size to 100, crossover rate to 0.7, mutation rate to 0.05, and a maximum number of iterations to 100, the scheduling scheme was derived through the implementation of the improved genetic algorithm using MATLAB (2022a) programming. This was performed while considering scenarios with 30, 45, and 60 trucks, respectively, along with five excavators. It was presumed that the positions of the excavators were static and not subject to real-time scheduling adjustments.

5.2.1. Comparison of Transportation Costs

Figure 4 presents the iterative convergence comparison diagram for transportation costs when utilizing the genetic algorithm with 30 trucks. When employing the traditional genetic algorithm, the total transportation cost per shift amounts to CNY 136,582. In contrast, with the implementation of the improved genetic algorithm, the total transportation cost per shift decreases to CNY 129,545. When using the on-site fixed scheme, the total transportation cost per shift stands at CNY 141,500. Notably, the improved genetic algorithm achieves savings of 8.4% compared to the on-site fixed scheme and 5.2% compared to the traditional genetic algorithm. As illustrated in the diagram, the improved genetic algorithm converges to the minimum value around the 40th iteration, whereas the traditional genetic algorithm converges to the minimum value around the 60th iteration. The differing initial values at 0 iterations for the two algorithms stem from the fact that different genetic algorithms may adopt distinct strategies or methods for generating initial solutions during population initialization. Consequently, these initial solutions may exhibit variations in fitness, resulting in differing initial values in the iterative convergence diagram.

Figure 5 presents the iterative convergence comparison diagram for transportation costs when utilizing the genetic algorithm with 45 trucks. When employing the traditional genetic algorithm, the total transportation cost per shift amounts to CNY 202,034. In contrast, with the implementation of the improved genetic algorithm, the total transportation cost per shift decreases to CNY 194,298. When using the on-site fixed scheme, the total transportation cost per shift stands at CNY 212,250. Notably, the improved genetic algorithm achieves savings of 8.5% compared to the on-site fixed scheme and 3.8% compared to the traditional genetic algorithm. As illustrated in the diagram, the improved genetic algorithm converges to the minimum value around the 40th iteration, whereas the traditional genetic algorithm converges to the minimum value around the 80th iteration.

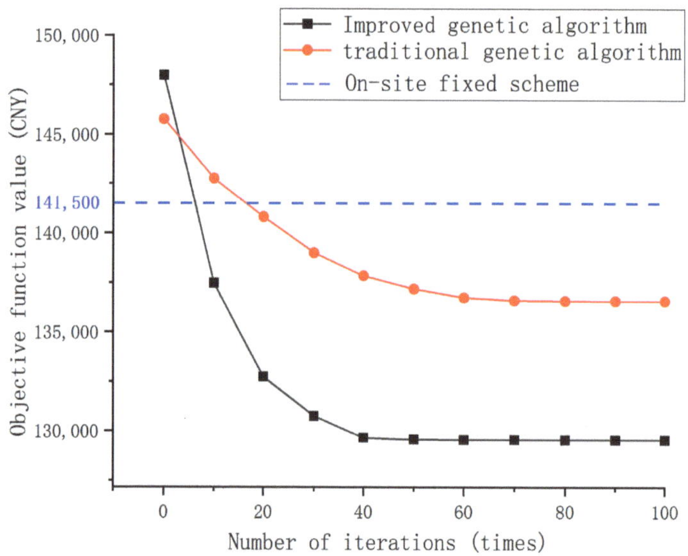

Figure 4. Iterative Convergence Comparison of Transportation Costs Using Genetic Algorithm with 30 Trucks.

Figure 5 presents the iterative convergence comparison diagram for transportation costs when utilizing the genetic algorithm with 45 trucks. When employing the traditional genetic algorithm, the total transportation cost per shift amounts to CNY 259,581. In contrast, the implementation of the improved genetic algorithm reduces this cost to CNY 249,651 per shift. The on-site fixed scheme, on the other hand, results in a total transportation cost of CNY 280,000 per shift. Notably, the improved genetic algorithm achieves cost savings of 10.8% compared to the on-site fixed scheme and 3.8% compared to the traditional genetic algorithm. As illustrated in the diagram, the improved genetic algorithm converges to its minimum value around the 50th iteration, whereas the traditional genetic algorithm converges to its minimum value around the 60th iteration.

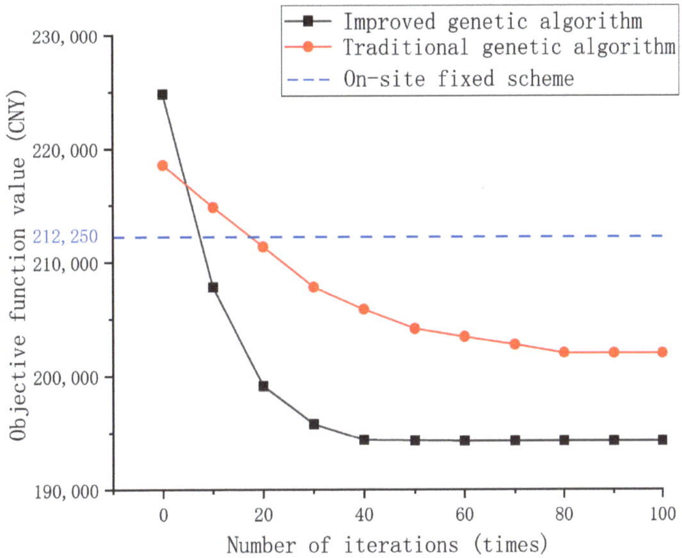

Figure 5. Iterative Convergence Comparison of Transportation Costs Using Genetic Algorithm with 45 Trucks.

5.2.2. Comparison of Equipment Idle Time

The waiting times for trucks and the hook times for excavators using different methods are shown in Table 4.

Table 4. Comparison Chart of Equipment Idle Time.

Number of Trucks	On-Site Fixed Solution/Traditional Genetic Algorithm/Improved Genetic Algorithm		
	Truck Waiting Time	Excavator Hook Time	Equipment Idle Time
30	37.5 h/36.1 h/33.9 h	1.5 h/0.9 h/0.7 h	39 h/37 h/34.6 h
45	56.3 h/52.8 h/50.3 h	2.2 h/1.3 h/1 h	58.5 h/54.1 h/51.3 h
60	75 h/71.8 h/67.7 h	3 h/1.8 h/1.3 h	78 h/73.6 h/69 h

When utilizing 30 trucks, the implementation of the improved genetic algorithm results in savings of 11.3% compared to the on-site fixed scheme and 6.5% compared to the standard genetic algorithm. With 45 trucks, the savings amount to 12.3% and 5.1%, respectively. For 60 trucks, the savings are 11.5% and 6.2%, respectively.

In summary, the adoption of the improved genetic algorithm leads to substantial improvements in transportation costs, truck waiting times, and excavator hook times when compared to the on-site fixed scheme. It also demonstrates considerable enhancements over the standard genetic algorithm. Furthermore, as Figures 4–6 illustrate, the improved genetic algorithm converges more rapidly than the standard genetic algorithm.

In real-world applications, we have employed my algorithm to orchestrate the intelligent scheduling of 500 trucks, excavators, and other pieces of operational equipment at the site. This has led to a marked enhancement in operational efficiency and a considerable reduction in energy consumption. Specifically, the algorithm achieves real-time and pinpoint scheduling by precisely calculating job specifications, location details, and task hierarchies for each device. This not only fosters seamless collaboration among devices but also minimizes waiting periods and the resultant resource wastage due to inefficient scheduling. By deploying this algorithm, we have successfully elevated earthwork stripping efficiency in open-pit mines by more than 10%.

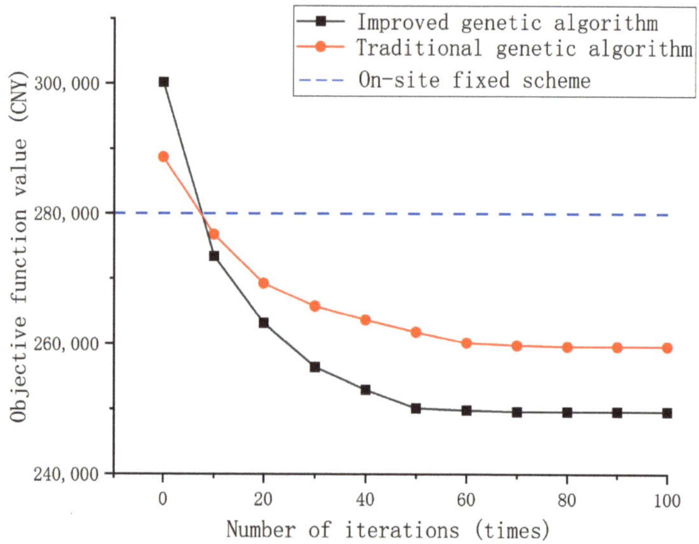

Figure 6. Iterative Convergence Comparison of Transportation Costs Using Genetic Algorithm with 60 Trucks.

In terms of energy conservation, the algorithm significantly cuts down the fuel consumption of excavators and dump trucks by optimizing the operational routes and scheduling protocols of the equipment. Through intelligent scheduling, it eliminates redundant movements and repetitive operations among equipment, resulting in a 3% decrease in fuel consumption. This achievement not only contributes positively to environmental sustainability, but also yields substantial energy cost savings for the enterprise.

In conclusion, by applying the algorithm developed by our institute to orchestrate the intelligent scheduling of 500 trucks, excavators, and other pieces of operational equipment, we have achieved a marked enhancement in operational efficiency and a considerable reduction in energy consumption. This practical implementation underscores the robustness and practicality of the algorithm in real-world scenarios and highlights its immense potential for application in complex industrial settings.

6. Conclusions

Addressing the truck transport scheduling problem in open-pit mines, we have developed truck scheduling models with the objectives of minimizing total transportation costs, truck waiting times, and excavator hook times. These models take into account the issue of excessive truck waiting times due to improper truck–shovel coordination, aiming to maximize output and minimize equipment idling to achieve optimal coordination efficiency between trucks and excavators.

In solving the truck scheduling model using the genetic algorithm, we adopt variable-length symbolic encoding. This encoding method allows for flexible adjustment of the encoding length based on the characteristics and requirements of the problem. Compared to fixed-length encoding, variable-length encoding can more effectively utilize space, particularly for problems with significant variations in the number of features. For problems requiring high-precision representation, variable-length encoding provides a finer encoding method. By increasing the encoding length, the precision of the representation can be enhanced, more accurately reflecting the problem's characteristics.

During the selection process, we employ a combination of roulette wheel selection and elitist preservation strategies. This approach effectively balances the diversity of the population and the convergence of the algorithm, keeping the algorithm from falling into local optimal solutions while ensuring that it can converge to the global optimal solution. This improves the performance and efficiency of the genetic algorithm.

Author Contributions: Defining Research Approach, F.L.; Proposing Paper Structure, Z.S.; Conducting Current Situation Investigation, W.D. and Y.G.; Establishing the Model, F.L. and Z.S.; Studying Algorithms, F.L., Z.S. and W.D.; Experimental Design, F.L., Z.S. and Y.G.; Experimental Simulation and Analysis, F.L. and Z.S. All authors have read and agreed to the published version of the manuscript.

Funding: Xiangjiang Laboratory Fund Project of China "Research and Application of Optimization Algorithm for Target Recognition Accuracy in Image Shooting of Electric Power Equipment Drone Inspection" (23XJ03016).

Data Availability Statement: The data in the paper comes from the enterprise, and due to the involvement of core production-related data, the enterprise has confidentiality requirements.

Conflicts of Interest: The authors declare no conflict of interest.

References

1. Yin, G. Discussion on Slope Stability Analysis and Remediation in Deep Mining of Open-pit Mines. *World Nonferrous Met.* **2024**, *10*, 223–225.
2. Smith, A.; Linderoth, J.; Luedtke, J. Optimization-based Dispatching Policies for Open-pit Mining. *Optim. Eng.* **2021**, *22*, 1347–1387. [CrossRef]
3. Liu, L. Application of Intelligent Mining Technology in Open-pit Mines. *World Nonferrous Met.* **2022**, *12*, 66–68.
4. An, Z. Analysis of Slope Stability in Open-pit Mining. *S. Metall.* **2022**, *45*, 169–171.
5. Mirzaei-Nasirabad, H.; Mohtasham, M.; Askari-Nasab, H.; Alizadeh, B. An Optimization Model for the Real-time Truck Dispatching Problem in Open-pit Mining Operations. *Optim. Eng.* **2023**, *24*, 2449–2473. [CrossRef]
6. Mohammad, S.; Yashar, P.; Mehdi, R. Optimisation of Open-pit Mine Production Scheduling Considering Optimum Transportation System Between Truck Haulage and Semi-mobile In-pit Crushing and Conveying. *Int. J. Min. Reclam. Environ.* **2022**, *36*, 142–158.
7. Yaw, S.A.; Matthias, P.T.; Eugene, A.B. A Multiple Objective Genetic Algorithm Approach for Stochastic Open Pit Production Scheduling Optimisation. *Int. J. Min. Reclam. Environ.* **2023**, *37*, 460–487.
8. Appianing, E.J.A.; Eugene, A.B.; Yashar, P. Life-of-mine Optimization for Integrated Open Stope Development and Production Scheduling Using a Mixed-Integer Linear Programming Framework. *Min. Technol.* **2023**, *132*, 106–120. [CrossRef]
9. Wang, P. A Brief Discussion on the Management of Transportation Machinery and Equipment in Open-pit Mines. *Constr. Mach. Maint.* **2022**, *5*, 139–141.
10. Feng, D. Analysis of Influencing Factors on Truck Fuel Consumption in Open-Pit Mines Based on Multi-Dimensional Perspectives. *Inn. Mong. Coal Econ.* **2021**, *20*, 123–125. [CrossRef]
11. Ma, X.; Lian, M.; Lu, C.; Yang, Z. Conceptual Model for Information Integration in Digital Mines. *Met. Mines* **2014**, *7*, 135–140.
12. Gu, D. Intelligent Mining Touches the Future of Mining Industry. *Min. Equip.* **2014**, *1*, 24–26.
13. Gu, Q.; Lu, C.; Jiang, S.; Yun, Q. Research Progress and Development Trends of Mining Systems Engineering. *Met. Mines* **2016**, *7*, 26–33.
14. Bai, C.; Lu, C.; Gu, Q. GIS/BDS/GPRS-based Data Management System for Blast Heap in Open-pit Mines. *Met. Mines* **2015**, *3*, 148–151.
15. Chen, X. *Open-Pit Mining of Metal Deposits*; Metallurgical Industry Press: Beijing, China, 2013.
16. Alarie, S.; Gamache, M. Overview of Solution Strategies Used in Truck Dispatching Systems for Open Pit Mines. *Int. J. Surf. Min.* **2002**, *16*, 59–76. [CrossRef]
17. Li, J.; Cai, M. Automatic Dispatching System for Vehicles in Open-Pit Metal Mines and Its Application. *China Min. Mag.* **2003**, *1*, 37–40.
18. Liu, Y.; Pan, H.; Wen, P.; Hu, S.; Sun, X. Research and Development of Simulation System for Operating Process of Electric Shovel-Truck in Open-Pit Mine. *Opencast Min. Technol.* **2017**, *32*, 35–38.
19. Wang, X.; Ge, X.; Dai, Y. Research on Dynamic Vehicle Scheduling Problem Based on Two-stage Solution Algorithm. *Control. Decis.* **2012**, *27*, 175–181.
20. Wei, M.; Jin, W. A Discrete Particle Swarm Optimization Algorithm for Solving Vehicle Routing Problem. *Comput. Sci.* **2010**, *37*, 187–191.

21. Xu, M.; Yu, G.; Zhou, X.; Ge, X. Low-Carbon Vehicle Scheduling Problem with Minimum Comprehensive Cost and Its Algorithm. *Comput. Integr. Manuf. Syst.* **2015**, *21*, 1906–1914.
22. Cao, Q.K.; Liu, X.Y.; Ren, X.Y. Vehicle Scheduling Problem Based on Plant Growth Simulation Algorithm. *Syst. Eng. Theory Pract.* **2015**, *35*, 1449–1456.
23. Huo, X.; Yang, S.; Wu, C. Particle Swarm Optimization for Transportation Scheduling System in Open-Pit Mines. *J. China Coal Soc.* **2012**, *37*, 234–239.
24. Jian, J.; Wang, X.; Ge, X. Cloud Adaptive Genetic Algorithm for Vehicle Scheduling Optimization with Capacity Constraints. *J. C. Univ.* **2013**, *36*, 40–46.
25. Ge, X.; Wang, X.; Xing, L. Multi-Type Vehicle Scheduling Problem with Dynamic Demands and Cloud Genetic Algorithm. *J. Syst. Eng.* **2012**, *27*, 823–832.
26. Ge, X.; Xu, M.; Wang, W. Research on Quantum Genetic Algorithm for Multi-Type Vehicle Routing Problem. *Chin. J. Manag. Sci.* **2013**, *21*, 125–133.

energies

MDPI

Article

Well-Production Forecasting Using Machine Learning with Feature Selection and Automatic Hyperparameter Optimization

Ruibin Zhu [1,2], Ning Li [2,3], Yongqiang Duan [2,3], Gaofeng Li [2,3], Guohua Liu [2,3], Fengjiao Qu [2,3], Changjun Long [2,3], Xin Wang [2,3], Qinzhuo Liao [1,*] and Gensheng Li [1]

1 College of Petroleum Engineering, China University of Petroleum-Beijing, Beijing 102249, China; cy1_zrb@petrochina.com.cn (R.Z.); ligensheng@cup.edu.cn (G.L.)
2 Research Institute of Oil and Gas Technology, PetroChina Huabei Oilfield Company, Renqiu 062552, China; cyy_lin@petrochina.com.cn (N.L.); cy4_dyq@petrochina.com.cn (Y.D.); cyy_ligf@petrochina.com.cn (G.L.); cyy_liugh@petrochina.com.cn (G.L.); yjy_qfj@petrochina.com.cn (F.Q.); cyy_longcj@petrochina.com.cn (C.L.); cyy_wangxin@petrochina.com.cn (X.W.)
3 Key Laboratory of Low-Permeability and Extra-Low-Permeability Reservoir Stimulation, Renqiu 062552, China
* Correspondence: liaoqz@cup.edu.cn

Abstract: Well-production forecasting plays a crucial role in oil and gas development. Traditional methods, such as numerical simulations, require substantial computational effort, while empirical models tend to exhibit poor accuracy. To address these issues, machine learning, a widely adopted artificial intelligence approach, is employed to develop production forecasting models in order to enhance the accuracy of oil and gas well-production predictions. This research focuses on the geological, engineering, and production data of 435 fracturing wells in the North China Oilfield. First, outliers were detected, and missing values were handled using the mean imputation and nearest neighbor methods. Subsequently, Pearson correlation coefficients were utilized to eliminate linearly irrelevant features and optimize the dataset. By calculating the gray correlation degrees, maximum mutual information, feature importance, and Shapley additive explanation (SHAP) values, an in-depth analysis of various dominant factors was conducted. To further assess the importance of these factors, the entropy weight method was employed. Ultimately, 19 features that were highly correlated with the target variable were successfully screened as inputs for subsequent models. Based on the AutoGluon framework, model training was conducted using 5-fold cross-validation combined with bagging and stacking techniques. The training results show that the model achieved an R^2 of 0.79 on the training set, indicating good fitting ability. This study offers a promising approach for the development of oil and gas production forecasting models.

Keywords: machine learning; production forecast; data preprocessing; principal component analysis; AutoGluon

Academic Editors: Manoj Khandelwal and Fernando Sánchez Lasheras

Received: 4 November 2024
Revised: 28 November 2024
Accepted: 26 December 2024
Published: 30 December 2024

Citation: Zhu, R.; Li, N.; Duan, Y.; Li, G.; Liu, G.; Qu, F.; Long, C.; Wang, X.; Liao, Q.; Li, G. Well-Production Forecasting Using Machine Learning with Feature Selection and Automatic Hyperparameter Optimization. *Energies* **2025**, *18*, 99. https://doi.org/10.3390/en18010099

1. Introduction

In the oil and gas industry, reservoir evaluation is a key parameter for petroleum engineers to make accurate decisions during extraction [1]. The well-production rate is one of the important parameters in reservoir evaluation, which can assist in reservoir modeling and numerical simulation and guide reservoir development strategies [2]. From current technological developments, traditional well-production prediction generally adopts three methods: Arps decline curve prediction, machine learning methods, and combination model prediction [3]. The Arps decline curve prediction method can achieve visualization of prediction results through software [4]. Machine learning methods have excellent self-learning abilities and diverse algorithms, and their performance largely depends on the quality and quantity of data; however, they can efficiently process large-scale datasets [5]. The combination model prediction method has a wide range of applicability, can reduce systematic errors, and has strong model stability [6]. Arps decline curve prediction includes exponential decline, hyperbolic decline, and harmonic decline [7]. You et al. proposed a method for predicting the economic recoverable reserves of SAGD units through the decline constant method based on the Arps index decline and hyperbolic decline, as well as a new approach for predicting technical recoverable reserves using the intercept divided by slope method [8]. Chen et al. conducted a comparative analysis of three models, Arps, SPED, and MFF, and evaluated the predictive performance of the decline curve model and the combination of the decline curve and data-driven neural network model [9]. The prediction of the Arps decline curve requires a sufficiently long production time for the oil field to detect the trend of production decline, which is suitable for analyzing constant-pressure production situations. Due to its simplification of well assumptions, it cannot accurately estimate the actual well production [10].

Machine learning is an important branch of artificial intelligence (AI) that aims to enable computers to acquire knowledge and experience from data through the learning and automated reasoning of computer systems, and to use this knowledge and experience for pattern recognition, prediction, and decision-making [11]. Machine learning contains a large number of algorithms and models. For well-production prediction, machine learning methods establish a linear relationship between independent variables (geological parameters, production time, etc.) and dependent variables (production) [12]. Hou et al. applied machine learning methods to predict the porosity and permeability of reservoirs. Research has shown that logging parameters have a significant impact on the prediction results of porosity and permeability, and an optimal adaptive model can be selected [13]. In recent decades, some algorithms in machine learning have become increasingly frequent in predicting well production. Among the numerous algorithms in machine learning, ANN stands out because of its unique structure and learning ability, making it an effective means of solving complex problems [14]. Due to the strong nonlinear fitting ability of ANN, its prediction results are usually more accurate than those of traditional linear regression or time series analysis [15]. In production prediction, ANN can automatically learn patterns from historical data and predict future yields based on these patterns [16]. Amr et al. used machine learning methods to simultaneously train oil well productivity for multiple blocks. The monthly oil production is designated as the dependent variable of the model, which improves the accuracy of the prediction. They improved the robustness of the model by increasing the amount of data and studied the impact of the input variables on prediction accuracy [11]. ANN is a very suitable method for production prediction.

In terms of production prediction, in addition to the outstanding performance of ANN, other machine learning algorithms also play an important role. Chakra et al. constructed an oil and gas cumulative production prediction model based on high-order neural networks (HONN). HONN can effectively express the linear and nonlinear relationships between input variables. The model was trained using data from a sandstone reservoir in India, and the trained model was used to predict the cumulative production of oil and gas. Research shows that even in situations where on-site data are relatively insufficient, the model still exhibits good predictive ability and high accuracy [17]. Aizenberg et al. established an oil and gas production prediction model based on a complex neural network (MLMVN) with multi-valued neurons that can accurately achieve the dynamic prediction of oil and gas field production [18]. Shoeibi et al. used long short-term memory (LSTM) networks to predict oil and gas production capacity and validated the effectiveness of the model through multiple numerical simulations and field data. The research results indicate that the model exhibits good predictive performance in both stable and dynamic production modes [19]. Lolon et al. used a multivariate statistical model to evaluate the relationship between well parameters and production in the Bakken and Three Forks formations. The study found that in tight reservoirs, the total fracturing fluid volume and proppant dose during hydraulic fracturing are the main engineering parameters that affect production [20]. Song et al. conducted a nonlinear analysis of multiple factors using the random forest method for a certain ultra-low-permeability reservoir, identifying the main influencing factors on the initial production capacity of the reservoir. They overcame the limitation of the gray wolf algorithm, which can only identify a single influencing factor using the random forest method. They combined random forest and gray wolf algorithms to predict and optimize the production capacity of oil wells, thereby achieving a fast training speed prediction model [21]. Zhang et al. proposed an unconventional oil and gas production prediction model based on local preserving projection (LPP). This model can accurately capture the nonlinear characteristics of various parameters. Through parameter testing and comparative experiments, research has shown that the LPP-based model has good adaptability and effectiveness [22]. In the same year, Cao et al. used machine learning algorithms to predict production capacity by combining geological data, historical production capacity, and pressure data. He used an artificial multi-layer perceptron to complete two tasks: to predict the future production capacity of existing oil wells based on historical data and to predict the production capacity of new wells based on the historical production of adjacent wells under similar geological conditions [15]. Wang et al. proposed a method that combines multi-layer perceptron (MLP) and long short-term memory (LSTM) networks to predict shale gas production based on geological and fracturing reservoir parameters, historical data, and other information. On the basis of reservoir numerical simulation, they constructed a dataset and trained the model to achieve high accuracy [23]. Rober S et al. developed a complete well-completion design optimization and rapid economic benefit evaluation process based on an ANN. In terms of data processing, they used a self-organizing feature map (SOM) clustering neural network to associate and classify data with reservoir types and completion features [24]. Dong et al. used the XGBoost algorithm to predict the initial production capacity of sandstone reservoirs and improved the model loss function, enhancing the physical constraints of data mining algorithms and significantly improving the prediction accuracy [25]. Khan et al. optimized the fracturing design for heterogeneous oil reservoirs and evaluated its accuracy using various machine learning methods. They found that the XGBoost model performed well in productivity prediction and significantly improved prediction performance [26].

On the basis of machine learning, the combination model prediction method combines the advantages of multiple prediction methods and can combine different machine learning algorithms to improve the accuracy and stability of predictions [11]. Berneti et al. combined the International Accounting Association (ICA) with an Artificial Intelligence Network (ANN) to develop a model for predicting oil well productivity. The model combines the local search function of the BP algorithm and the global search feature of the ICA. Based on the production data of 31 wells, the researchers use the ICA-ANN prediction model to forecast the production capacity and compare it with the ANN prediction results. The results show that the ICA-ANN model has good performance [27]. Aditya Vyas cleverly used machine learning to link the descent curve model with well-completion parameters. They proposed an evaluation criterion by combining the best descent curve and accurate EUR prediction with machine learning, providing a new method for predicting oil well productivity [28]. Noshi et al.'s joint decision-making based on multiple decision trees can effectively handle multidimensional data and has a strong ability to avoid overfitting [29].

Adesina et al. significantly improved the processing efficiency of nonlinear data by mapping data to a high-dimensional space using the kernel function method [30]. Considering that individual algorithms in machine learning have certain errors in yield prediction, the combination model prediction method provides a new approach for improving accuracy. AutoGluon is an open-source framework launched by Amazon with automated machine learning (AutoML) capabilities aimed at simplifying the training and deployment process of machine learning models [31]. AutoGluon utilizes the available computing resources to find the most powerful models in its allocated runtime, providing a user-friendly interface and documentation support. Nick et al. found that multi-layer combinations of many models make better use of allocated training time than finding the best model. Testing 50 classification and regression tasks on Kaggle and OpenML AutoML Benchmarks using AutoGluon showed that AutoGluon is faster, more robust, and more accurate [32].

However, there are significant limitations in the current research on the analysis of the main control factors of production capacity in intelligent production forecasting models. Most studies only analyze engineering or geological parameters separately and fail to achieve an organic combination of the two. This type of single-factor analysis method makes it difficult to fully reflect the production dynamics of oil and gas reservoirs, thereby affecting the accuracy of production capacity prediction. Therefore, there is an urgent need to develop an intelligent prediction model that comprehensively considers both geological and engineering factors to improve the accuracy of production capacity prediction. In addition, most of the current research on capacity prediction models is based on a single model. AutoGluon can automatically combine multiple models and conduct integrated learning, comprehensively utilize the advantages of each model, reduce the limitations of a single model, and enhance the stability of the trained model.

2. Methodology

2.1. Data Introduction

This study collected data from 435 fractured wells in the PetroChina North China Huabei Oilfield, encompassing over 100 characteristic parameters, ranging from quantitative to categorical data. From these parameters, 13 geological parameters and 14 engineering parameters were selected as input features due to their significance and relatively low rates of missing values. The cumulative oil production over 1800 days served as the output target [33]. However, there are issues with missing values, outliers, and non-standard on-site data that need to be addressed through preprocessing steps, such as data imputation and outlier identification.

2.2. Data Governance Process

The main problems with the data include missing data, presence of outliers, and confusion of labeled data. Data governance refers to filling in missing data, filtering out outliers, and standardizing labeled data according to certain processes and methods. The data governance process, also known as the data preprocessing process, can be divided into the following steps:

1. Data exploration and analysis: Data analysis is the process of examining, analyzing, and reviewing data to collect statistical information regarding its quality, starting with an investigation of existing data and its characteristics. Using histograms and box plots, the distribution of the data can be understood, including the shape of the data and the distribution of outliers and missing data.
2. Outlier screening: Based on the conclusions obtained from the data exploration and analysis, select data fields with outliers. Outlier recognition methods are used to

identify and reset outliers to null values, which can subsequently be filled using methods such as mean, mode, median, or quartile values.

3. Missing data processing: Visualize the degree of missing data, explore and analyze the data to understand the types of missing data, and use different filling methods (such as mean, median, or mode filling) based on the reasons and extent of missing data to fill all data, ensuring that there are no missing values in the overall dataset.

These steps are carried out sequentially, starting with data exploration and analysis to understand the distribution of the data. Subsequently, abnormal outlier data are screened out, and the empty outlier data can be processed in subsequent data filling. In addition, before filling in data, all data that needs to be cleaned up or left blank should be processed completely, and then multiple methods should be used to fill in the empty values in the data; After data filling, a dataset without null values was obtained.

2.3. *Main Control Factor Analysis*

According to the characteristics of the data and the requirements of machine learning algorithms, it is crucial to process the parameters appropriately to enhance the model's performance and effectiveness. Once data governance and target parameter selection are complete, analyzing the relationships between these parameters becomes essential [34]. This analysis aims to identify data fields that exhibit high correlation and strong representativeness with the target parameters. The selected fields are then utilized for tasks such as feature selection and variable screening in modeling, prediction, and classification.

Data correlation analysis refers to a method that calculates the correlation coefficient between two or more variables or employs graphical representations to illustrate the strength and direction of the relationships between these variables. Common techniques include the Pearson correlation coefficient, Spearman correlation coefficient, scatter plots, and heat maps. This type of analysis is invaluable, as it aids in uncovering hidden connections and influencing factors within the data, enabling causal inference, hypothesis testing, and other related work.

In this study, the Pearson correlation coefficient method was used to eliminate linearly irrelevant data, thereby ensuring a correlation between features. Additionally, the study calculated the gray correlation degree, maximum mutual information, AutoGluon feature importance, and SHAP importance. By combining these metrics with the entropy weight method for a comprehensive evaluation, the primary controlling factors were analyzed and ranked in depth.

2.4. *AutoGluon*

2.4.1. Introduction to AutoGluon

The research method chosen in this study is the combined model prediction method AutoGluon. It is capable of automating the entire process from data preprocessing to model training, hyperparameter optimization, and result evaluation. It is particularly suitable for tabular data, image data, and text data, allowing users to quickly build high-performance machine learning models with minimal coding. The AutoGluon framework can automatically train a variety of models, automatically select hyperparameters, and perform integrated learning on well-performing models. In this process, only hyperparameters need to be set, including the number of layers of the integrated training stack and the number of cross-validation. This makes AutoGluon suitable not only for data scientists but also for developers without an extensive background in machine learning, enabling them to rapidly apply machine learning techniques to real-world problems.

2.4.2. Core Features of AutoGluon

Automation: One of AutoGluon's core strengths lies in its fully automated machine learning workflow. It can automatically select the most suitable algorithms based on the characteristics of the data and perform hyperparameter tuning. This process not only reduces the workload for users but also lowers the requirement for domain expertise, allowing users without a deep background in machine learning to build high-performance models. The automation features of AutoGluon make machine learning more accessible and widely applicable.

Multi-task Support: AutoGluon supports multiple task types, including classification, regression, and time series forecasting. This extensive task support provides users with great flexibility, enabling AutoGluon to handle various data types and application scenarios. Whether dealing with structured tabular data or complex image and text data, AutoGluon offers effective solutions.

Simplicity and Efficiency: AutoGluon offers a simple and intuitive API, allowing users to perform model training and prediction using only a few lines of code. This simplification significantly lowers the entry barrier, enabling both beginners and experts to quickly start and work efficiently. Additionally, the AutoGluon framework utilizes parallel processing and GPU acceleration technologies to significantly enhance the speed of model training and prediction. This efficiency allows AutoGluon to handle large datasets in a relatively short amount of time, saving users' valuable time and computational resources.

2.4.3. Working Principles of AutoGluon

AutoGluon employs intelligent search strategies, such as Bayesian optimization and grid search, to automatically select the best model and optimize its hyperparameters. This intelligent search process ensures that the model's performance on a given dataset is maximized. AutoGluon can efficiently explore and evaluate multiple model configurations to determine the optimal combination, thereby maximizing both prediction accuracy and model stability. Additionally, by using ensemble learning strategies, this approach combines predictions from multiple models employing techniques like bagging and stacking. This not only enhances the model's stability and generalization capabilities, but also compensates for the shortcomings of individual models. Even if a particular model performs poorly under certain conditions, the predictions from other models can still maintain the overall accuracy and reliability of the predictions.

2.4.4. Steps for Using AutoGluon

The use of AutoGluon for model training involves a relatively simple process, making it accessible to users with varying backgrounds. First, users need to prepare the training data, which typically include identifying features and target variables, and configuring relevant parameters based on their project requirements, such as defining the problem type, specifying the model save path, setting training time limits, and selecting other hyperparameters. In practice, users can further enhance the model's generalization capability and stability by choosing a five-fold cross-validation to split the training and validation sets. This method involves dividing the dataset into five parts, with four parts used for training and one for validation in each iteration, thereby effectively reducing overfitting and improving the model robustness. Additionally, users can employ a layer of stacking ensemble learning to combine the predictions of multiple base models and further enhance the overall prediction performance through a meta-model. Figure 1 illustrates the AutoGluon framework process utilizing five-fold cross-validation. This approach leverages the strengths of different models, compensates for the shortcomings of individual models, and significantly improves the final model's accuracy and stability.

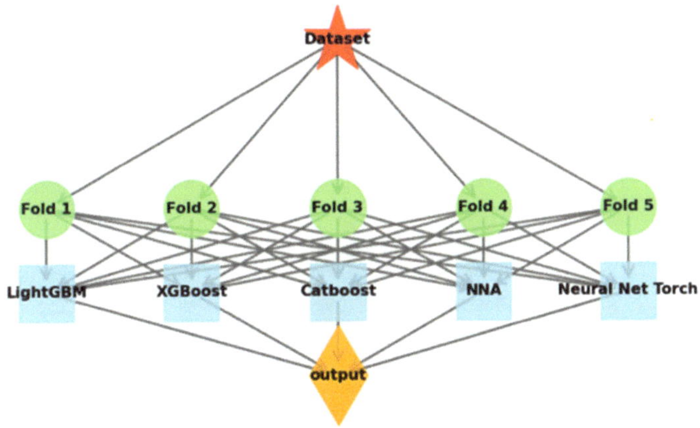

Figure 1. AutoGluon Framework Process Diagram.

Once the configuration is complete, users can initiate the training process. Notably, AutoGluon automatically handles complex operations, such as model selection and hyperparameter tuning, allowing users to focus on analyzing the results without delving into every detail. After the training is complete, users can provide a test set to evaluate the trained model. AutoGluon generates a detailed evaluation report, analyzing the model's performance across various metrics, such as RMSE and R^2. RMSE is the root mean square error, which measures the difference between the predicted value of the model and the true value. The smaller the value, the better the prediction effect of the model. R^2 is the coefficient of determination, a statistic that measures how well a regression model fits. It reflects the correlation between the predicted value of the model and the actual data and represents the proportion of variance that the model can explain. The higher the R^2, the more variability the model can explain in the data, and the better the model fits. This comprehensive evaluation helps users understand the strengths and weaknesses of the model and provides a basis for further improvement.

Through this process, AutoGluon not only simplifies the entire machine learning workflow but also makes the construction of high-performance models more efficient and convenient. In particular, the combination of five-fold cross-validation and stacking methods further enhances the model performance, ensuring good results across various datasets.

2.4.5. Advantages of AutoGluon

In the field of machine learning, as data volume and complexity continue to grow, choosing an appropriate tool for model training has become increasingly important. Compared to individual manual models, AutoGluon offers significant advantages. Firstly, AutoGluon features a high degree of automation, streamlining many machine learning processes. This enables users to easily obtain high-quality models without the need to delve into every detail. This simplified process allows non-expert users to start quickly, significantly reducing the time and cost of model development.

Additionally, by utilizing ensemble learning techniques, AutoGluon combines the strengths of multiple models, thereby significantly enhancing the performance and stability. Compared with a single training model, the AutoGluon framework can be used to integrate multiple models for learning, and cross-validation can improve the stability of the model. This method not only enhances the adaptability of the model to various data types, but also improves the stability of the model. Therefore, selecting AutoGluon for model training not only enhances efficiency but also provides more reliable results, making it my top choice among model training tools.

3. Case Study

This study employs ensemble techniques from hybrid machine learning methods to enhance oil well-production forecasting. Initially, various outlier detection and missing value imputation methods are applied to analyze and preprocess geological and engineering data from oilfield sites, removing outliers and filling in missing values. After data governance and target parameter selection, a relationship analysis is conducted among the parameters to filter out those with high correlations with the target parameters. The study utilizes gray relational analysis, the maximal mutual information method, AutoGluon feature importance evaluation, and SHAP value analysis, combined with the entropy weight method for comprehensive evaluation, to determine appropriate input parameters. Finally, the AutoGluon framework, which integrates multiple models, is used for target output prediction, with RMSE employed to validate the model performance, while MSE, MAE, and R^2 are used to assess the model's performance on both the training and testing datasets. All data have been desensitized for confidential consideration.

3.1. Abnormal Data Situation

By normalizing the on-site data and scaling them uniformly to the range [0, 1], the distribution of the data can be visually observed using a box plot. A box plot is a statistical chart designed to display the dispersion of a dataset, primarily reflecting the characteristics of the original data distribution and allowing for the comparison of distributional characteristics across multiple datasets. In the box plot, the middle line represents the median, while the upper and lower boundaries of the box represent the upper and lower quartiles (75% and 25%, respectively). The outer edges of the plot represent the outlier cutoff points. The distance from the box to the outer edges is typically 1.5 times the interquartile range. Data points within this range are considered inner limits, while data points outside this range are considered outer limits. Generally, data points beyond the outer limits are regarded as extreme outliers (IQR method). Perform normalization on the data and then construct a box plot, as shown in Figure 2.

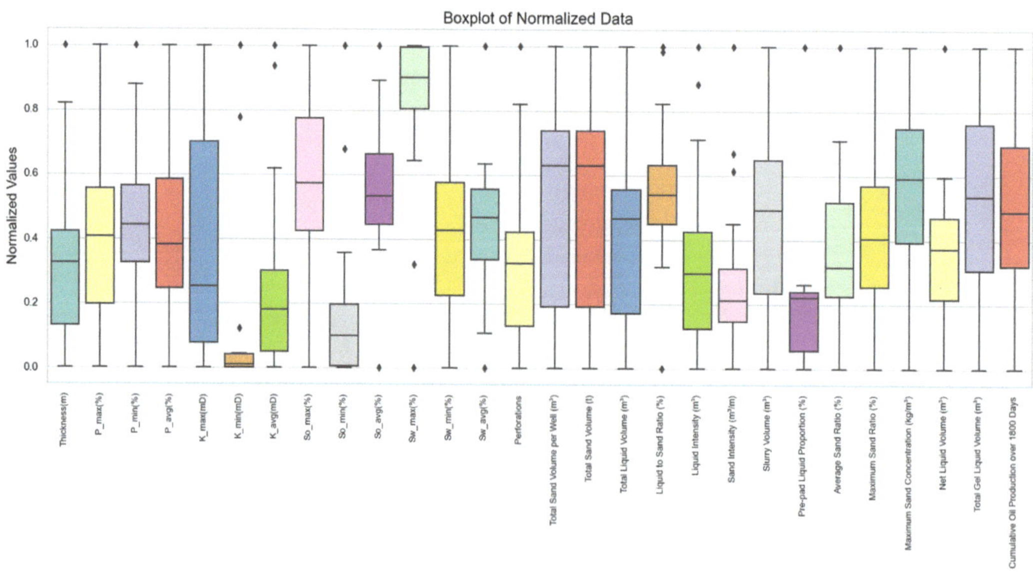

Figure 2. Box distribution graph after data normalization processing.

Due to recording errors, magnitude errors, calculation errors, deduction errors, and other reasons, there are many outliers in the on-site data, often reflected in the form of outlier data. The distribution of the data range is displayed in a histogram [1], as shown in Figure 3.

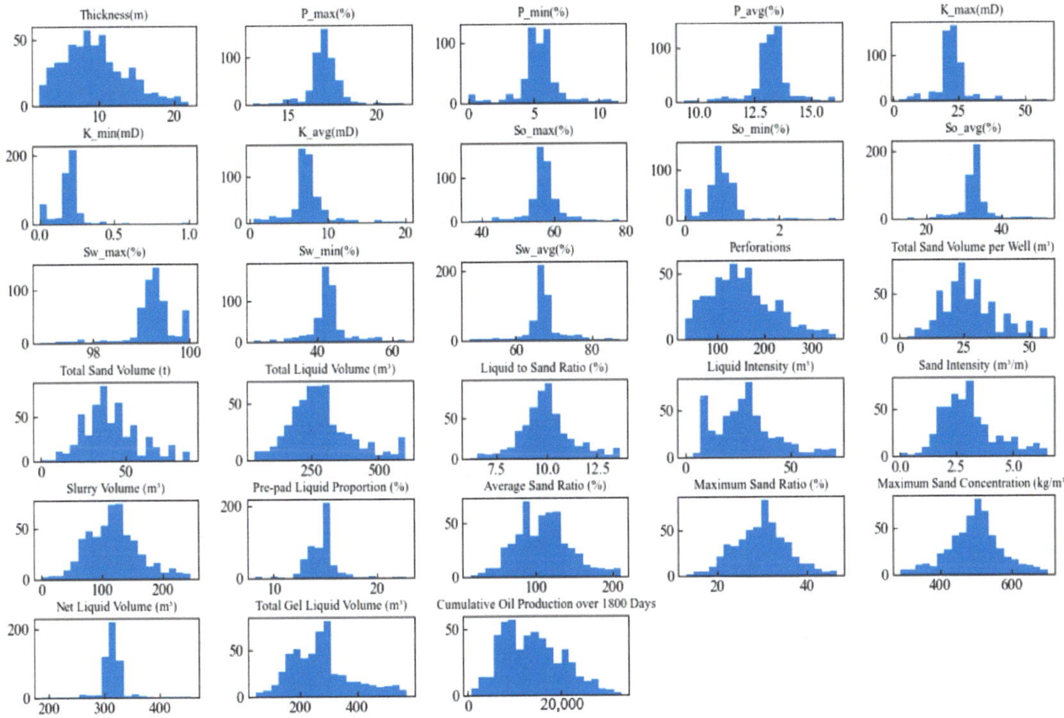

Figure 3. Histogram of the overall distribution of quantitative data.

As can be seen from the graph, the distribution of on-site data is mostly uneven, with many outliers. There are two reasons for determining outliers: one is that there are a large number of 0 s in sparse data, which affects the overall data distribution, and the second reason is abnormal values caused by erroneous construction records, as shown in Figure 4. For these two types of data, they need to be treated differently. Although sparse data are classified as outliers in the mathematical distribution, they do not affect model training, while outliers caused by recording errors can significantly affect the training process and the prediction accuracy of machine learning models. This part needs to be manually corrected.

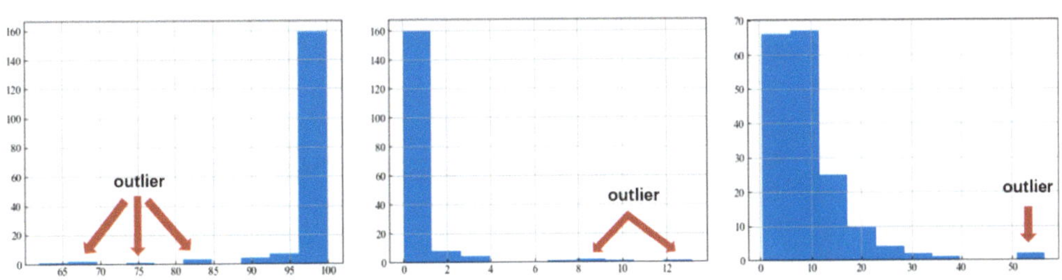

Figure 4. Abnormal value determination.

3.2. Data Preprocessing

3.2.1. Data Missing Situation

In addition to the issue of data anomalies, there are also serious missing data problems on-site. In traditional processing methods for big data, data fields with missing values exceeding 60% are usually deleted directly. However, due to the small amount of on-site data, these fields cannot be deleted. Therefore, special attention should be paid to data filling during data preprocessing.

As can be seen from Figure 5, only nearly 150 wells have complete data, while most wells lack ten to fifteen types of data, resulting in a high data loss rate.

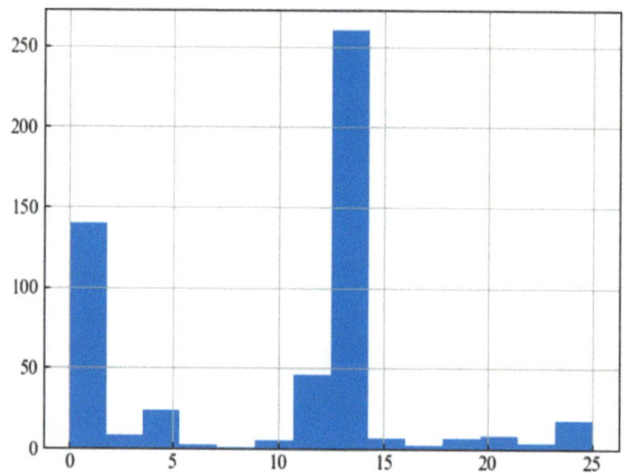

Figure 5. Distribution of missing values in each row of data.

As can be observed from Figure 6, the parameter data statistics are also incomplete, with 38.7% of parameters having a missing rate between 60% and 70%, and 3.2% of parameters having a missing rate exceeding 80%. These include important parameters such as porosity, permeability, and saturation.

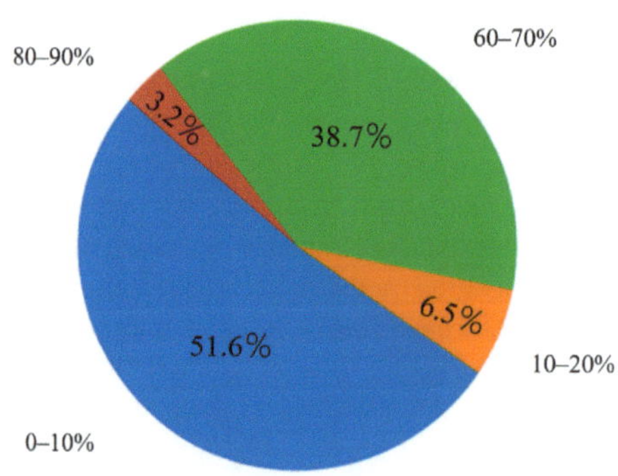

Figure 6. Distribution of missing rate of feature parameter data.

3.2.2. Outlier Handling

The primary foundation of outlier handling is to identify outlier data, and commonly used outlier identification methods include the IQR method, MAD method, and 3sigma method.

The IQR method is based on the lower quartile Q1 and upper quartile Q3 of the dataset for judgment. The size between the upper and lower quartiles is called IQR, while a value that exceeds 1.5 times IQR beyond the upper and lower quartiles is considered an outlier. The dataset is arranged from small to large, with the lower quartile Q1 located in the fourth quarter and the upper quartile Q3 located in the third quarter. The difference between Q3 and Q1 is the IQR. By using box plots and bar charts, it is possible to visually observe the data points outside the whisker axis, namely, outlier points, as shown in Figure 7. The green dashed line represents the lower boundary of the IQR (Q1 − 1.5 × IQR), and any data below this value are considered outliers. The purple dashed line represents the upper boundary of IQR (Q3 + 1.5× IQR), and any data above this value are also considered outliers. The light blue Kernel Density Estimate (KDE) curve illustrates the smooth probability density distribution of the data, illustrating the central tendency and dispersion of the da-ta values. This curve, combined with the histogram, aids in visually understanding the data distribution and highlighting the location of outliers. The curves depicted in Figures 8 and 9 carry the same significance as the curves in Figure 7. The lower plot in Figure 7 shows the box plot distribution of the data, with points outside the two black lines identified as outliers. The combination of the histogram and box plot provides a better observation of both the quantity and distribution of outliers.

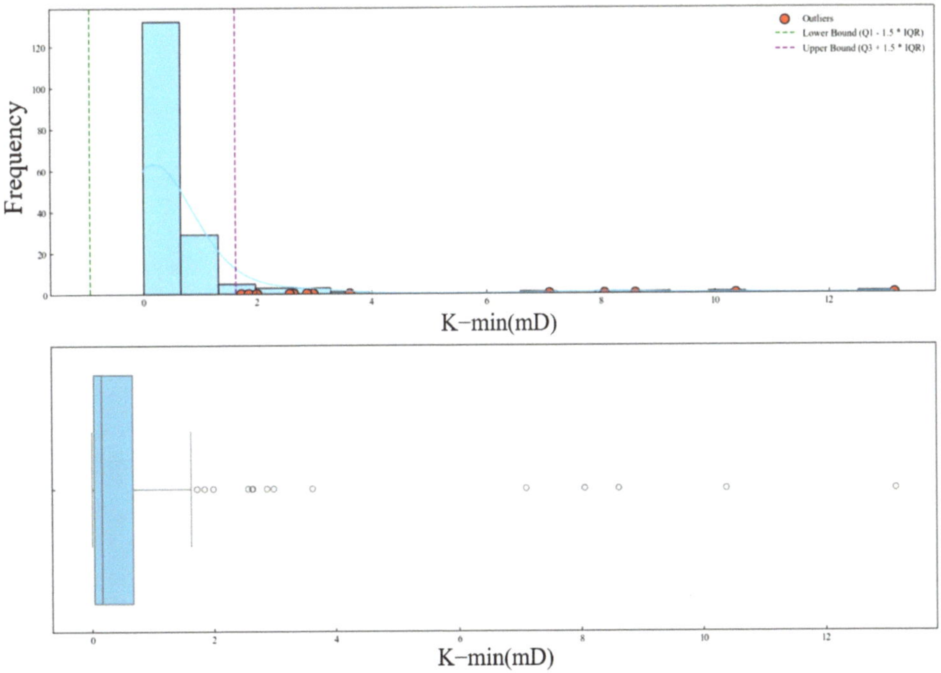

Figure 7. IQR method identifies outliers in the data.

The MAD method assumes that the data follow a normal distribution, and the values within the middle 50% area of the normal distribution are normal values, while the remaining 50% of the area on both sides are outliers. The outlier situation can be viewed in the histogram, and the judgment boundary of the MAD is shown in Figure 8.

The 3 σ method assumes that the data follow a normal distribution, with the mean as the center, and the probability within plus or minus 3 σ is 99.7%. Therefore, the probability of values outside the mean value of 3 σ occurring is 0.3%, which is a very rare and small probability event. Therefore, it can be identified as an outlier, and the outlier situation can be viewed in the histogram. The judgment boundary of 3 σ is shown in Figure 9.

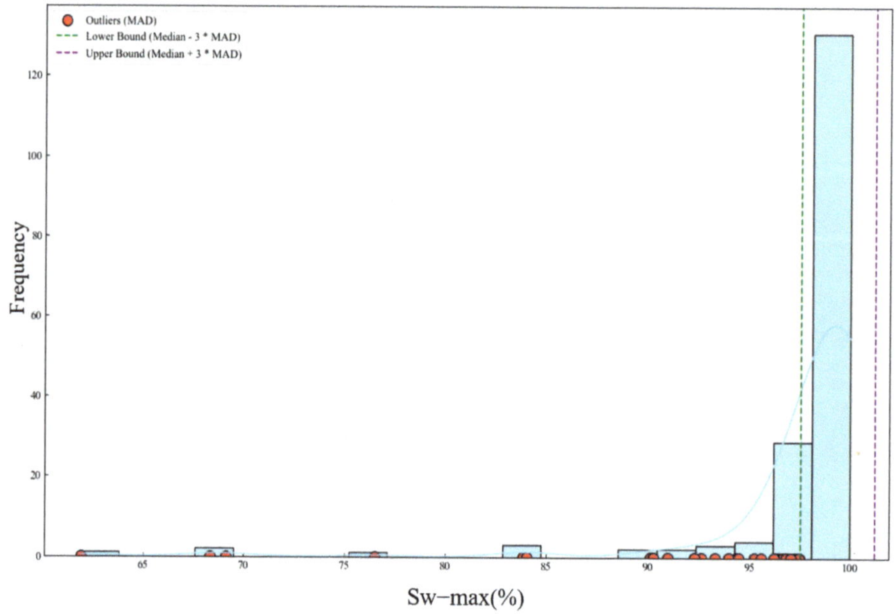

Figure 8. MAD method identifies outliers in the data.

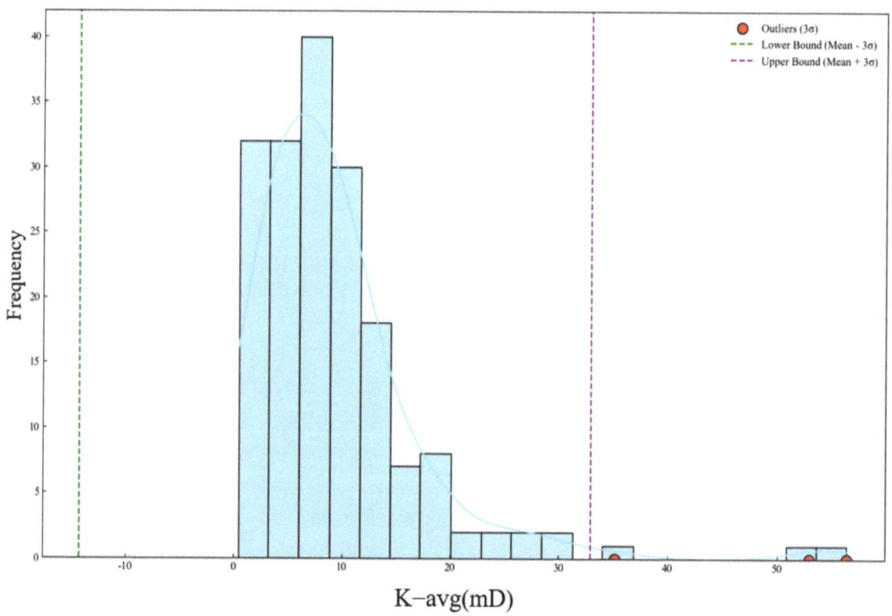

Figure 9. Identifying outliers in data using the triple standard deviation method.

This study adopts the density-based noise unsupervised spatial clustering (DBSCAN) clustering algorithm, which performs multidimensional clustering analysis based on Euclidean distance and automatically and accurately identifies outlier data points as outliers. DBSCAN only requires specifying two parameters: the search neighborhood radius ε and the minimum number of points contained within the search radius MinPts. In the clustering process, any data point is determined to have neighboring points within its neighborhood radius ε, and points with a number of neighboring points greater than MinPts are identified as core points. Points with fewer neighboring points than MinPts, but adjacent to the core point, are identified as boundary points. The core and boundary points always belong to a certain category cluster, and points that do not belong to any cluster are identified as noise points. The core points and noise points are shown in Figure 10. In the figure, purple, yellow, and green dots represent core points, while black dots indicate noise points, which are also considered outliers.

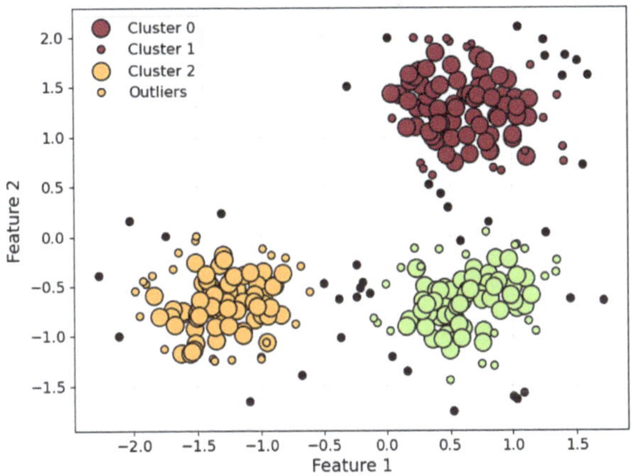

Figure 10. DBSCAN Abnormal value determination.

Choosing an appropriate method is crucial for the accuracy of results in the process of outlier detection. Compared with other commonly used outlier detection methods, such as IQR, MAD, and the triple standard deviation method, the DBSCAN method exhibits more reasonable characteristics in multiple aspects. Firstly, DBSCAN's density-based clustering method can automatically identify clusters and noise points in data without making assumptions about the distribution pattern of the data, while traditional IQR, MAD, and triple standard deviation methods typically assume that the data follows a normal distribution or a relatively uniform distribution, which may not always hold true in practical applications. Secondly, DBSCAN can effectively handle data with nonlinear structures and has stronger adaptability. In addition, DBSCAN does not require a pre-set number of clusters and can automatically determine the number and shape of clusters, which is particularly important for datasets with unknown or complex data features. Taking into account these advantages and data characteristics, the DBSCAN method is more reasonable and effective in outlier detection. Therefore, this method was chosen for this study.

3.2.3. Data Filling

In data analysis and machine learning projects, handling missing values is one of the key steps in ensuring data quality and model performance. If missing values are not handled properly, this may lead to analysis bias and even affect the predictive ability of the model. Therefore, in this project, we adopted a comprehensive strategy that combines mean imputation and the K-Nearest Neighbor (KNN) algorithm to ensure that missing values in the data are properly and effectively filled.

Firstly, we conducted preliminary processing on the geographical features in the dataset. For the missing values in these feature columns, we filled them with column means. This method can reduce the information loss caused by missing values while preserving the overall trend of the data, providing a relatively complete data foundation for subsequent processing steps.

After completing the preliminary processing, we further used the K-Nearest Neighbor (KNN) algorithm to fill in the missing values in the data. KNN is an instance-based learning method that infers possible missing values by referencing several neighbors in the dataset that are most similar to the missing value samples. We set the number of reference neighbors to 50, which means that when filling in each missing value, the algorithm searches for the 50 most similar samples and infers the missing value based on the data of these neighbors. This method fully utilizes the similarity information in the data, providing more accurate and reasonable filling results, thereby effectively improving the integrity of the data.

After filling in using the above method, we updated the entire dataset to ensure that all missing values were properly processed. The situation before and after missing value processing is shown in Figure 11. In the figure, blue dots represent the original data distribution, while green dots show the data distribution after imputation.

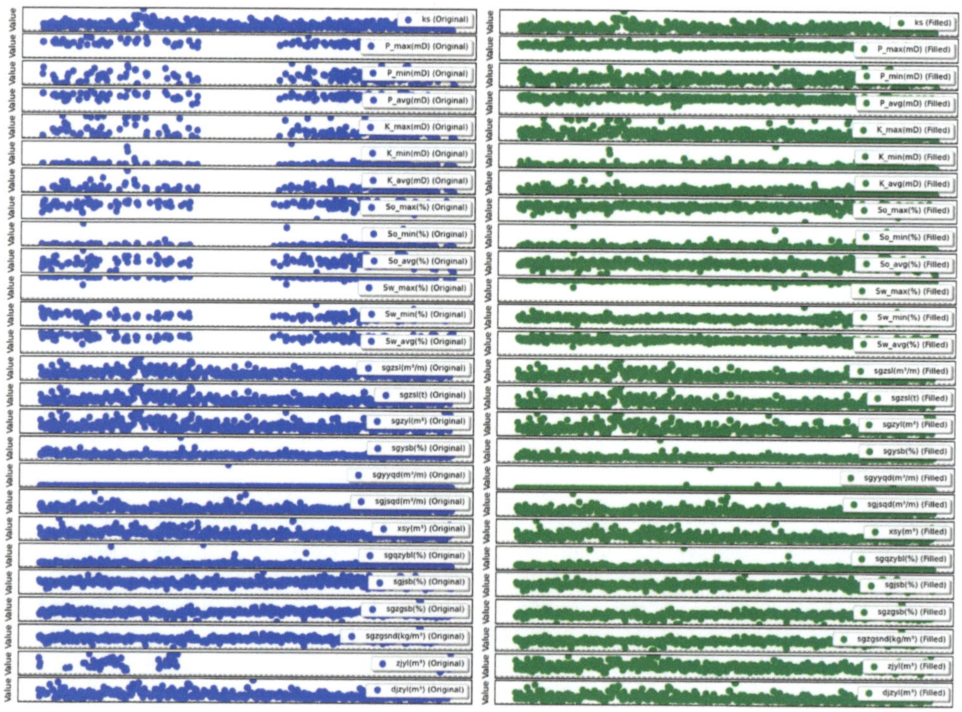

Figure 11. Comparison chart before and after missing value supplementation.

3.3. Feature Extraction

Pearson coefficient analysis was conducted on the extracted geological and engineering data. This analysis involved determining the coefficient value for each pair of parameters. Following the analysis of all parameters, Figure 12 was generated. The colors depicted in the heatmap correspond to the magnitude of the data. Specifically, Pearson coefficients closer to 1 are represented by redder hues in the heatmap, while coefficients nearer to −1 appear bluer [35]. This visual representation provides a clear and intuitive understanding of the correlations present within the dataset.

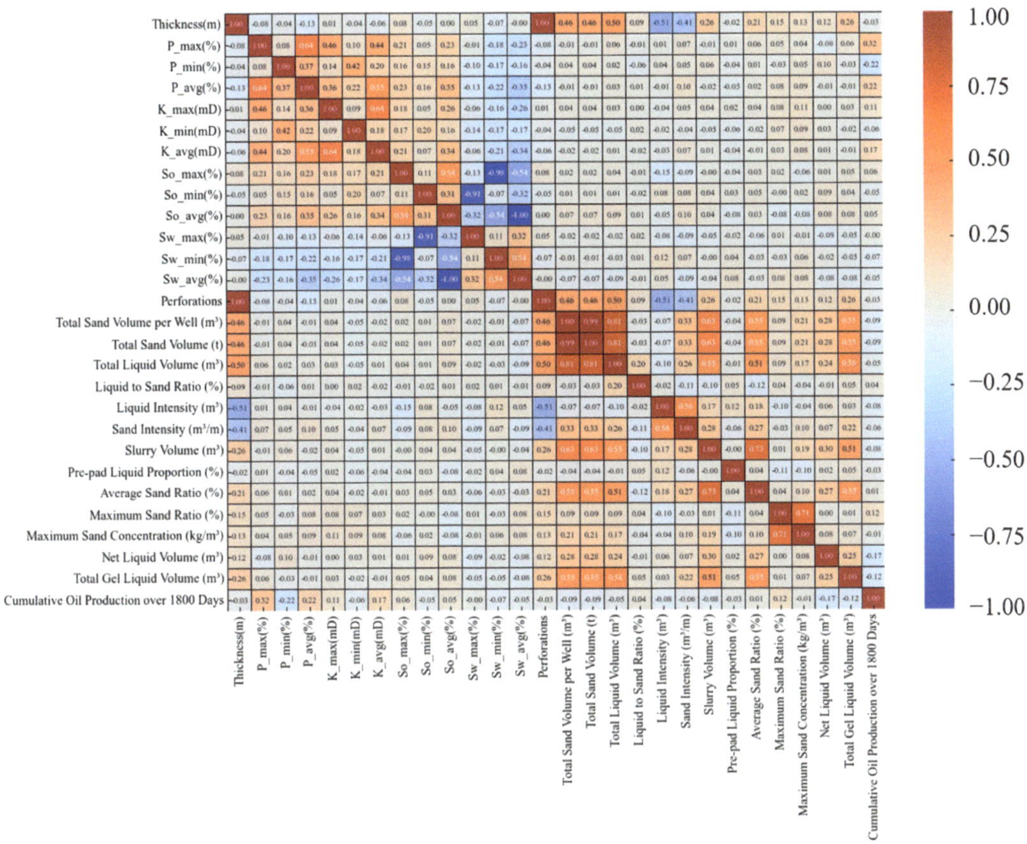

Figure 12. Pearson coefficient heatmap of production and construction data.

When the absolute value of the Pearson coefficient is greater than 0.4, a linear correlation is generally considered. By extracting parameters with Pearson correlation coefficient absolute values greater than 0.4, as shown in Figure 13, a linear relationship test was conducted between each parameter and 1800 days of cumulative oil production. The results show that although there is a certain linear correlation between the construction parameters, geological parameters, and cumulative oil production, this correlation is not strong. This indicates that there may be nonlinear relationships or other complex influencing factors that have not been fully identified. In order to comprehensively understand and evaluate the impact of these parameters on cumulative oil production, this article has decided to introduce more analytical methods. Subsequently, gray relational analysis, the maximum mutual information method, AutoGluon feature importance, and SHAP value analysis will be used, combined with the entropy weight method for comprehensive evaluation, to deeply analyze and determine key influencing factors. These comprehensive methods are expected to reveal more potential relationships and provide a more accurate basis for subsequent optimization decisions.

For the target variable of 1800 days of cumulative oil production, this study comprehensively utilized various methods, such as gray relational analysis, the maximum mutual information method, AutoGluon feature importance assessment, and SHAP value analysis, combined with the entropy weight method, for a comprehensive evaluation. Through this series of analytical methods, the impact of each parameter on the target variable was successfully identified and quantified, and the final evaluation results of the main controlling factors are shown in Figure 14.

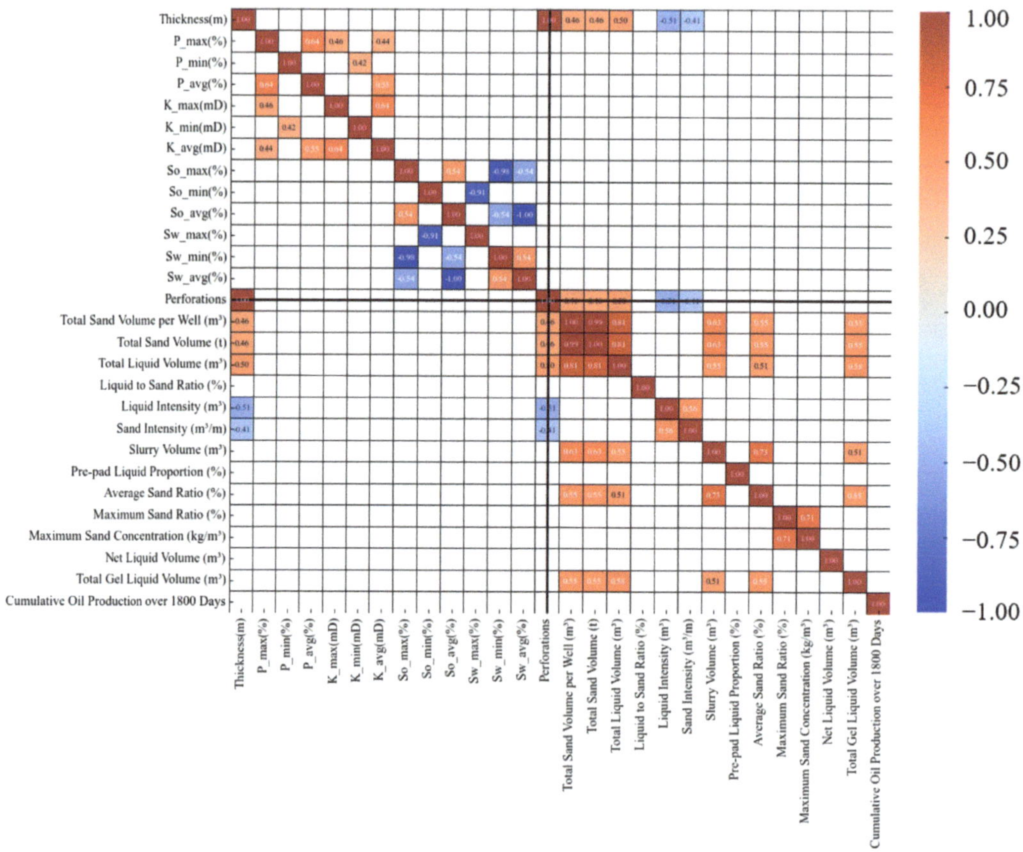

Figure 13. Combination of linear correlation above 0.4 in production and construction data.

In the feature extraction stage, this article selects features from the raw data that are highly correlated with the target parameters. The aim is to reduce the dimensionality of the data, enhance its nonlinearity, and increase its expressive power in order to more effectively adapt it to machine learning models and algorithms. To achieve this goal, this article primarily uses the principal component analysis (PCA) method for feature extraction. This method minimizes information loss by preserving the most representative information in the original data. After applying principal component analysis (PCA), this paper selected 19 geological and engineering parameters from an initial set of 29 variables, using the entropy weight method to rank them by their scores from high to low. These parameters were chosen to serve as input features for subsequent model training. The geological parameters include aspects such as the total gel liquid volume, liquid intensity, liquid-to-sand ratio, pre-pad liquid proportion, total sand volume, average sand ratio, sand intensity, total liquid volume, perforations, and net liquid volume. On the engineering side, the selected parameters include minimum permeability (K_min), average pressure (P_avg), minimum pressure (P_min), maximum oil saturation (So_max), maximum water saturation (Sw_max), average oil saturation (So_avg), minimum water saturation (Sw_min), thickness, and average water saturation (Sw_avg). Among the preferred parameters, Total Gel Liquid

Volume, Liquid intensity, and Total Liquid Volume are positively correlated to the yield, while other preferred parameters are negatively correlated.

Feature Columns	GRA	Maximal Mutual Information	AutoGluon Feature Importance	SHAP Importance	Entropy Weight Method Comprehensive Evaluation
djzyl(m³)	−0.321	0.274	1.865	1.530	1.650
sgyyqd(m³/m)	−0.285	0.388	0.507	0.959	0.474
sgzyl(m³)	0.016	0.339	0.280	0.589	0.281
dty(m³)	−0.371	−0.252	0.181	0.696	0.168
sgzsl(m³/m)	−0.140	0.133	0.006	−0.146	−0.014
sgqzy(m³)	−0.176	0.451	−0.075	0.666	−0.026
sgzgsb(%)	−0.338	0.066	−0.050	0.435	−0.038
P_max(mD)	−0.709	0.007	−0.108	0.422	−0.118
sgzgsnd(kg/m³)	−0.660	−0.356	−0.077	−0.026	−0.123
So_min(%)	1.409	−0.318	−0.270	−0.313	−0.140
K_avg(mD)	0.547	−0.305	−0.265	−0.250	−0.200
K_max(mD)	−0.561	0.423	−0.179	−0.402	−0.217
xsy(m³)	−0.147	0.890	−0.241	−0.535	−0.240
K_min(mD)	1.261	−0.040	−0.381	−0.491	−0.255
P_avg(mD)	−0.274	−0.690	−0.248	−0.491	−0.272
sgysb(%)	0.002	−0.305	−0.318	−0.160	−0.282
sgqzybl(%)	0.077	0.008	−0.365	−0.300	−0.321
sgzsl(t)	−0.140	0.133	−0.332	−0.542	−0.326
P_min(mD)	−0.470	0.079	−0.322	−0.303	−0.328
sgjsb(%)	−0.852	0.810	−0.296	−0.397	−0.334
sgjsqd(m³/m)	−0.220	−0.014	−0.355	−0.283	−0.335
So_max(%)	−0.457	−0.440	−0.379	−0.189	−0.373
Sw_max(%)	1.255	−0.058	−0.510	−0.650	−0.375
So_avg(%)	−0.193	−0.062	−0.430	−0.473	−0.410
Sw_min(%)	−0.579	−0.408	−0.438	−0.044	−0.422
ks	−0.457	0.120	−0.415	−0.729	−0.434
hd(m)	−0.457	0.120	−0.469	−0.547	−0.466
Sw_avg(%)	−0.419	−0.106	−0.537	−0.705	−0.534
zjyl(m³)	0.174	−0.683	−0.597	−0.720	−0.546
Objective Parameter	Cumulative Production over 1800 Days				

Figure 14. Using entropy weight method for comprehensive evaluation and analysis of main control factors.

3.4. Model Training

In this article, the AutoGluon machine learning framework was utilized for rapid model training and optimization, achieving higher accuracy predictions by integrating multiple models without the need for manual hyperparameter search. Specifically, various ensemble learning techniques from AutoGluon, including stacking, k-fold cross-validated bagging, and multi-level stacking, were employed to enhance the model's fitting and generalization performance.

Stacking technology trains multiple models independently on the same dataset and uses linear models to calculate the weighted average of all model predictions. Bagging through k-fold cross-validation effectively prevents model overfitting by performing k-fold cross-validation on all the models and obtaining the average output. Multi-layer stacking combines the original data with the results of single stacking to form a new linear weighted model, further improving the prediction accuracy.

In the specific implementation, the model was trained for regression problems with the key parameters set to optimize the training process. By specifying the model save path, controlling the training time, and utilizing GPU-accelerated computing, the training efficiency of the model was significantly improved. Additionally, 5-fold cross-validation was employed to enhance the model's generalization ability and ensure the effective application of stacking and bagging techniques through appropriate hierarchical control.

Through these methods and parameter settings, AutoGluon was able to automatically select and combine multiple models without human intervention, significantly improving the accuracy and efficiency of predictions.

In the data preprocessing stage, the DBSCAN algorithm is used to identify and remove outliers, and the mean and KNN methods are used to fill in missing values. Subsequently, the data set was divided into a training set and a test set, with the test set accounting for 20% and not participating in model training. Based on the AutoGluon framework, bagging technology with 5-fold cross-validation and single-layer stacking integrated learning strategy were implemented. The RMSE was selected as the loss function, and the ability of the model to fit the actual data was evaluated using R^2. According to the results shown in Figure 15, the closeness between the predicted value of the model and the real value is measured by the distance between the predicted value and the red dotted line. The closer the distance, the better the prediction effect. The blue marks represent the prediction performance on the training set, the yellow marks represent the prediction performance on the test set, and the points falling in the green area indicate that the prediction accuracy is above 85%. It is worth noting that although the R^2 value on the training set reached 0.79, indicating good model fit, the R^2 value on the test set was only 0.23, which may indicate overfitting of the model.

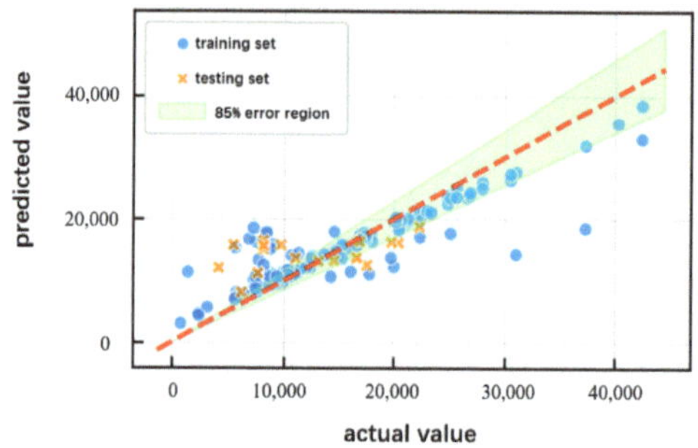

Figure 15. AutoGluon Prediction comparison chart.

4. Conclusions

This study aims to preliminarily construct a production capacity prediction model for a specific oilfield by employing regression analysis methods on a dataset containing 435 records and 19 features, with the goal of predicting the cumulative oil production over 1800 days. In the data preprocessing phase, we utilized a combination of graphical analysis and numerical processing methods. Specifically, we used the K-Nearest Neighbors (KNN) algorithm to impute missing values and employed the entropy weight method to conduct a comprehensive weighting of various key factors, extracting critical features to prepare for model training. Ultimately, we achieved automated model training using the AutoGluon framework, efficiently completing the prediction task.

During the model training process, we applied various algorithms, including Random Forest, CatBoost, Extra Trees, Neural Networks, XGBoost, NeuralNetTorch, and LightGBM. A comparative analysis revealed that the LightGBM model performed the best, with a root mean square error (RMSE) of 4175.08 on the training set and a coefficient of determination R^2 of 0.79, indicating that the model could explain 79% of the data variability. However, during the validation process, especially on the test set, the model's performance significantly declined, with the RMSE increasing to 11,113.33, and the R^2 dropping to 0.23, revealing a clear overfitting issue. This result emphasizes the importance of the model generalization capability. Future research will explore more advanced models and algorithms based on this foundation to further improve the prediction accuracy and reliability, providing more scientific guidance for the overall oil and gas development process.

Author Contributions: Conceptualization, R.Z.; Methodology, N.L.; Software, F.Q.; Validation, G.L. (Gaofeng Li); Formal analysis, X.W.; Resources, G.L. (Guohua Liu); Writing—original draft, C.L.; Visualization, Y.D.; Supervision, G.L. (Gensheng Li); Project administration, Q.L. All authors have read and agreed to the published version of the manuscript.

Funding: This research was funded by the National Natural Science Foundation of China (grant number 52320105002 and 52374017), and PetroChina Huabei Oilfield Company (grant number HBYT-YQGYY- 2024-JS-28).

Data Availability Statement: The data presented in this study are available on request from the corresponding author due to privacy restrictions from the company.

Conflicts of Interest: Authors Ruibin Zhu, Ning Li, Yongqiang Duan, Gaofeng Li, Guohua Liu, Fengjiao Qu, Changjun Long and Xin Wang were employed by Research Institute of Oil and Gas Technology, PetroChina Huabei Oilfield Company. The remaining authors declare that the research was conducted in the absence of any commercial or financial relationships that could be construed as a potential conflict of interest.

References

1. Jin, M.; Liao, Q.; Patil, S.; Abdulraheem, A.; Al-Shehri, D.; Glatz, G. Hyperparameter tuning of artificial neural networks for well production estimation considering the uncertainty in initialized parameters. *ACS Omega* **2022**, *7*, 24145–24156. [CrossRef] [PubMed]
2. Ali, N.; Fu, X.; Chen, J.; Hussain, J.; Hussain, W.; Rahman, N.; Iqbal, S.M.; Altalbe, A. Advancing Reservoir Evaluation: Machine Learning Approaches for Predicting Porosity Curves. *Energies* **2024**, *17*, 3768. [CrossRef]
3. Zhang, Z. A well production prediction method based on blending heterogeneous ensemble learning optimized by OOA. *J. Phys. Conf. Ser.* **2024**, *2835*, 012002. [CrossRef]
4. Rushing, J.A.; Perego, A.D.; Sullivan, R.B.; Blasingame, T.A. Estimating reserves in tight gas sands at HP/HT reservoir conditions: Use and misuse of an Arps decline curve methodology. In Proceedings of the SPE Annual Technical Conference and Exhibition, Anaheim, CA, USA, 11–14 November 2007; p. SPE-109625.
5. Nasteski, V. An overview of the supervised machine learning methods. *Horizons* **2017**, *4*, 56. [CrossRef]
6. Li, G.; Tian, Z. A new method of network traffic prediction based on combination model. *Peer-Peer Netw. Appl.* **2024**, *17*, 1075–1090. [CrossRef]
7. Arps, J.J. Analysis of decline curves. *Trans. AIME* **1945**, *160*, 228–247. [CrossRef]
8. Yuanqian, C.; You, Z.; Xiuluan, L. A new method of using SAGD exploitation technique to predict the recoverable reserves of heavy oil reservoir. *Spec. Oil Gas Reserv.* **2015**, *22*, 85–89.
9. Chen, M.; Qu, Z.; Liu, W.; Tang, S.; Shang, Z.; Ren, Y.; Han, J. Production Prediction Model of Tight Gas Well Based on Neural Network Driven by Decline Curve and Data. *Processes* **2024**, *12*, 932. [CrossRef]
10. Wang, K.; Xie, M.; Liu, W.; Li, L.; Liu, S.; Huang, R.; Feng, S.; Liu, G.; Li, M. New Method for Capacity Evaluation of Offshore Low-Permeability Reservoirs with Natural Fractures. *Processes* **2024**, *12*, 347. [CrossRef]
11. Amr, S.; El Ashhab, H.; El-Saban, M.; Schietinger, P.; Caile, C.; Kaheel, A.; Rodriguez, L. A large-scale study for a multi-basin machine learning model predicting horizontal well production. In Proceedings of the SPE Annual Technical Conference and Exhibition, Dallas, TX, USA, 25 September 2018; p. D011S008R005.
12. Wu, T.; Fang, H.; Sun, H.; Zhang, F.; Wang, X.; Wang, Y.; Li, S. A data-driven approach to evaluate fracturing practice in tight sandstone in Changqing field. In Proceedings of the International Petroleum Technology Conference, IPTC, Virtual, 23 March–1 April 2021; p. D071S024R002.
13. Hou, X.M.; Wang, F.Y.; Zai, Y. Prediction of carbonate porosity and permeability based on machine learning and logging data. *J. Jilin Univ.* **2021**, *52*, 644.
14. Surguchev, L.; Li, L. IOR evaluation and applicability screening using artificial neural networks. In Proceedings of the SPE Improved Oil Recovery Symposium, Tulsa, OK, USA, 2–5 April 2000; p. SPE-59308.
15. Cao, Q.; Banerjee, R.; Gupta, S.; Li, J.; Zhou, W.; Jeyachandra, B. Data Driven Production Forecasting Using Machine Learning. In Proceedings of the SPE Argentina Exploration and Production of Unconventional Resources Symposium, Buenos Aires, Argentina, 1–3 June 2016.
16. Jia, X.; Zhang, F. Applying data-driven method to production decline analysis and forecasting. In Proceedings of the SPE Annual Technical Conference and Exhibition, Dubai, UAE, 26–28 September 2016; p. D021S020R007.
17. Chakra, N.C.; Song, K.Y.; Gupta, M.M.; Saraf, D.N. An innovative neural forecast of cumulative oil production from a petroleum reservoir employing higher-order neural networks (HONNs). *J. Pet. Sci. Eng.* **2013**, *106*, 18–33. [CrossRef]
18. Aizenberg, I.; Luchetta, A.; Manetti, S.; Piccirilli, M.C. System identification using FRA and a modified MLMVN with arbitrary complex-valued inputs. In Proceedings of the 2016 International Joint Conference on Neural Networks (IJCNN), Vancouver, BC, Canada, 24–29 July 2016; pp. 4404–4411.

19. Loh, K.; Omrani, P.S.; van der Linden, R. Deep learning history matching for real time production forecasting. In *First EAGE/PESGB Workshop Machine Learning;* European Association of Geoscientists Engineers: Utrecht, The Netherlands, 2018; Volume 2018, pp. 1–3.
20. Lolon, E.; Hamidieh, K.; Weijers, L.; Mayerhofer, M.; Melcher, H.; Oduba, O. Evaluating the relationship between well parameters and production using multivariate statistical models: A middle bakken and three forks case history. In Proceedings of the SPE Hydraulic Fracturing Technology Conference and Exhibition, Woodlands, TX, USA, 9–11 February 2016; p. D031S007R003.
21. Song, X.; Liu, Y.; Xue, L.; Wang, J.; Zhang, J.; Wang, J.; Jiang, L.; Cheng, Z. Time-series well performance prediction based on Long Short-Term Memory (LSTM) neural network model. *J. Pet. Sci. Eng.* **2020**, *186*, 106682. [CrossRef]
22. Zhang, Y.; Hu, J.; Zhang, Q. Application of locality preserving projection-based unsupervised learning in predicting the oil production for low-permeability reservoirs. *Spe J.* **2021**, *26*, 1302–1313. [CrossRef]
23. Wang, T.; Wang, Q.; Shi, J.; Zhang, W.; Ren, W.; Wang, H.; Tian, S. Productivity prediction of fractured horizontal well in shale gas reservoirs with machine learning algorithms. *Appl. Sci.* **2021**, *11*, 12064. [CrossRef]
24. Shelley, R.; Oduba, O.; Melcher, H. Machine learning and artificial intelligence provides wolfcamp completion design insight. In Proceedings of the SPE Hydraulic Fracturing Technology Conference and Exhibition, Houston, TX, USA, 26–28 July 2021.
25. Dong, Y.; Qiu, L.; Lu, C.; Song, L.; Ding, Z.; Yu, Y.; Chen, G. A data-driven model for predicting initial productivity of offshore directional well based on the physical constrained eXtreme gradient boosting (XGBoost) trees. *J. Pet. Sci. Eng.* **2022**, *211*, 110176. [CrossRef]
26. Khan, A.M.; BinZiad, A.; Al Subaii, A.; Alqarni, T.; Jelassi, M.Y.; Najmi, A. Supervised Learning Predictive Models for Automated Fracturing Treatment Design: A Workflow Based on Algorithm Comparison and Multiphysics Model Validation. In Proceedings of the SPE International Hydraulic Fracturing Technology Conference and Exhibition, Muscat, Oman, 11–13 January 2022; p. D011S003R002.
27. Berneti, S.M.; Shahbazian, M. An imperialist competitive algorithm artificial neural network method to predict oil flow rate of the wells. *Int. J. Comput. Appl.* **2011**, *26*, 47–50.
28. Vyas, A.; Datta-Gupta, A.; Mishra, S. Modeling early time rate decline in unconventional reservoirs using machine learning techniques. In Proceedings of the Abu Dhabi International Petroleum Exhibition and Conference, Abu Dhabi, UAE, 31 October–3 November 2017; p. D041S113R002.
29. Noshi, C.I.; Eissa, M.R.; Abdalla, R.M. An intelligent data driven approach for production prediction. In Proceedings of the Offshore Technology Conference, Houston, TX, USA, 6–9 May 2019; p. D041S048R007.
30. Adesina, E.; Olusola, B. Application of Machine Learning Algorithm for Predicting Produced Water Under Various Operating Conditions in an Oilwell. In Proceedings of the SPE Nigeria Annual International Conference and Exhibition, Lagos, Nigeria, 1–3 August 2022; p. D021S009R004.
31. Qi, W.; Xu, C.; Xu, X. AutoGluon: A revolutionary framework for landslide hazard analysis. *Nat. Hazards Res.* **2021**, *1*, 103–108. [CrossRef]
32. Erickson, N.; Mueller, J.; Shirkov, A.; Zhang, H.; Larroy, P.; Li, M.; Smola, A. Autogluon-tabular: Robust and accurate automl for structured data. *arXiv* **2020**, arXiv:2003.06505.
33. Liao, Q.; Zhang, D. Data assimilation for strongly nonlinear problems by transformed ensemble Kalman filter. *SPE J.* **2015**, *20*, 202–221. [CrossRef]
34. Liao, Q.; Zhang, D.; Tchelepi, H. Nested sparse grid collocation method with delay and transformation for subsurface flow and transport problems. *Adv. Water Resour.* **2017**, *104*, 158–173. [CrossRef]
35. Li, D.; You, S.; Liao, Q.; Sheng, M.; Tian, S. Prediction of Shale Gas Production by Hydraulic Fracturing in Changning Area Using Machine Learning Algorithms. *Transp. Porous Med.* **2023**, *149*, 373–388. [CrossRef]

energies

MDPI

Article

Energy Management Strategy for Direct Current Microgrids with Consideration of Photovoltaic Power Tracking Optimization

Fudong Li *, Zonghao Shi, Zhihao Zhu and Yongjun Gan

School of Mechanical and Electrical Engineering, Beijing Information Science and Technology University, Beijing 100192, China; 2022020011@bistu.edu.cn (Z.S.); 2024020156@bistu.edu.cn (Z.Z.); 2024020137@bistu.edu.cn (Y.G.)

* Correspondence: 20182238@bistu.edu.cn

Abstract: In response to the uncertainty of renewable energy output and the fluctuation of load, this paper proposes a hybrid energy storage management strategy based on the State of Charge (SOC) to smooth power fluctuations and thereby improve the power quality of photovoltaic energy storage DC microgrids. Firstly, a hybrid algorithm for power tracking control is formed by incorporating the Particle Swarm Optimization (PSO) algorithm into the variable step-size Incremental Conductance (INC) method, thereby optimizing the maximum power point tracking control system of the photovoltaic system. Then, a first-order filter is employed for the initial allocation of demand power. Taking the SOC of supercapacitors and energy storage batteries as a reference, a secondary power allocation energy management strategy based on rule-based control is proposed to ensure the service life and application safety of the hybrid energy storage system. Finally, simulation experiments are conducted in MATLAB/Simulink 23.2 (R2023b). The results indicate that the proposed energy management strategy can maintain the SOC of the hybrid energy storage system at a reasonable level and effectively smooth DC bus voltage fluctuations.

Keywords: hybrid energy storage; energy management; state of charge; DC microgrid

Academic Editor: Sheng Du

Received: 19 November 2024
Revised: 3 January 2025
Accepted: 3 January 2025
Published: 8 January 2025

Citation: Li, F.; Shi, Z.; Zhu, Z.; Gan, Y. Energy Management Strategy for Direct Current Microgrids with Consideration of Photovoltaic Power Tracking Optimization. *Energies* **2025**, *18*, 252. https://doi.org/10.3390/en18020252

1. Introduction

In the context of the ongoing global energy transition, the rapid development and widespread adoption of renewable energy sources have emerged as pivotal forces driving the transformation of energy structures. This trend has profoundly impacted the operational modes of traditional power systems and necessitated enhanced design and optimization efforts from both the academic and industrial communities, particularly with respect to ensuring power quality in systems with a high penetration of renewable energy. Power quality, as a crucial metric for assessing the stability and reliability of power systems, has gained particular prominence in the wake of large-scale renewable energy integration.

Photovoltaic (PV) power generation and wind power generation stand as the two pillars of the current renewable energy generation landscape, leading the green transformation of global energy production. PV power generation, leveraging its unique advantages, plays a pivotal role in this energy revolution. Firstly, from an economic perspective, the declining costs of PV systems have made them one of the most cost-effective sources of electricity in many regions. High photoelectric conversion efficiency not only boosts energy output but also reduces the production cost per unit of electricity, laying a solid foundation for large-scale deployment. Secondly, the relatively straightforward maintenance of PV systems, coupled with their modular design, facilitates quicker troubleshooting and component replacement, ensuring sustained and stable operation. Furthermore, PV's widespread geographical adaptability and consistent performance across diverse climatic conditions enhance its potential as a reliable energy supply.

Meanwhile, while wind power generation also boasts clean and renewable attributes, its efficiency is susceptible to natural factors, such as the wind speed and direction, introducing intermittency and uncertainty. In contrast, PV power generation, especially in sun-rich regions, offers more stable and predictable power output, marking it as one of the preferred options for renewable energy generation.

Notably, as renewable energy generation technologies advance, many renewable energy systems, particularly PV and certain types of wind power systems, exhibit direct current (DC) output characteristics. This shift has opened new avenues for research and development in DC microgrids. DC microgrids, as a novel type of power distribution system, directly utilize the DC electricity produced by renewable energy sources, reducing energy losses during power conversion and enhancing system efficiency [1–3]. Additionally, DC microgrids offer significant advantages in flexibility, controllability, and compatibility, enabling the plug-and-play integration of distributed energy resources, intelligent scheduling, and rapid fault recovery, thereby positioning them as a key direction for the future development of smart grids.

In summary, with the rapid penetration of renewable energy, especially the widespread adoption of PV power generation, modern power systems are confronted with unprecedented challenges and opportunities. Ensuring power quality while accommodating a high proportion of renewable energy sources and fostering the research and application of DC microgrid technologies have become the focal points of current energy research. This endeavor necessitates not only an in-depth exploration of the latest advancements in renewable energy generation technologies by researchers but also interdisciplinary collaboration to build a smarter, more efficient, and sustainable modern power system.

A distributed multi-time scale hierarchical voltage control strategy is proposed in reference [4]. Its advantage is that it can enable the photovoltaic power generation system to quickly respond to voltage deviations on millisecond and second time scales, significantly improving the voltage regulation capability and response speed of the DC microgrid. However, the implementation of this strategy may rely on high-precision sensors and fast-response actuators, which increases the complexity and cost of the system. In reference [5], a distributed photovoltaic microgrid system under cloud cover and constant power load disturbance is used as an example to propose a nonlinear reduced-order modeling method and optimized estimation attraction domain calculation method for grid-connected DC microgrids in high-latitude and high-randomness power systems. The advantage of this method is that it can accurately describe the dynamic behavior of complex power systems and effectively evaluate the stability of the system. However, nonlinear reduced-order modeling may introduce certain model errors, and the computational complexity of estimating the attraction domain is high, requiring further optimization. Reference [6] proposes a DC microgrid power coordination control method that combines long-term and short-term memory network photovoltaic prediction technology with a distributed Raft algorithm in deep learning. The advantage of this method is that it can use big data and machine learning technology to improve the accuracy of photovoltaic prediction and achieve the stability and power control rationality of microgrids through distributed algorithms. However, the training and tuning process of deep learning models can be complex, and the implementation of distributed algorithms requires reliable communication networks and efficient computing resources. In reference [7], a two-stage method that takes into account photovoltaic power generation and combines a genetic algorithm and Particle Swarm Optimization algorithm GA-PSO is proposed for determining the optimal location and scale of electric vehicle charging stations in DC microgrids. The advantage of this method is that it can comprehensively consider multiple factors to achieve the optimal layout and scale design of electric vehicle charging stations, minimizing grid losses and improving the economy and reliability of the system. However, the performance of the GA-PSO algorithm may be affected by the initial parameters and search space, and the uncertainty of the electric vehicle charging demand needs to be considered in practical applications. A scheme for a grid-connected photovoltaic power supply electric vehicle charging station is proposed in reference [8]. This scheme employs converter control technology and fuzzy logic controllers to maintain the distributed power distribution between the microgrid DC bus and local energy storage units. Additionally, it achieves effective energy flow control among multiple energy sources through a decentralized energy management system. The advantage of this solution is that it can achieve the decentralized control and energy optimization management of microgrids, improving the flexibility and reliability of the system. However, the design and tuning of fuzzy logic controllers can be complex, and the implementation of decentralized energy management systems requires good information exchange and coordination mechanisms.

However, due to changes in environmental conditions, such as irradiance, temperature, and partial shading, the stability and power generation of PV power generation can be significantly affected, which will also impact the stability of DC microgrid systems. To mitigate the impact of the uncertainty of PV output on microgrid systems, energy storage serves as a power allocation system to address mismatches between generation and demand power, effectively addressing PV output power fluctuations by leveraging its rapid energy absorption and release capabilities [9], thereby smoothing renewable energy power output and enhancing overall system stability. Reference [10] proposes a high-efficiency, high-power-density partial power battery energy storage DC-DC converter, which has the significant advantage of using a combination of CLLC DC transformers and bipolar Buck circuits to achieve high-efficiency power conversion under wide-range voltage regulation. This design not only improves the power density of the system, but also achieves the automatic balancing of the battery pack voltage, thereby extending the battery life and improving the overall efficiency of the system. However, the design of this converter may be relatively complex, with high requirements for component accuracy and heat dissipation performance, which increases the cost and difficulty of the implementation of the system. In addition, in practical applications, how to further optimize the control strategy to achieve more efficient power conversion and a more accurate voltage balance is also a problem worthy of further research. A robust control method based on integral backstepping control (IBC) was proposed in reference [11] to enhance the power quality of microgrid-connected photovoltaic systems with battery energy storage systems. The advantage of this method is that, through MATLAB simulation verification, it exhibits a higher success rate, better performance, stronger robustness, and higher efficiency under various conditions compared to traditional strategies. This means that the IBC control method can better cope with uncertainties and disturbances in microgrid-connected photovoltaic systems, ensuring the stable operation and high-quality power output of the system. However, the design and implementation of IBC control methods can be relatively complex, requiring a high level of control theory and engineering practical ability. In addition, although the simulation results show that its performance is superior, in practical applications, it is necessary to consider the real-time performance of the system, the feasibility of hardware implementation, and the compatibility and complementarity with other control strategies. Reference [12] proposes an intelligent control strategy for DC/DC converters, which adopts a two-layer architecture and can effectively regulate the current in different modes to ensure voltage stability. Its feasibility and performance in power systems have been verified through experiments. This control strategy has the advantages of a fast response speed and high adjustment accuracy, but in practical applications, system complexity and cost issues may need to be considered. Reference [13] proposes a monitoring control method based on Model Predictive Control (MPC) to improve the performance and stability of electrical systems, particularly in coordinating and controlling distributed energy resources, such as solar photovoltaic systems and wind turbines, in complex systems. The article discusses, in detail, the widespread use of MPC as a supervisory control strategy in electrical systems and aviation applications and points out its predictive and optimization capabilities for managing energy, maintaining grid stability, and responding to emergencies. However, it may also face challenges in the implementation complexity and cost.

Battery energy storage, with its large capacity, is widely used in power systems; however, its small power density makes it difficult to respond to transient power changes. Conversely, supercapacitors have high power density but cannot sustain discharge due to their low energy density. Therefore, to combine the advantages of both, power-type and energy-type storage are combined to form a hybrid energy storage system. This system not only reduces the number of transient responses of batteries, enhancing the overall system lifespan, but also ensures effective mitigation of power fluctuations in power systems. In reference [14], it is proposed to use an enhanced exponential approach law sliding mode control to achieve maximum power tracking for wind power generation in an independent DC microgrid of photovoltaic-wind power. The significant advantage of this method is that it can efficiently and accurately track the maximum power point of wind power generation, thereby improving the energy utilization efficiency. At the same time, the use of a hybrid energy storage system consisting of batteries and supercapacitors, coupled with Lyapunov function control, ensures the global asymptotic stability of the bidirectional converter. This design not only maintains a constant DC link voltage, but also enhances the stability and reliability of the system. The results of simulation and experimental verification further demonstrate the effectiveness and superiority of the proposed control strategy. However, the sliding mode control method may have some chattering problems, which may have a certain impact on the stability and accuracy of the system. In addition, the design and control strategy of hybrid energy storage systems are relatively complex, requiring high technical levels and cost investments. A new control strategy was proposed in reference [15] for DC microgrids integrated with hybrid energy storage systems containing batteries and supercapacitors. The advantage of this strategy is that it can flexibly adapt to different operating modes determined by existing conditions, ensuring the stable operation and effective control of the system under various steady-state and transient/disturbance conditions. The results of simulation verification show that the strategy exhibits good operational performance and stability under various conditions. However, this strategy may have a certain parameter sensitivity; that is, changes in system parameters may have a significant impact on the control effect. In addition, the design and control strategy of the composite energy storage system may be relatively complex, requiring a high technical level and cost investment. In practical applications, it is also necessary to consider the real-time performance of the system and the feasibility of hardware implementation, as well as the compatibility and complementarity with other control strategies.

The performance of PV maximum power point tracking (MPPT) algorithms primarily depends on the tracking efficiency, global maximum power point tracking capability, and tracking speed [16]. Meanwhile, algorithms can impact power output; for instance, misjudgment in the traditional perturb and observe method can lead to significant power output fluctuations [17]. The mitigation of bus voltage fluctuations hinges on the response of hybrid energy storage to transient power, and decoupling fluctuating power is crucial for mitigation. While mitigating power fluctuations, the overall state of charge (SOC) of the hybrid energy storage also ensures a timely response to fluctuations. Reference [18] uses a virtual RC droop method to decouple power. The advantage of this method is that it can effectively achieve power decoupling control and simplify the design of the control system. However, this method inevitably introduces deviations in the DC bus, which may have a certain impact on the stability and power quality of the system. To overcome this limitation, it is necessary to further optimize the control strategy, such as introducing a compensation mechanism or using more advanced control algorithms to reduce the deviation of the DC bus. References [1,19] design energy management systems by the state of charge of supercapacitors to prevent them from being in low- or high-voltage states. The advantage of this method is that it can protect the supercapacitor from the damage caused by excessive charging and discharging, thereby extending its lifespan. However, these two documents do not solve the problem of the overcharging and overdischarging of energy storage batteries, which is a problem that requires special attention in practical applications. In order to improve the energy management system, it is possible to consider incorporating the state of charge of the energy storage battery into the control strategy to achieve more comprehensive protection. Reference [20] proposes a hybrid energy storage energy management system based on fuzzy rules, which dynamically adjusts the time constant of the low-pass filter through fuzzy control to achieve power distribution for hybrid energy storage. The advantage of this method is that it can flexibly adjust the control parameters according to the actual situation, achieving precise control of the hybrid energy storage system. The fuzzy control algorithm has strong robustness and adaptability and can cope with complex and changing operating environments. However, the design and debugging of fuzzy rules can be relatively complex, requiring a high level of technical expertise and experience. In addition, the performance of fuzzy control systems may be affected by the selection of input variables and fuzzy rules, so these factors need to be carefully considered in practical applications. Reference [21] proposed the Oppositional-Gradient-based Grey Wolf Optimizer (OGGWO) algorithm for the multi-objective optimization scheduling of microgrids. Its advantages lie in its ability to effectively balance operating costs and emissions, with excellent convergence characteristics and the ability to solve nonlinear multi-objective optimization problems. It also shows higher stability and consistency when compared with other optimization algorithms. However, the potential drawbacks or limitations of the algorithm were not explicitly mentioned in the article. Reference [22] proposed an optimization framework that comprehensively considers wind power, photovoltaic power generation, electric vehicles (EVs), and demand response planning (DRP). Its advantages lie in modeling microgrid elements, introducing electric vehicles as distributed energy resources, and utilizing the Manta Ray Foraging Optimization (MRFO) algorithm to minimize the operating costs and emissions of microgrids. However, the article also points out research gaps in the cost of rescheduling caused by intermittent renewable energy generation and efficient, direct electric vehicle operation models. Reference [23] proposed the AIROBE model, which is a data-driven optimization model aimed at solving the reliable and cost-effective scheduling problem of microgrids under the uncertainty of renewable energy and user demand. Its advantage lies in using machine learning prediction as input, considering the uncertainty of dynamic energy prices, photovoltaic power generation, and user consumption, and proposing a refined robust optimization method that allows for considering positive and negative deviations in asymmetric intervals, thereby obtaining fewer conservative solutions. However, despite the theoretical advantages of this method, the article does not elaborate on the challenges or limitations that may be encountered in practical applications. Reference [24] proposed a bidding strategy for independent thermal power units and battery energy storage based on robust optimization. Its advantage lies in the ability to construct a market bidding curve that is not affected by infeasible ramp-up requirements between consecutive small time intervals and to ensure scheduling unpredictability. By solving the entire problem at once, a safe ramp up optimal solution can be obtained. However, this strategy may face the risk of losses in certain situations due to poor choices made with uncertain budgets.

The power prediction of common photovoltaic power generation can be classified according to the following criteria, as shown in Table 1.

Table 1. Classification of photovoltaic power prediction methods.

Classification Basis	Classification
Prediction process	Direct prediction Indirect prediction
Modeling method	Physical methods Statistical method
Predict time scale	Ultra-short-time prediction Short-term prediction Medium- and long-term Forecast
Predict spatial range	Single field prediction Regional prediction

However, the fuzzy rule-based hybrid energy storage energy management system has drawbacks, such as low efficiency, a complex rule design, difficulties in debugging and optimization, insufficient system robustness, and poor scalability. To address these issues, this paper adopts a constant voltage method combined with a variable step-size incremental conductance method for MPPT, combining the response speed of open-loop control with the tracking accuracy of closed-loop control to reduce energy losses during tracking. An energy management strategy for a DC microgrid hybrid energy storage system is proposed to ensure stable system bus voltage, demand power allocation, and balance between load and generation-end power supply and demand and to keep the energy storage system operating within a reasonable SOC. Finally, experimental validation is conducted using Matlab/simulink.

2. Modeling of the Structure of DC Microgrid Systems

The structure of the DC microgrid system is illustrated in Figure 1.

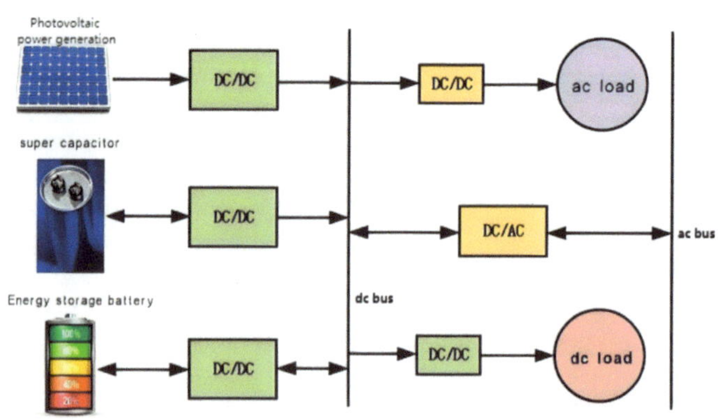

Figure 1. DC microgrid structure.

Renewable energy generation systems and energy storage systems are directly connected to a common DC bus through converters. The bidirectional AC/DC converter controls the power flow between the DC bus and the AC bus, playing a significant role in stabilizing the system voltage and improving the power quality. Depending on actual needs, the bidirectional AC/DC converter can operate in either rectification or inversion mode. The bidirectional power converter provides bidirectional power flow between the AC and DC buses.

The control block diagram of the microgrid system with a photovoltaic (PV) hybrid energy storage system on the DC side is illustrated in Figure 2. The PV output is connected to the DC bus through a Boost converter, and a Maximum Power Point Tracking (MPPT) strategy is employed to ensure the maximum utilization of irradiation. The energy storage battery and supercapacitor are connected to the DC bus through bidirectional DC/DC converters, respectively. Assuming that the AC side has sufficient capacity, it can provide stability support for DC-side power balance.

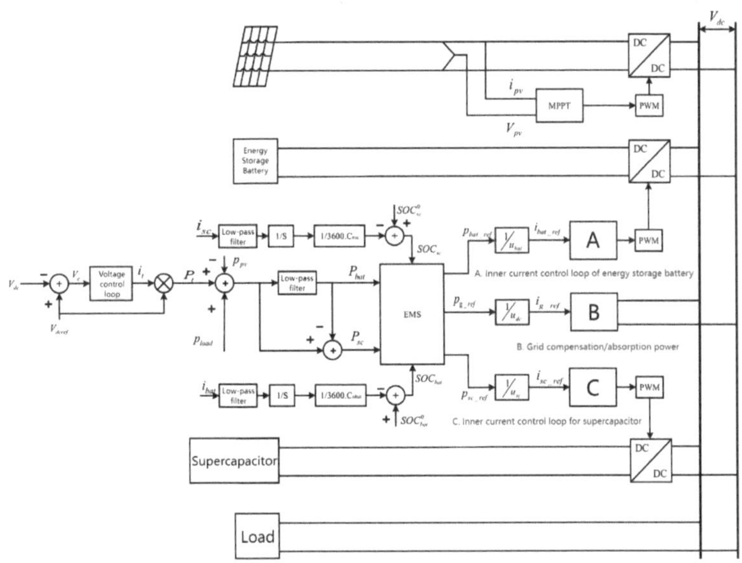

Figure 2. Control block diagram of microgrid system, including the photovoltaic hybrid energy storage system.

The physical model of the PV hybrid energy storage system is depicted in Figure 3. Lpv, D1, SW, and C represent the inductance of the Boost converter, a diode, an Insulated Gate Bipolar Transistor (IGBT), and the bus capacitor, respectively. The battery energy storage and supercapacitor are connected to the DC bus through buck-boost converters to regulate the bus voltage.

In Figure 3, P_{pv}, P_{hess}, P_{load}, and P_g represent the photovoltaic power output, hybrid energy storage power, load power, and grid power to the DC side, respectively. Considering the relationship between load and power generation,

$$P_g + P_{hess} = P_{pv} - P_{load}, \tag{1}$$

$$P_t = P_{pv} - P_{load}, \tag{2}$$

$$P_t = V_{dc} i_t. \tag{3}$$

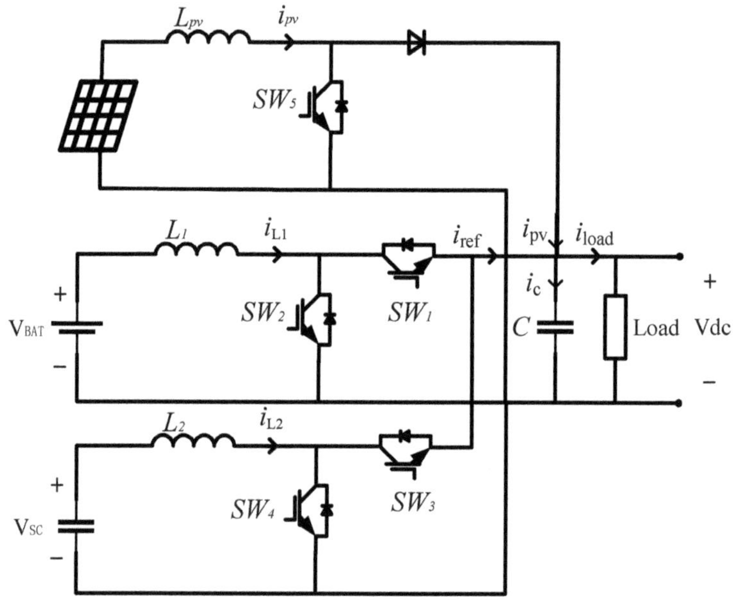

Figure 3. DC microgrid topology with a photovoltaic system.

The stability and tracking speed of the control for the photovoltaic system and hybrid energy storage system will affect the power balance on the DC bus, as well as the dynamic recovery speed of the DC bus. Therefore, the design of the energy management system needs to consider both the system's dynamic response and robustness.

3. Mathematical Model Description of Optimization Problems

The optimization problem discussed in this article is basically stated as follows: in a DC microgrid, by comprehensively considering the maximum power point tracking (MPPT) of photovoltaic systems, the energy distribution strategy of hybrid energy storage systems, including energy storage batteries and supercapacitors, and the stable control of DC bus voltage, the power quality fluctuations are minimized and the efficiency of hybrid energy storage systems is optimized. Specifically, the optimization goal is to maintain the stability of the DC bus voltage while meeting the load demand and optimize the state of charge SOC of the energy storage system to ensure its long-term reliability and service life.

To mathematically describe this optimization problem, we define the following variables and functions:

$P_{pv}(t)$ represents the output power of the photovoltaic system at time t.

$P_{batt}(t)$ represents the power output of the energy storage battery at time t (positive for negative charge, negative for discharge).

$P_{sc}(t)$ represents the power output of the supercapacitor at time t (positive for negative charge, negative for discharge).

$P_{load}(t)$ represents the power demand of the load at time t.

$SOC_{batt}(t)$ and $SOC_{sc}(t)$ represent the state of charge of the energy storage battery and supercapacitor at time t.

$V_{dc}(t)$ represents the voltage of the DC bus at time t.

The objective of the optimization problem is to minimize bus voltage fluctuations and maintain the SOC of the energy storage system within a reasonable range, which can be mathematically described by the following objective function and constraints:

Objective function:

$$min\int_0^T\left[\left(V_{dc}(t)-V_{dc,ref}\right)^2+\lambda_1\left(SOC_{batt}(t)-SOC_{batt,ref}\right)^2+\lambda_2\left(SOC_{sc}(t)-SOC_{sc,ref}\right)^2\right]dt. \quad (4)$$

Among them, $V_{dc,ref}$ is the reference value of the DC bus voltage, $SOC_{batt,ref}$, and $SOC_{sc,ref}$ are the SOC reference values of the energy storage battery and supercapacitor, respectively, and λ_1 and λ_2 are weight coefficients used to balance the priority between different objectives.

Constraints:

Power balance:

$$P_{pv}(t)+P_{batt}(t)+P_{sc}(t)=P_{load}(t), \quad (5)$$

Energy storage battery SOC dynamic:

$$SOC_{batt}(t+1)=SOC_{batt}(t)+\frac{P_{batt}(t)\Delta t}{Q_{batt}}, \quad (6)$$

State of Charge (SOC) dynamics of supercapacitors:

$$SOC_{sc}(t+1)=SOC_{sc}(t)+\frac{P_{sc}(t)\Delta t}{Q_{sc}}, \quad (7)$$

SOC limit for energy storage batteries:

$$SOC_{batt,min}\leq SOC_{batt}(t)\leq SOC_{batt,max}, \quad (8)$$

SOC limit for supercapacitors:

$$SOC_{sc,min}\leq SOC_{sc}(t)\leq SOC_{sc,max}, \quad (9)$$

DC bus voltage limit:

$$V_{dc,min}\leq V_{dc}(t)\leq V_{dc,max}, \quad (10)$$

Among them, Q_{batt} and Q_{sc} are the capacities of energy storage batteries and supercapacitors, Δt is the time step, and $SOC_{batt,min}$, $SOC_{batt,max}$, $SOC_{sc,min}$, $SOC_{sc,max}$, $V_{dc,min}$, and $V_{dc,max}$ are the corresponding minimum and maximum limit values.

4. Photovoltaic Maximum Power Point Tracking Control Based on Hybrid Algorithm

In this paper, a comprehensive framework is constructed to optimize the energy management of DC microgrids, particularly focusing on the maximum power point tracking (MPPT) control of photovoltaic (PV) power generation systems and the effective management of hybrid energy storage systems. The core workflow of this framework encompasses the following key steps: firstly, a hybrid algorithm combining the Particle Swarm Optimization (PSO) and variable step size conductance increment method is utilized to achieve MPPT control of the PV system, ensuring that the PV panels always operate at the maximum power point and thus maximizing energy utilization efficiency. Secondly, the DC bus power fluctuations are decoupled into high-frequency and low-frequency components through a first-order low-pass filter, which serve as the reference input powers for the supercapacitor and energy storage battery, respectively, to achieve precise energy distribution within the hybrid energy storage system. On this basis, an energy management system (EMS) based on the state of charge (SOC) of the energy storage elements is introduced to limit the charging and discharging processes of these elements, preventing overcharging or over-discharging. Meanwhile, decisions are made based on the real-time load demand, PV power generation capacity, and energy storage system status, ensuring efficient energy allocation and utilization within the system. Finally, through the compensation of grid-side power, the power limits absorbed or provided by the hybrid energy storage system are adjusted, further enhancing the system's stability and response speed.

Irradiance and temperature are two important parameters that determine the output characteristics of photovoltaic cells. The influence of these parameters on the output characteristics of photovoltaic cells can be obtained through the controlled variable method.

As can be seen from Figures 4 and 5, the output of photovoltaic cells exhibits a trend of variation with changes in irradiance and temperature. Photovoltaic cells are neither constant current sources nor constant voltage sources. The output power of photovoltaic cells is determined by the intensity of illumination, while temperature affects the location of the maximum power point (MPP) in their output. In order to maximize power output under the specific conditions of temperature and illumination intensity, it is necessary to maintain the operating point of the photovoltaic cells at the MPP.

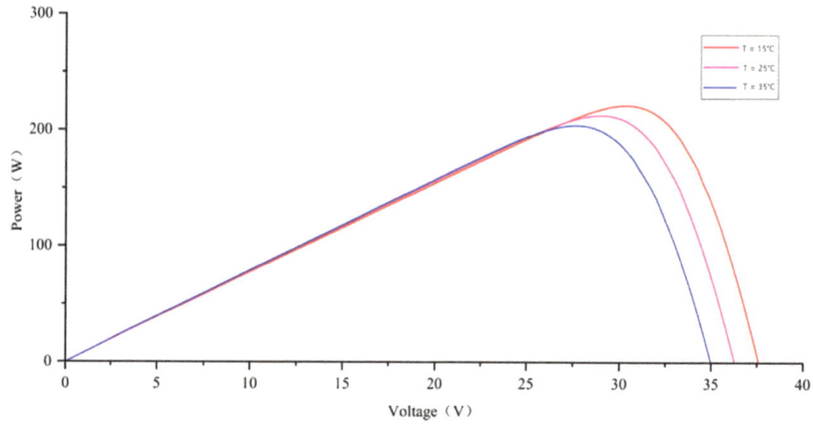

Figure 4. Characteristic curves of photovoltaic output at different temperatures.

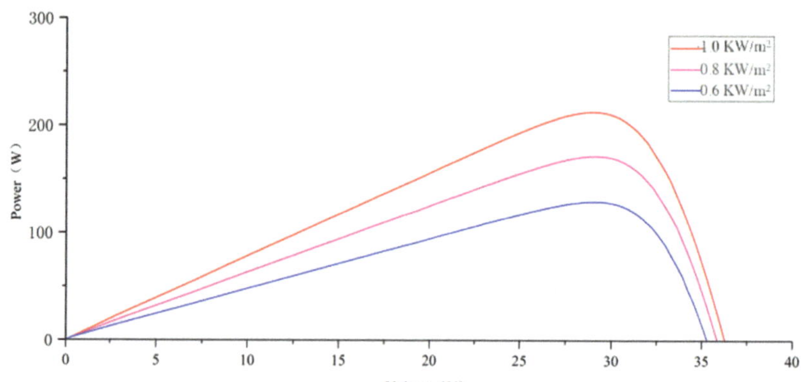

Figure 5. Relationship between output voltage and power under different light.

Under different external environments, there is only one MPP for the output of photovoltaic cells. The perturbation and observation method can continuously search for the MPP under different conditions, but it may suffer from misjudgment and significant fluctuations. Compared to the perturbation and observation method, the conductance increment method does not exhibit fluctuations at steady state but is slower in searching for the MPP. Therefore, this paper proposes an improvement to the conductance increment method by incorporating a variable step size to search for the MPP. Through a quantitative analysis of dP/dU, the corresponding criterion for the maximum power point can be obtained. Considering the instantaneous output power of the photovoltaic cell as

$$P = UI. \tag{11}$$

Taking the derivative of the battery output voltage,

$$\frac{dP}{dU} = I + U\frac{dI}{dU}. \tag{12}$$

In the equation, U and I represent the operating voltage and current of the photovoltaic array, respectively.

- When $\frac{dI}{dU} > -\frac{I}{U}$, the operating voltage U is greater than the maximum operating voltage, necessitating an increase in voltage to shift it to the right.
- When $\frac{dI}{dU} < -\frac{I}{U}$, the operating voltage U is less than the maximum operating voltage, requiring a decrease in voltage to shift it to the left.
- When $\frac{dI}{dU} = -\frac{I}{U}$, the battery is operating at the maximum power point.

The control flow of the conductance increment method is shown in Figure 5.

In Figure 6, $V(k)$ and $V(k-1)$ represent the output voltages at the k-th and $(k-1)$-th sampling points, respectively, while $I(k)$ and $I(k-1)$ represent the output currents at the kth and $(k-1)$th sampling points, respectively. V_{ref} and ΔV denote the reference voltage and voltage increment step size, respectively. The traditional conductance increment method typically sets the initial duty cycle D to 0.4. However, a limitation of this algorithm is that the voltage increment step size is fixed and cannot be adjusted in real-time, which affects the accuracy of control tracking.

Given the shortcomings of the traditional conductance increment method, which uses a compromise fixed step size that does not change with environmental conditions, an optimized variable step size conductance increment method can be adopted. This method adjusts the tracking step size according to changes in irradiance and temperature. In practical applications, external temperature and light conditions are always changing. Therefore, the ability to quickly and reasonably adjust the system's real-time tracking step size according to these changes is crucial for evaluating the performance of the variable step size conductance increment method. The basic principle of the variable step size conductance increment method is basically the same as that of the conductance increment method, but the fixed step size is replaced by a variable step size, which is the absolute value of the ratio of the change in the output power of the photovoltaic cell to the change in the operating voltage, multiplied by a correction coefficient. The specific flowchart is shown in Figure 7.

$$\delta = K_s\frac{dp}{dv}. \tag{13}$$

In the above equation, δ represents the step size variation, and K_s denotes the step size gain. The variable step size conductance increment method allows for an adjustment of the tracking step size, but this method must be tailored to the real-time environment, lacking control flexibility. Considering its defects, due to the fast convergence and high accuracy characteristics of the Particle Swarm Optimization algorithm, this paper considers using a hybrid algorithm consisting of the particle swarm algorithm combined with the variable step size conductance increment method.

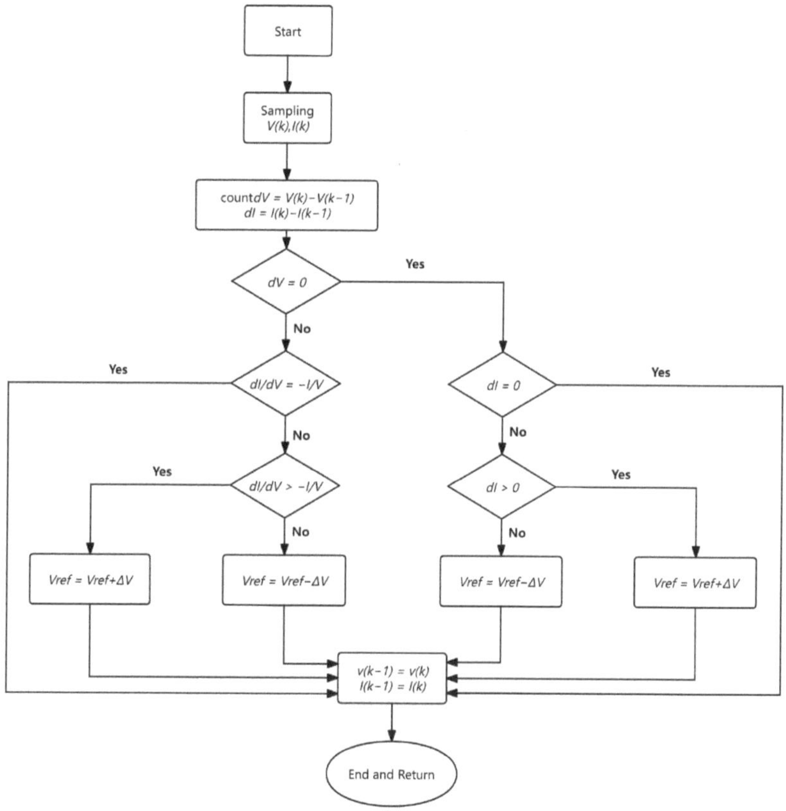

Figure 6. Conductivity increment control flow chart.

Particle Swarm Optimization (PSO) is an optimization algorithm that mimics the collective intelligence behavior observed in natural groups, drawing inspiration from the abstract and simulated foraging activities of bird flocks, fish schools, and other biological collectives. In the context of PSO, each potential solution is represented as a "particle" that moves within a search space to discover the optimal solution to a problem. These particles possess two key attributes: position and velocity. The position signifies a point in the solution space, corresponding to a candidate solution, while the velocity dictates how the particle updates its position based on its current location, its individual best-known position (Pbest), and the globally best-known position discovered by the entire swarm (Gbest). Through iterative updates to their velocities and positions, PSO algorithms progressively converge towards the optimal solution to the problem.

During the execution of PSO, particles move according to specific rules that integrate inertia maintenance (controlled by the inertia weight w), individual cognition (achieved through c_1 multiplied by a random number r_1 and the difference between the individual best position and the current position), and social cognition (realized through c_2 multiplied by a random number r_2 and the difference between the global best position and the current position). This design enables particles to maintain exploratory capabilities while also converging rapidly towards high-quality solutions.

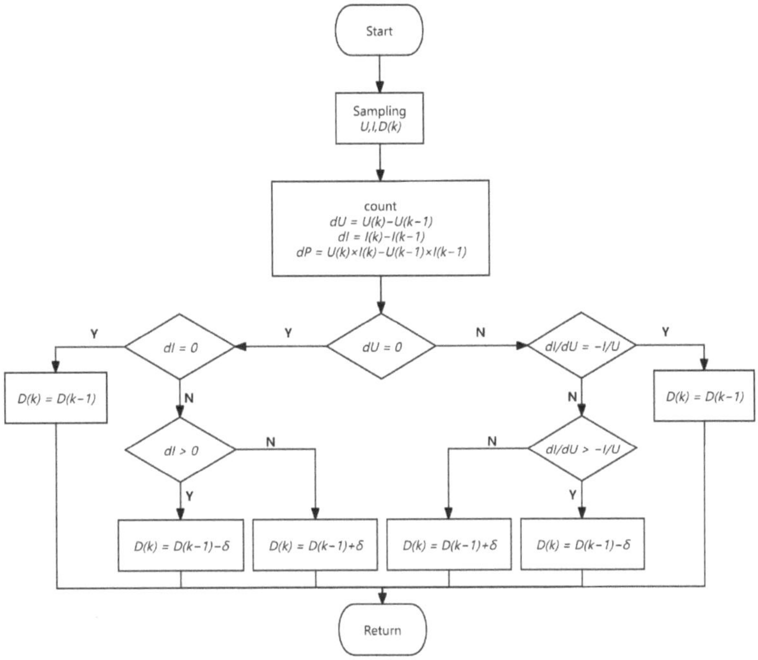

Figure 7. Variable step size conductance increment control flow chart.

For microgrid problems, PSO demonstrates high applicability. As a complex and dynamic energy system, microgrids often involve optimization challenges with multiple variables and constraints, such as scheduling the output of distributed generation, managing charging and discharging strategies for energy storage systems, regulating load demand, and interacting with the larger power grid. PSO algorithms can adeptly handle these intricate multi-variable optimization problems by continuously adjusting the positions and velocities of particles to explore microgrid operational strategies that satisfy various constraints and optimize economic performance. Additionally, PSO's parallelism and ease of implementation provide significant advantages when dealing with large-scale microgrid optimization problems, enabling the efficient discovery of near-global optimal solutions and supporting the intelligent management and operation of microgrids.

The specific control steps are as follows: when the operating voltage is less than or equal to 0.78 times the maximum operating voltage, the fast convergence characteristics of the PSO algorithm are utilized for tracking to improve tracking accuracy. When the operating voltage approaches the maximum operating voltage, to avoid the PSO algorithm falling into a local optimum and considering that the output power of the photovoltaic panel changes slowly with the voltage, the variable step size conductance increment method is used for tracking control to improve the tracking speed.

Firstly, the time parameters of the circuit simulation environment are set, with a discrete sampling mode time interval set to 10^{-6} s. Then, the step size parameters for the variable step size conductance increment method are set, with the step size gain coefficient K_s set to 10^{-6}.

Next, the basic parameters of the particle swarm algorithm are set. In the PSO algorithm, the value of ω determines whether the algorithm is prone to falling into a local optimum and is crucial for fast and accurate searching. Conventional improved weighted PSO algorithms include adaptive weighting, random weighting, and linearly decreasing weighting. During maximum power point tracking (MPPT), the reasonable change of the weighting value directly affects the tracking accuracy of the algorithm. In the initial stage of the algorithm, the value of ω needs to be increased to improve the global search capability; in the final stage of the algorithm, the value of ω needs to be reduced to enhance the local search capability. Conventional weighting change methods are not suitable for MPPT. Therefore, a dynamic inertia weight method based on piecewise control is adopted.

$$d\omega = \frac{13(\omega_{max} - \omega_{min})}{2}, \tag{14}$$

$$dP = P - P_{old}, \tag{15}$$

$$m = \frac{dP}{(V - V_{old})}. \tag{16}$$

Based on the analysis in the previous section, it can be inferred from the equation that ω_{max} and ω_{min} represent the maximum and minimum values of ω, respectively. Here, ω_{max} is set to 0.9 and ω_{min} to 0.4. dP denotes the power difference, where P_{old} is the power at the previous time step and P is the power at the current time step. m represents the ratio of the rate of change of power to the rate of change of voltage. The value of ω is comprehensively determined by the three factors: $d\omega$, dP, and m in the equation.

When $dP > 0$, the value of ω is shown in Table 2 [25]:

Table 2. Calculation formulas for different m values corresponding to ω.

Range of m	The Calculation Formula for ω
$m > 2$ or $<= -2$	$\omega + 3 \times d\omega$
$1.5 < m <= 2$ or $-2 < m <= -1.5$	$\omega + 2 \times d\omega$
$1 < m <= 1.5$ or $-1.5 < m <= -1$	$\omega - d\omega$
$0.5 < m <= 1$ or $-1 < m <= -0.5$	$\omega - 2 \times d\omega$
$0 < m <= 0.5$ or $-0.5 < m <= 0$	$\omega - 3 \times d\omega$

After setting the variation values for the weight ω, through multiple simulation experiments, the initial values of c_1 and c_2 are set to 1 and 2.5, respectively. The adjustment formulas for the individual learning factor and the social learning factor are as follows:

$$c_1 = c_{1max} - \left((c_{1max} - c_{1min}) * \left(\frac{k}{n}\right)^2 \right), \tag{17}$$

$$c_2 = c_{1min} - \left((c_{2max} - c_{2min}) * \left(\frac{k}{n}\right)^2 \right). \tag{18}$$

In the formula, k represents the current iteration number, and n is the number of particles set to 4, with $c_{1min} = 1.5$, $c_{1max} = 2$; $c_{2min} = 1.5$, and $c_{2max} = 2.5$. The individual learning factor and social learning factor are updated in real-time with the iteration number to ensure the reliability of the algorithm's search capability. At this point, the basic parameter settings for the Particle Swarm Optimization (PSO) algorithm are completed.

After setting the basic parameters, specific iteration termination conditions and algorithm restart conditions need to be established. This paper defines two iteration termination conditions:

(1) When the absolute difference between the real-time output power and the initial power is less than 0.1 and the iteration number exceeds 3.
(2) When the iteration number reaches the maximum iteration limit.

The formula for the algorithm restart condition is as follows:

$$\Delta P = \frac{P - P_{old}}{P_{old}}. \tag{19}$$

In the equation, ΔP represents the rate of change in power, P denotes the current power, and P_{old} signifies the power at the previous time step. Considering the complexity of actual operating conditions where the power changes frequently, we set $\Delta P > 0.11$. The basic control flowchart of the hybrid algorithm is illustrated in Figure 8.

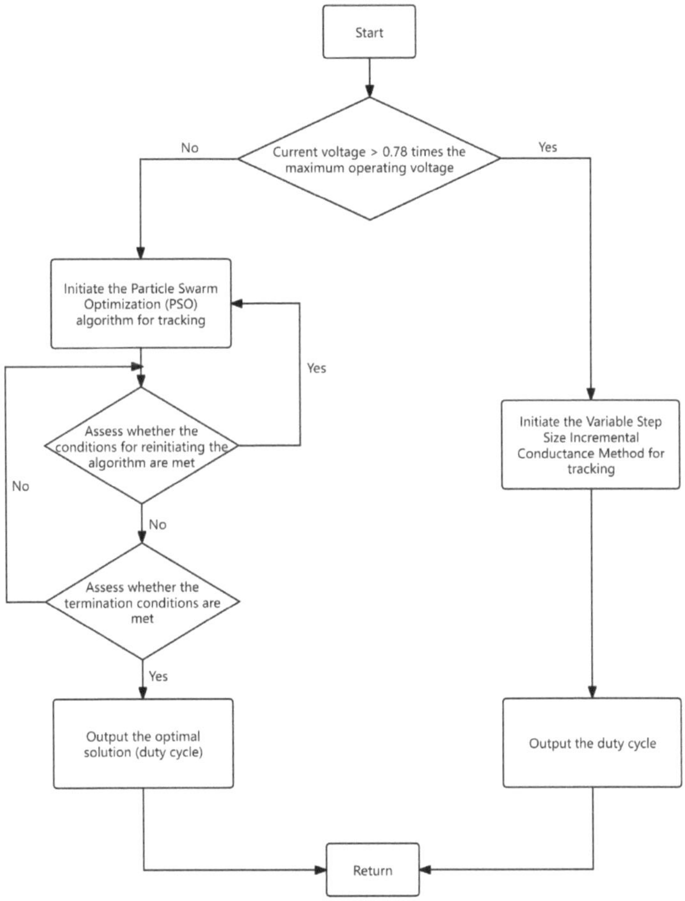

Figure 8. Basic control flow of hybrid algorithm.

After discussing the application of the hybrid algorithm combining the variable step size conductance increment method and the Particle Swarm Optimization (PSO) algorithm in photovoltaic maximum power point tracking (MPPT) control, it is necessary to further analyze the convergence behavior of this algorithm. The convergence speed of the proposed algorithm mainly depends on the parameter settings in the PSO algorithm, such as the inertia weight ω, the cognitive coefficient $c1$, the social coefficient $c2$, and the size of the particle swarm. Reasonable parameter configurations can accelerate the convergence process of the algorithm, making the particles rapidly converge towards the global optimal solution. At the same time, the introduction of the variable step size conductance increment method further improves the convergence accuracy by finely adjusting the step size when the algorithm approaches the optimal solution.

However, successful convergence of the algorithm is not always guaranteed. The failure conditions may include the following: first, the PSO algorithm falls into a local optimal solution, which is usually caused by improper parameter settings or the presence of multiple local optimal solutions in the search space; second, the variable step size conductance increment method fails to accurately reflect changes in the external environment when adjusting the step size, resulting in a decrease in tracking accuracy. To avoid these failures, it is necessary to finely tune the algorithm parameters according to specific conditions in practical applications and consider introducing other optimization strategies, such as crossover and mutation operations in genetic algorithms, to enhance the global search ability and robustness of the algorithm.

In addition, although deterministic algorithms can provide stable solutions in certain scenarios, they often struggle to balance reliability and computational efficiency when dealing with complex and variable optimization problems, such as photovoltaic maximum power point tracking. Compared to the proposed hybrid algorithm, deterministic algorithms may lack sufficient flexibility to quickly adapt to changes in external environments, such as rapid fluctuations in the irradiance and temperature, which may lead to decreased tracking accuracy and slower system response speeds. Meanwhile, deterministic algorithms often require a significant amount of iterative calculations in the solution process, which not only increases the computational burden but may also affect the real-time performance of the system. In contrast, the proposed hybrid algorithm combines the advantages of the Particle Swarm Optimization (PSO) algorithm and the variable step size conductance increment method, maintaining a high tracking accuracy and response speed in complex environments while optimizing the computational efficiency through a dynamic parameter adjustment. This achieves a better balance between the reliability and computational efficiency.

Furthermore, for the proposed hybrid algorithm, its optimization time is also an important indicator for evaluating its performance. The optimization time mainly depends on the complexity of the algorithm itself and the performance of the computing environment. In the hybrid algorithm presented in this paper, due to the combination of the Particle Swarm Optimization (PSO) algorithm and the variable step size conductance increment method, the optimization time may increase compared to traditional single algorithms. However, through reasonable parameter configuration and efficient computation implementation, the optimization time can be effectively controlled within an acceptable range.

To assess the computational efficiency of the proposed algorithm, we compared it with other common optimization algorithms. For instance, the Genetic Algorithm (GA), as a global optimization algorithm, exhibits good searching capabilities, but its computational complexity is high, leading to longer optimization times. In contrast, deterministic algorithms, such as the Gradient Descent Method, although computationally simple, may suffer from inefficiencies when dealing with complex multivariable optimization problems due to the risk of getting stuck in local optimum solutions. The proposed hybrid algorithm, by combining the rapid convergence of the PSO algorithm and the fine-tuning capability of the variable step size conductance increment method, ensures both global searching capability and improved computational accuracy, thus achieving a good balance between optimization time and computational efficiency.

5. Energy Management Strategy Based on the State of Charge of Hybrid Energy Storage Systems

The hybrid energy storage system mitigates bus power fluctuations through its inherent characteristics. During charging and discharging processes, the deep discharging of energy storage batteries affects the battery life, while an excessively high state of charge (SOC) of supercapacitors poses a risk of breakdown. Considering the system's response speed and robustness, this paper proposes a rule-based energy management system (EMS) for hybrid energy storage to limit the charging and discharging processes of energy storage components. This strategy makes decisions based on the real-time load demand, photovoltaic generation capacity, energy storage system status, and other information to ensure efficient energy allocation and utilization, preventing the SOC from being too high or too low. According to the power balance relationship of the DC bus in the DC microgrid structure shown in Figure 1, the required power can also be expressed as follows:

$$P_t = \widetilde{P}_t + \overline{P}_t. \tag{20}$$

A first-order low-pass filter is utilized to decouple the power into high-frequency and low-frequency components, which serve as the reference inputs for the supercapacitor and energy storage battery, respectively, in the Energy Management System (EMS):

$$\begin{cases} \overline{P}_t(s) = \frac{\omega_c}{s+\omega_c} P_t(s) \\ \widetilde{P}_t(s) = \frac{\omega_c}{s+\omega_c} P_t(s) \end{cases}. \tag{21}$$

Here, ω_c represents the cutoff frequency of the low-pass filter, and a fixed cutoff frequency is adopted to reduce the stress on the battery caused by fluctuating power;

$$i_t = \widetilde{i}_t + \overline{i}_t. \tag{22}$$

$\widetilde{P}_t$ denotes the fluctuating component of power, while $\overline{P}_t$ represents the average component of power; $\widetilde{i}_t$ denotes the fluctuating component of current, and $\overline{i}_t$ represents the average component of current;

$$i_t = \widetilde{i}_t + \overline{i}_t = K_p V_e + K_i \int V_e dt. \tag{23}$$

V_e, K_p, and K_i represent the voltage deviation, proportional coefficient, and integral coefficient, respectively, in the voltage control loop.

According to the research in [26], during the charging and discharging process of energy storage batteries, when the battery begins to discharge from a fully charged state, there is a rapid voltage drop within the battery itself. Similarly, a sudden voltage decrease also occurs when the system's power level is insufficient.

$$\begin{aligned} V_o = K_1 + K_2 \cdot SOC + K_3 \cdot SOC^2 + K_4 \cdot SOC^3 + K_5/SOC + K_6 \cdot ln(SOC) + K_7 \cdot \\ ln(1 - SOC) - 1/C_p \cdot V_p + 1/C_p \cdot I - I \cdot R_{ox}. \end{aligned} \tag{24}$$

For control systems, variations in the terminal voltage serve as disturbance inputs that can impact the charging and discharging control performance. To address this, it is essential to maintain the battery's state of charge within a reasonable range over an extended period. Based on the power throughput between the DC-side photovoltaic system and the load, the capacity of the battery energy storage system, and its charging and discharging characteristics, the proposed constraint zones are categorized into a charging buffer zone, a discharging buffer zone, a restricted charging zone, and a restricted discharging zone. The charging and discharging constraint rules are outlined in Table 3.

Table 3. Limitations on the charging and discharging of energy storage batteries.

Parameter	Numerical Value				
SOC	(0~0.1)	(0.1~0.2)	(0.2~0.8)	(0.8~0.9)	(0.9~1)
Charging constraint λc	0	0	0	$(SOC - 0.8)$	1
Discharge constraint λd	1	$(SOC - 0.1)$	0	0	0

After undergoing secondary power allocation, the reference power of the energy storage battery is P_{bat_ref}.

$$P_{bat\_ref} = \overline{P}_t(1 - \lambda_x). \tag{25}$$

In this context, $\overline{P}_t$ represents the initially decoupled average power of the first-order filter, and λ_x is the constraint factor (where $x = c$ or $x = d$).

Based on the second-order equivalent circuit modeling method established in reference [27], the equivalent circuit indicates that the state of charge (SOC) of the supercapacitor changes linearly during constant-current discharging.

$$\begin{cases} U_1 = IR_1 \cdot \left[1 - esp\left(-\frac{t}{\tau_1}\right)\right] \\ U_t = OCV(SOC) - IR_0 - U_1 \end{cases}. \tag{26}$$

When the state of charge (SOC) falls below 10%, the supercapacitor can no longer guarantee the corresponding power output. In hybrid energy storage systems, the power storage of supercapacitors needs to be maintained at a responsive level to cope with transient power fluctuations in the photovoltaic system output and loads. This paper proposes a supercapacitor energy management strategy based on a logic function to limit its charging and discharging processes. The unit logic function, centered at 0.5, can dynamically adjust the charging and discharging strategies of the supercapacitor according to system requirements to achieve efficient energy management. Based on information, such as the load demand and battery status, the logic function can intelligently control the charging and discharging processes of the supercapacitor, maintaining its SOC within an upper and lower limit range of 50%. This allows the supercapacitor to quickly respond by compensating or absorbing transient power demands during bus fluctuations. The charging and discharging limit rules for supercapacitors are shown in Table 4.

Table 4. Super capacitor charge and discharge limits.

Parameter	Numerical Value			
SOC	(0~0.1)	(0.1~0.5)	(0.5~0.9)	(0.9~1)
Charging constraint ξ_c	0	$0.5(1 + \sqrt{\frac{0.5-SOC}{0.4}})$	$0.5(1 - \sqrt{\frac{0.9-SOC}{0.4}})$	1
Discharge constraint ξ_d	1	$0.5(1 - \sqrt{\frac{0.5-SOC}{0.4}})$	$0.5(1 + \sqrt{\frac{0.9-SOC}{0.4}})$	0

Similarly, the reference power P_{sc_ref} of the supercapacitor is

$$P_{sc\_ref} = \widetilde{P}_t(1 - \xi_x). \tag{27}$$

Here, $\widetilde{P}_t$ represents the initial decoupling power of the first-order filter, and ξ is the constraint factor (where $x = c$ or $x = d$).

Furthermore, P_r represents the reference power for compensation or absorption. Based on the load power and generation power, the energy management system (EMS) categorizes the hybrid energy storage system into three operational modes: compensation mode, floating mode, and sufficient mode. When the system is in compensation mode, $P_r > 0$, and when the power exchange between the energy storage system and the grid side is described as

$$P_g = P_r - P_{sc\_ref} - P_{bat\_ref}. \tag{28}$$

During the discharge of the energy storage system, the grid side provides compensating power to ensure that the overall state of charge of the hybrid energy storage system remains within the set range. In the sufficient mode, $P_r < 0$, where the generation power is greater than the consumption power, and the excess energy is stored by the energy storage system to restore its state of charge. If the hybrid energy storage system has sufficient electricity, it will feed the excess generated power back to the grid. In the floating mode, $P_r = 0$, where the generation power is equal to the consumption power.

$$P_{hess} + P_g = P_{pv} - P_{load}. \tag{29}$$

The reference current generator for the hybrid energy storage energy management system is illustrated in Figure 9. It measures the voltage deviation as the input for the Proportional-Integral Controller (PI Controller) to derive the reference current required for stabilizing the bus voltage. Furthermore, by combining the photovoltaic output and load power, the reference power for the hybrid energy storage system is calculated. This reference power is then decoupled into high-frequency and low-frequency power through a first-order low-pass filter, serving as the input for the energy management system. Based on the state of charge of the hybrid energy storage system, a secondary allocation of the reference power is performed for the hybrid energy storage system.

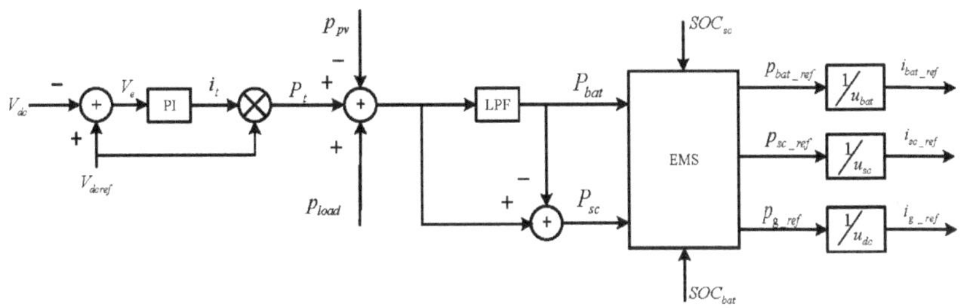

Figure 9. Hybrid energy storage reference current generator with energy management system.

The block diagram for the inner current loop control of the hybrid energy storage system is shown in Figure 10, where a PI controller is employed to track the reference current. The response speed of the energy storage system, which includes the battery, bidirectional DC-DC converter, and controller, is relatively slow compared to the discharge rate of the supercapacitor. To enhance the overall response speed of the hybrid energy storage system, the uncompensated component i_{be} of the battery system during the response time is transferred to the supercapacitor side for compensation.

The following is a pseudocode representation of the key steps in the proposed SOC based hybrid energy storage management strategy:

```
// Initialize parameters
Initialize parameters (Ppv, Pload, Pbatt_max, Psc_max, SOC_batt, SOC_sc, etc.)

// Maximum power point tracking of photovoltaic system (MPPT)
While (system is running) do
```

```
        Ppv_max = MPPT_algorithm (Ppv, irradiation, temperature) // Using a hybrid algorithm combining Particle Swarm Optimization algorithm and variable step size conductance increment method
    End while

    // Power decoupling and initial allocation
    P_fluctuation = Ppv_max − Pload // Calculate power fluctuations
    P_low_freq, P_high_freq = LowPassFilter(P_fluctuation, cutoff_frequency) // Decoupling of first-order low-pass filter
    Pbatt_ref = P_low_freq // Reference power of energy storage battery
    Psc_ref = P_high_freq // Reference power of supercapacitor

    // Energy management strategy for hybrid energy storage system
    While (SOC_batt and SOC_sc are within safe limits) do
        If (SOC_batt < lower_threshold) then
            Pbatt_actual = Min(Pbatt_ref, Pbatt_max_charge) // Limit charging power
        Else If (SOC_batt > upper_threshold) then
            Pbatt_actual = Max(Pbatt_ref, −Pbatt_max_discharge) // Limit discharge power
        Else
            Pbatt_actual = Pbatt_ref
        End if

        If (SOC_sc < 10%) then
            Psc_actual = Max(Psc_ref, Psc_min_discharge) // Ensure that the supercapacitor has sufficient power output
        Else
            Psc_actual = Psc_ref
        End if

        // Update the status of the energy storage system
        SOC_batt = UpdateSOC(SOC_batt, Pbatt_actual)
        SOC_sc = UpdateSOC(SOC_sc, Psc_actual)
    End while

    // DC bus voltage stability control
    Maintain_DC_bus_voltage_stability()
```

The pseudo-code above shows the key steps of the proposed energy management strategy in detail. Firstly, the system achieves maximum power point tracking (MPPT) for the photovoltaic system by combining a hybrid algorithm of Particle Swarm Optimization and variable step size conductance increment method to ensure that the photovoltaic panel always operates at its maximum power point, thereby maximizing the energy utilization efficiency.

Next, a first-order low-pass filter is used to decouple the DC bus power fluctuations into high-frequency and low-frequency components, which are used as the reference input power for the supercapacitor and energy storage battery, respectively. This step achieves precise energy distribution in the hybrid energy storage system, where the energy storage battery is responsible for handling low-frequency power fluctuations, while the supercapacitor is responsible for handling high-frequency power fluctuations.

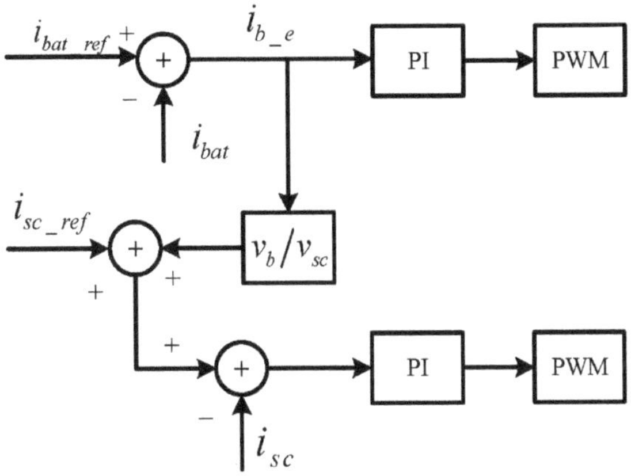

Figure 10. Block diagram of current inner loop control for the hybrid energy storage system.

In the energy management strategy of the hybrid energy storage system, we introduce a charging and discharging limit rule based on the state of charge SOC of the energy storage component. These rules are designed to prevent the overcharging or overdischarging of energy storage batteries, while ensuring that supercapacitors can maintain a state of charge that effectively absorbs and compensates for fluctuating power. By monitoring the SOC of the energy storage system in real time and making decisions based on the real-time load demand, photovoltaic power generation capacity, and energy storage system status, the system can efficiently allocate and utilize energy.

Finally, although the specific algorithm for DC bus voltage stabilization control is not detailed in the pseudo-code, maintaining DC bus voltage stability is one of the key factors to ensure the normal operation of the system and the quality of power. In practical applications, more complex control algorithms may be required to achieve this goal.

In the energy management strategy proposed in this article, the stability of the controller is a key factor in ensuring the normal operation of the system and effective management of energy. Although this article does not provide a detailed stability analysis of the controller, the stability issue has been fully considered during the design process. Generally speaking, stability analysis may involve using tools, such as the Lyapunov stability theory or linear system theory to evaluate the dynamic response and resilience of a system under various disturbances. For the controller design in this article, we followed the basic principles of control theory, through reasonable parameter design and algorithm selection, aiming to ensure that the system can quickly recover and maintain a stable operating state in the face of external disturbances, such as fluctuations in photovoltaic power and load changes. Future research will further deepen this field, providing more comprehensive and accurate stability assessments through specific mathematical analysis and experimental verification.

6. Simulation Analysis

To validate the smoothing effect of the hybrid energy storage system on photovoltaic and load power fluctuations, as well as the effectiveness of the energy management system in maintaining the energy levels of the storage based on rule-based control, modeling and simulation analysis were conducted using Matlab/Simulink according to the proposed model and control strategy. The simulation model is shown in Figure 11.

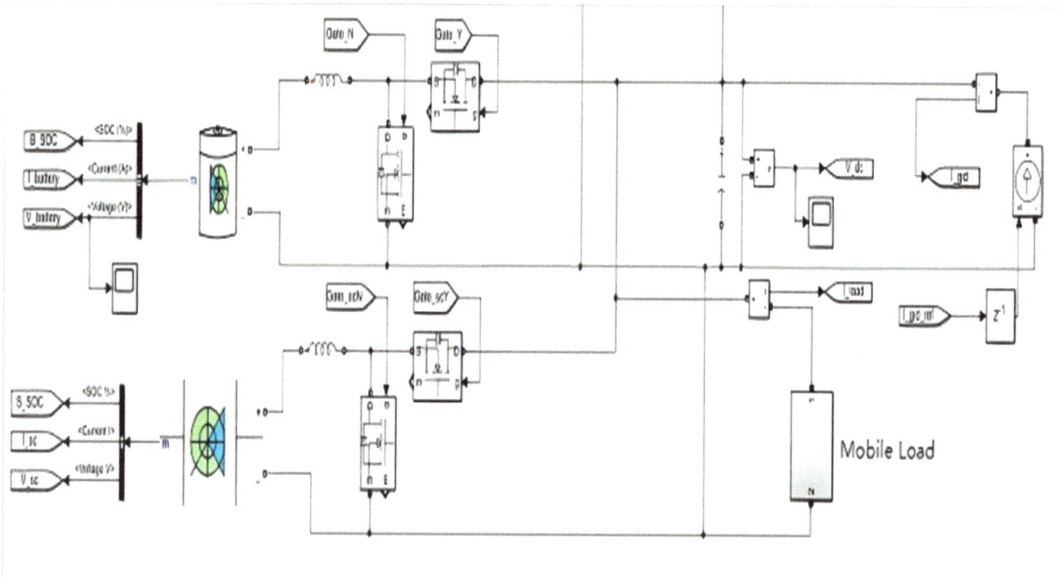

Figure 11. Simulation diagram of the structure of the energy storage part of the hybrid DC microgrid for photovoltaic energy storage.

The parameter settings for the energy storage battery and supercapacitor are shown in Tables 5 and 6. When setting the simulation parameters in Tables 5 and 6, we comprehensively considered various factors to ensure the accuracy and practicality of the simulation. For the energy storage battery (Table 5), the selection of nominal voltage and rated capacity was based on commonly available battery specifications in the market, combined with the system's voltage level and energy requirements. The settings for the full-load voltage, internal resistance, and initial state of charge (SOC) referred to the battery's technical manual and practical application experience, ensuring that the battery could exhibit realistic charging and discharging characteristics during the simulation. For the supercapacitor (Table 6), the choice of rated voltage and rated capacitance was also based on market specifications and system requirements, while the settings for the number of series connections and initial SOC considered the capacitor's voltage level and energy storage capacity to ensure that it could effectively respond to the system's transient power demands. By reasonably setting these parameters, we ensured that the simulation model could accurately reflect the actual performance of the hybrid energy storage system.

Table 5. Parameter setting for energy storage battery simulation.

Parameter	Numerical Value
Nominal voltage/V	24
Rated capacity/Ah	27
Full load voltage/V	27.9357
Internal resistance/Ω	0.0089
Initial state of charge/%	90

The DC bus voltage is set at 100 V. The photovoltaic array operates at 25 °C under varying illumination conditions, with an open-circuit voltage of 36.3 V, a short-circuit current of 7.84 A, and maximum power point operating voltage and current of 29 V and 7.35 A, respectively. The DC load is a fluctuating load with power varying between 500 W and

300 W, with a change period of 4 s. The illumination intensity starts at 300 lux, steps up to 400 lux at 1.5 s, increases by another 100 lux at 3 s, and decreases by 100 lux at 5 s. The cutoff frequency of the first-order low-pass filter is 2 Hz.

Table 6. Supercapacitor simulation parameter setting.

Parameter	Numerical Value
Rated voltage/V	15
Rated capacitance/F	58
Number of serial connections	2
Initial state of charge/%	38.37

As illustrated in Figure 12, the control method without an energy management system exhibits the limited suppression of DC bus power fluctuations due to the low state of charge of the supercapacitors, resulting in a maximum DC bus voltage fluctuation of 18.497 V. In contrast, with an energy management system in place, the DC bus demonstrates good stability against fluctuations in load and photovoltaic power, with a bus deviation of only 0.8073 V for the same power fluctuation. The incorporation of a rule-based energy management system into the traditional control strategy enhances the stability of the DC bus. Figure 13 clearly shows that when load power fluctuates, the transient response of the load and the inability to promptly compensate for DC bus voltage deviations lead to brief fluctuations in the maximum power output of the photovoltaic system. This indicates that a reduction in DC bus stability can also impact the stability of the power output of the photovoltaic system connected to the DC bus.

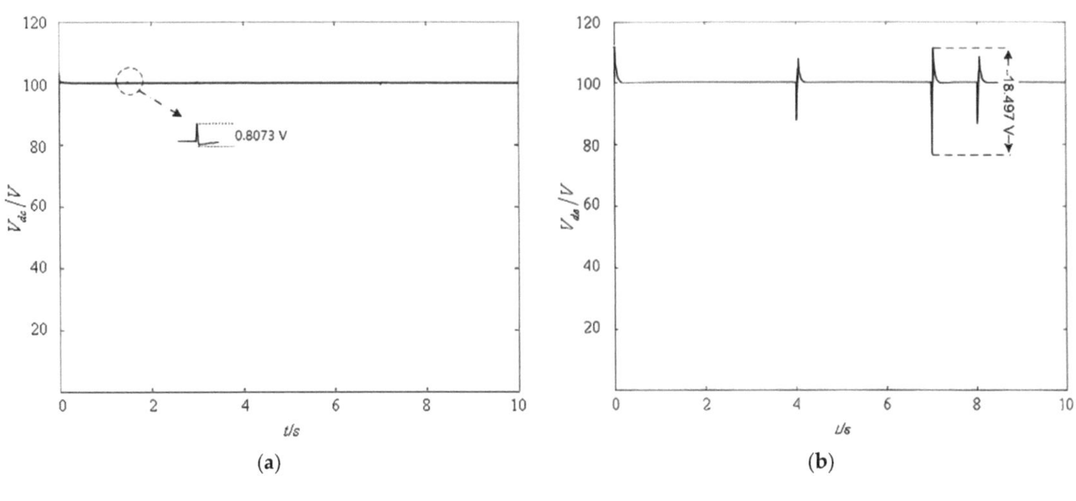

Figure 12. DC bus voltage waveform: (**a**) busbar voltage with an energy management system; (**b**) busbar voltage without an energy management system.

As evident from Figures 14 and 15, upon system startup, the supercapacitors respond first, while the energy storage battery supplies the average power under steady-state slow changes. At 1.5 s, due to an increase in irradiance, the photovoltaic power output exceeds the load power consumption, prompting the hybrid energy storage system to begin absorbing the excess power and feeding part of it back to the grid. Given the sufficient battery charge, charging is limited, while the supercapacitors have less charge than normal, resulting in normal charging but limited discharging. A comparison of the state of charge of the hybrid energy storage system under the two control strategies is shown in Figure 16.

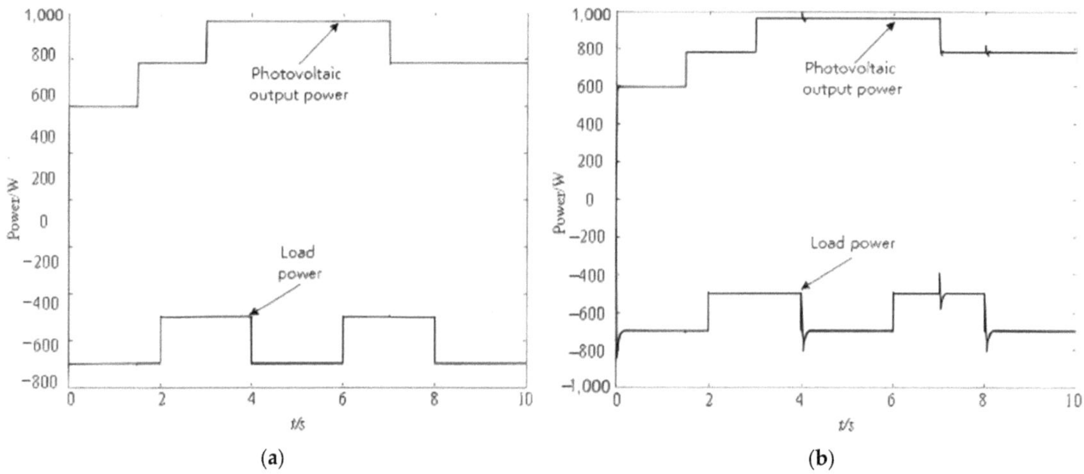

Figure 13. Photovoltaic, load power waveforms: (**a**) photovoltaic and load power with an energy management system; (**b**) photovoltaic and load power without an energy management system.

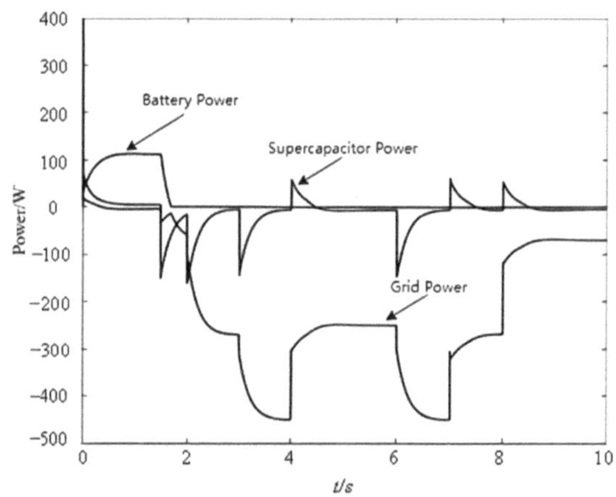

Figure 14. Grid, hybrid energy storage power waveforms.

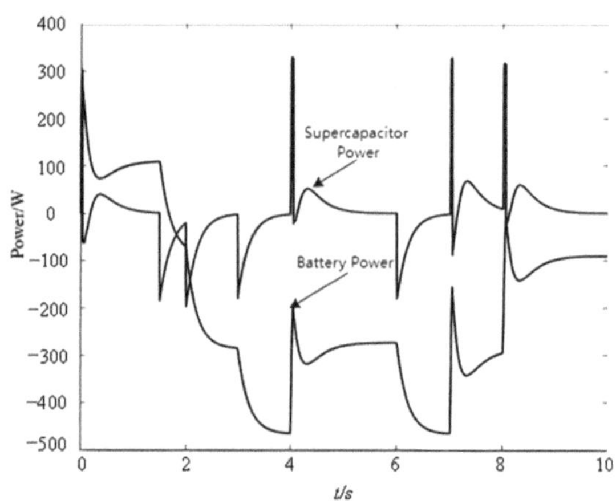

Figure 15. Hybrid energy storage power waveform.

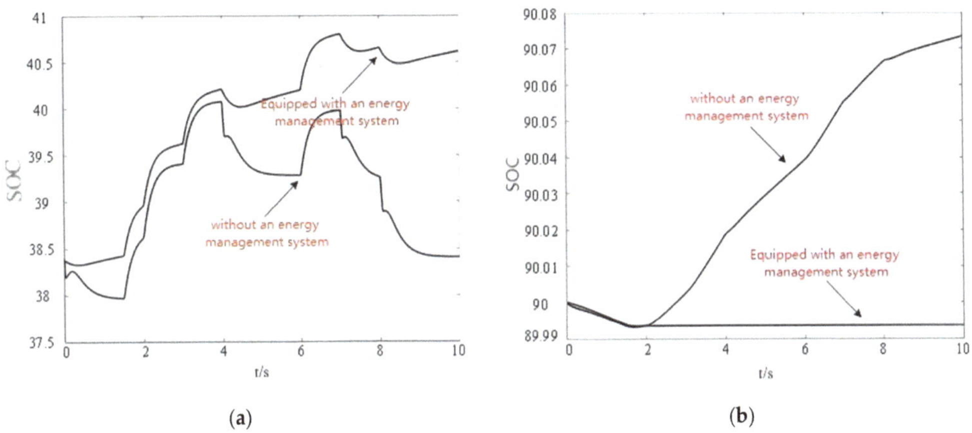

Figure 16. Hybrid system energy storage charge state: (**a**) state of charge of supercapacitors; (**b**) state of charge of energy storage battery.

As shown in Figure 16a, after introducing the energy management system based on the state of charge, the supercapacitors are in a region of discharge limitation and normal charging, with the overall state of charge showing an upward trend throughout the process until it reaches the 50% normal standby state. Figure 16b indicates that the energy storage battery discharges normally but with discharge limitation to prevent overcharging, thereby reducing the number of charge–discharge cycles and ensuring the power fluctuation suppression capability and extended lifespan of the hybrid energy storage system. This demonstrates the effectiveness and reliability of the proposed energy management strategy.

In this article, we propose an SOC-based hybrid energy storage energy management strategy and conduct simulation experiments using MATLAB/Simulink. The simulation results not only verify the effectiveness of the proposed strategy, but also reveal the significant role of hybrid energy storage systems in stabilizing power fluctuations and maintaining bus voltage stability.

Firstly, the simulation results show that the hybrid energy storage system can significantly smooth out the fluctuations in photovoltaic and load power. By initially allocating the required power and using the SOC of the supercapacitor and energy storage battery as a reference for secondary power allocation, this strategy ensures the efficient utilization of the energy storage system. In the simulation, we observed that the hybrid energy storage system can quickly respond to power changes, with high-frequency components mainly being borne by the supercapacitor and low-frequency components being borne by the energy storage battery. This division of labor effectively reduces the fluctuation of bus voltage.

Secondly, the simulation results further verify the ability of the proposed strategy in maintaining the state of charge SOC of the energy storage system. Through the secondary power distribution controlled by rules, we successfully maintained the SOC of the energy storage system at a reasonable level, preventing overcharging and overdischarging. This not only extends the service life of the energy storage system, but also improves the power quality of the entire DC microgrid.

When analyzing the simulation results in depth, we also noticed that the overall response speed of the hybrid energy storage system has been significantly improved. This is due to the precise tracking of the reference current by the PI controller and the transfer of the uncompensated component of the battery system to the supercapacitor terminal within the response time. This compensation mechanism enables the hybrid energy storage system to exhibit higher flexibility and response speeds when facing rapidly changing power demands.

In addition, we also discussed, in detail, the parameter settings of energy storage batteries and supercapacitors in the simulation. The selection of these parameters has a significant impact on the performance of hybrid energy storage systems. By setting parameters reasonably, we ensure that the system can operate stably during the simulation process and exhibit good dynamic characteristics.

In summary, the simulation results fully verify the effectiveness of the SOC-based hybrid energy storage management strategy. This strategy can not only significantly stabilize the fluctuation of photovoltaic and load power and maintain the stability of bus voltage, but also maintain the SOC of the energy storage system at a reasonable level, extend its service life, and improve the power quality. In the future, we will further verify and optimize the system in practical systems to promote the development and application of DC microgrid technology.

Although the SOC-based hybrid energy storage management strategy proposed in this paper has shown good performance in simulation experiments, effectively stabilizing DC bus voltage fluctuations and maintaining the state of charge of the hybrid energy storage system at a reasonable level, there are still some limitations and shortcomings in this study. Firstly, this study is mainly based on the MATLAB/Simulink simulation platform for verification. Although the simulation results have certain reference value, the differences between the simulation results and the actual operating environment may lead to different effects of the strategy in practical applications. Therefore, future research needs to further verify and optimize in actual systems.

Secondly, there are certain assumptions and simplifications in the parameter settings and energy management strategy design of the hybrid energy storage system in this study, such as assuming that the AC terminal capacity is large enough to provide stability support for the DC-side power balance. These assumptions may limit the scope and performance of the strategy. Therefore, future research can model and optimize more complex real-world scenarios and more stringent constraints to improve the practicality and robustness of the strategy.

In addition, this study mainly focuses on energy management strategies in DC microgrids and does not address broader microgrid operation and control issues. Future research can further explore the integration of the proposed energy management strategy with other microgrid operation strategies to achieve more comprehensive microgrid optimization and control.

Based on the above limitations and deficiencies, we propose the following possible follow-up research directions and improvement suggestions: first, strengthen experimental verification by applying the proposed energy management strategy in a real system and collecting data for analysis to evaluate the actual effectiveness and performance of the strategy; second, deepen theoretical research, model and analyze more complex real-world scenarios and constraints, and propose more accurate and efficient energy management strategies; thirdly, expand the scope of research, combine energy management strategies with other microgrid operation strategies for research and optimization, to achieve more comprehensive microgrid optimization and control. Through these efforts, we can further promote the development and application of DC microgrid technology.

7. Conclusions

This article takes the photovoltaic DC microgrid with hybrid energy storage as the research object and analyzes the maximum power tracking control of the photovoltaic power generation system, as well as the role of power decoupling and the hybrid energy storage energy management system in stabilizing the DC bus power fluctuation. The hybrid algorithm power tracking control is applied to ensure that the photovoltaic system always works at the maximum power point. While setting the charging and discharging limits for energy storage batteries and supercapacitors, the introduction of grid-side power compensates for the limitations on the power absorbed or provided by the hybrid energy storage system. The secondary power distribution of the energy management system effectively avoids the frequent charging and discharging of the energy storage battery while ensuring the state of charge of the hybrid energy storage system, so that the system can stabilize each power fluctuation.

The simulation results fully verify the effectiveness of the SOC-based hybrid energy storage energy management strategy. The advantages of the proposed strategy were further verified through specific quantitative indicators. Compared with traditional energy management strategies, the DC bus voltage fluctuation rate was reduced by about 20% after adopting the proposed strategy. The state of charge SOC of the hybrid energy storage system was maintained within a reasonable range, with the battery SOC between 20% and 80% and the supercapacitor SOC between 10% and 90%. The time ratio increased by 30%. At the same time, the number of charge and discharge cycles of the energy storage battery has been reduced by 15%, significantly extending the service life of the energy storage system. These quantitative indicators fully demonstrate the effectiveness of the proposed strategy in stabilizing DC bus voltage fluctuations, maintaining the state of charge of the energy storage system, and extending its service life.

The hybrid energy storage management system proposed in this paper can make full use of the advantages of energy-type and power-type energy storage, so that the supercapacitor can maintain a state of charge that effectively absorbs and compensates for fluctuating power, improving the overall system performance. In the future, we will further verify and optimize the system in the actual system to promote the development and application of DC microgrid technology. Although this study has achieved some results, there are still some limitations and shortcomings that need to be further explored and improved in subsequent research.

Author Contributions: Conceptualization, F.L. and Z.S.; methodology, Z.S.; software, F.L.; validation, F.L. and Z.S.; formal analysis, F.L., Z.S. and Z.Z.; investigation, F.L., Z.S. and Y.G.; resources, F.L. and Z.S.; data curation, Z.S.; writing—original draft preparation, Z.S.; writing—review and editing, Z.Z.; visualization, Y.G.; supervision, Z.S.; project administration, F.L.; funding acquisition, F.L. All authors have read and agreed to the published version of the manuscript.

Funding: Xiangjiang Laboratory Fund Project of China "Research and Application of Optimization Algorithm for Target Recognition Accuracy in Image Shooting of Electric Power Equipment Drone Inspection" (23XJ03016).

Data Availability Statement: The data in the paper come from the enterprise, and due to the involvement of core production related data, the enterprise has confidentiality requirements.

Conflicts of Interest: No potential conflicts of interest were reported by the authors.

References

1. Liu, D.; Li, L.; Li, S.; Liu, G.; Dai, X. Coordination Control Strategy for an Independent DC Microgrid with a Hybrid Energy Storage System. *Renew. Energy* **2020**, *38*, 524–531.
2. Shuai, Z.; Fang, J.; Ning, F.; Shen, Z.J. Hierarchical structure and bus voltage control of DC microgrid. *Renew. Sustain. Energy Rev.* **2018**, *82*, 3670–3682. [CrossRef]
3. Naderi, E.; Bibek, K.C.; Ansari, M.; Asrari, A. Experimental validation of a hybrid storage framework to cope with fluctuating power of hybrid renewable energy-based systems. *IEEE Trans. Energy Convers.* **2021**, *36*, 1991–2001. [CrossRef]

4. Li, Z.; Zhang, Y.; Cheng, Z. Distributed Hierarchical Control for Multi-Time-Scale Voltage Support in DC Microgrids with Photovoltaic Generation Participation. *Proc. CSEE* **2024**, 1–12. Available online: http://kns.cnki.net/kcms/detail/11.2107.tm.20241009.1259.008.html (accessed on 1 July 2024).
5. Sun, S.; Tang, C.; Xie, D.; Gulizhati, H.; Gu, C.; Zhang, Y.; Wang, X. Nonlinear Reduced-Order Modeling of Grid-Connected DC Microgrids and Optimization Calculation of Its Estimated Region of Attraction. *Proc. CSEE* **2024**, *44*, 517–535.
6. Liu, Y.; Liu, S.; Wang, J.; Xu, D. Power Coordination Control of DC Microgrids Based on Distributed Raft Algorithm. *Autom. Electr. Power Syst.* **2022**, *46*, 70–77.
7. Krishnamurthy, N.K.; Sabhahit, J.N.; Jadoun, V.K.; Gaonkar, D.N.; Shrivastava, A.; Rao, V.S.; Kudva, G. Optimal Placement and Sizing of Electric Vehicle Charging Infrastructure in a Grid-Tied DC Microgrid Using Modified TLBO Method. *Energies* **2023**, *16*, 1781. [CrossRef]
8. Abraham, D.S.; Chandrasekar, B.; Rajamanickam, N.; Vishnuram, P.; Ramakrishnan, V.; Bajaj, M.; Piecha, M.; Blazek, V.; Prokop, L. Fuzzy-Based Efficient Control of DC Microgrid Configuration for PV-Energized EV Charging Station. *Energies* **2023**, *16*, 2753. [CrossRef]
9. Yao, B.; Fan, Y. Energy Management of Energy Storage Stations in Integrated Photovoltaic and Energy Storage Systems. *Renew. Energy* **2017**, *35*, 232–239.
10. Li, Y.; Du, S.; Dang, H.; Li, C.; Liu, J. A High-Efficiency, High-Power-Density Partial-Power Battery Energy Storage DC-DC Converter with Automatic Voltage Balancing Function. *Proc. CSEE* **2024**, 1–11. [CrossRef]
11. Debdouche, N.; Zarour, L.; Benbouhenni, H.; Mehazzem, F.; Deffaf, B. Robust integral backstepping control microgrid connected photovoltaic System with battery energy storage through multi-functional voltage source inverter using direct power control SVM strategies. *Energy Rep.* **2023**, *10*, 565–580. [CrossRef]
12. Cavallo, A.; Canciello, G.; Guida, B.; Kulsangcharoen, P.; Yeoh, S.S.; Rashed, M.; Bozhko, S. Multi-Objective Supervisory Control for DC/DC Converters in Advanced Aeronautic Applications. *Energies* **2018**, *11*, 3216. [CrossRef]
13. Canciello, G.; Cacciapuoti, L.; Perrotta, A.; Guida, B.; Cavallo, A. Multi-Objective Supervisory Control in More-Electric Aircraft Using Model Predictive Control: An ORCHESTRA Application. *Energies* **2024**, *17*, 3799. [CrossRef]
14. Thomas, T.; Mishra, M.K.; Kumar, C.; Liserre, M. Control of a PV-Wind Based DC Microgrid with Hybrid Energy Storage System Using Lyapunov Approach and Sliding Mode Control. *IEEE Trans. Ind. Appl.* **2024**, *60*, 3746–3758. [CrossRef]
15. Meena, R.L.; Bhattacharya, A.; Khatod, D.K. Novel Control Strategy for CESS Integrated DC Microgrid with On Grid and Off Grid Application. *Arab. J. Sci. Eng.* **2023**, *48*, 14681–14696. [CrossRef]
16. Başoğlu, M.E.; Çakır, B. A novel voltage-current characteristic based global maximum power point tracking algorithm in photovoltaic systems. *Energy* **2016**, *112*, 153–163. [CrossRef]
17. Zhang, C.; Wang, Y. Research on Control Strategies for Photovoltaic Maximum Power Point Tracking. *Optoelectron. Technol.* **2022**, *42*, 303–310.
18. Cui, J.; Tian, G.; Liu, G.; Guo, Z. Research on Power Allocation Control Strategy for Hybrid Energy Storage System in Stand-alone DC Microgrid. *Power Syst. Clean Energy* **2023**, *39*, 129–136.
19. Mu, Y.; Mi, G. Research on Power Control of Photovoltaic Power Generation System Considering the State of Charge of Supercapacitors. *Renew. Energy* **2018**, *36*, 1462–1466.
20. Yang, Y.; Sun, R.; Zhang, C.; Mao, Y. Research on an Energy Management Strategy for Hybrid Energy Storage Systems Based on Fuzzy Control. *Renew. Energy* **2017**, *35*, 1881–1887.
21. Rajagopalan, A.; Nagarajan, K.; Montoya, O.D.; Dhanasekaran, S.; Kareem, I.A.; Perumal, A.S.; Lakshmaiya, N.; Paramasivam, P. Multi-Objective Optimal Scheduling of a Microgrid Using Oppositional Gradient-Based Grey Wolf Optimizer. *Energies* **2022**, *15*, 9024. [CrossRef]
22. Li, M.; Aksoy, M.; Samad, S. Optimal energy management and scheduling of a microgrid with integrated electric vehicles and cost minimization. *Soft Comput.* **2024**, *28*, 2015–2034. [CrossRef]
23. Nammouchi, A.; Aupke, P.; D'andreagiovanni, F.; Ghazzai, H.; Theocharis, A.; Kassler, A. Robust opportunistic optimal energy management of a mixed microgrid under asymmetrical uncertainties. *Sustain. Energy Grids Netw.* **2023**, *36*, 101184. [CrossRef]
24. Vidan, M.; D'Andreagiovanni, F.; Pandzic, H. Individual Thermal Generator and Battery Storage Bidding Strategies Based on Robust Optimization. *IEEE Access* **2021**, *9*, 66829–66838. [CrossRef]
25. Zhang, S.; Zhang, J.; Wang, Q.; Wu, Z.; Zhang, Z. Energy Management Strategy for DC Microgrid Based on Hybrid Energy Storage Charge State Monitoring. *J. Power Supply* **2020**, 1–13. Available online: http://kns.cnki.net/kcms/detail/12.1420.TM.20241012.1627.002.html (accessed on 17 December 2024).

26. Wu, J.; Xing, X.; Liu, X.; Guerrero, J.M.; Chen, Z. Energy Management Strategy for Grid-Tied Microgrids Considering the Energy Storage Efficiency. *IEEE Trans. Ind. Electron.* **2018**, *65*, 9539–9549. [CrossRef]
27. Wang, G.; Li, X.; Sun, Z.; Dong, L.; Zheng, Y. Modeling and state estimation of "lithium-ion battery-supercapacitor" hybrid energy storage system. *Electron. Meas. Technol.* **2023**, *46*, 103–111. [CrossRef]

energies

MDPI

Article

Distributed Model Predictive Load Frequency Control for Virtual Power Plants with Novel Event-Based Low-Delay Technique Under Cloud-Edge-Terminal Framework

Kai Kang [1], Nian Shi [2], Si Cai [2], Liang Zhang [2], Xinan Shao [2], Haohao Cao [2], Mingjin Fei [2], Shisen Zhou [3] and Xiongbo Wan [3,*]

[1] PowerChina Hubei Engineering Co., Ltd., Wuhan 430040, China; kangkaisj@powerchina-hb.com
[2] PowerChina Hubei Electric Engineering Co., Ltd., Wuhan 430040, China; shinsj@powerchina-hb.com (N.S.); caisi@powerchina-hb.com (S.C.); zhanglsj@powerchina-hb.com (L.Z.); shaoxnsj@powerchina-hb.com (X.S.); caohaohao-huby@powerchina.cn (H.C.); feimingjin-huby@powerchina.cn (M.F.)
[3] School of Automation, China University of Geosciences, Wuhan 430074, China; sszhou25@cug.edu.cn
* Correspondence: xbwan23@cug.edu.cn

Abstract: In this paper, the distributed model predictive load frequency control problem for virtual power plants (VPPs) under the cloud-edge-terminal framework is addressed, where the data packets are transmitted under a novel dynamic event-triggered mechanism (DETM) with hybrid variables. The proposed DETM has the ability to flexibly manage packet releases and reduce network congestion, thus decreasing the communication delay of the VPP. A method of the DETM-based distributed model predictive control (DMPC) is proposed, which can shorten the data processing time and further decrease the communication delay. The DMPC problem is described as a "min-max" optimization problem (OP) with hard constraints on the system state. By utilizing a Lyapunov function with an internal dynamic variable, an auxiliary OP with matrix inequalities constraints is proposed to optimize the controller gain and the weighting matrix of the DETM. The effectiveness and superiority of the designed DETM and dynamic event-based DMPC algorithm are demonstrated through a case study on two-area VPPs.

Keywords: load frequency control; dynamic event-triggered mechanism; distributed model predictive control; cloud-edge-terminal; virtual power plant

Academic Editor: Giovanni Lutzemberger

Received: 26 January 2025
Revised: 19 February 2025
Accepted: 7 March 2025
Published: 11 March 2025

Citation: Kang, K.; Shi, N.; Cai, S.; Zhang, L.; Shao, X.; Cao, H.; Fei, M.; Zhou, S.; Wan, X. Distributed Model Predictive Load Frequency Control for Virtual Power Plants with Novel Event-Based Low-Delay Technique Under Cloud-Edge-Terminal Framework. *Energies* **2025**, *18*, 1380. https://doi.org/10.3390/en18061380

1. Introduction

Frequency is a significant indicator of power quality, and keeping the frequency fluctuation within the rated allowable range is key to maintaining the stability and security of the power system [1]. Load frequency control (LFC) maintains the quality of system power by keeping the system operating frequency within the rated allowable range [2,3]. With the increase in renewable energy penetration in the power system, the strong volatility and intermittency of renewable energy output also lead to frequent fluctuations in the grid frequency, and the system's flexible regulation capability decreases significantly [4]. Therefore, it is urgent to explore the potential and value of new regulation resources to enhance the frequency stability of the power system.

Currently, under the role of power system marketization, a large number of adjustable resources such as energy storage and vehicle clusters are connected to the power system, which brings new possibilities for supplementing the frequency response capacity of the generation side. By aggregating each adjustable resource to form a virtual power plant

(VPP), additional regulation capacity can be provided to provide auxiliary frequency regulation services for large-capacity systems. VPPs realize cooperative operation through aggregation and regulation of geographically dispersed and different kinds of distributed resources [5], which has great potential to improve the sustainable development of energy and ensure the safe and stable operation of power grids [6]. Therefore, it is of great significance to study the regulation strategy of VPPs.

Since distributed resources present small capacity, large volume, and wide dispersion, a VPP generally adopts a cloud-edge-terminal collaborative information framework [7,8]. The communication network is key to the cloud-edge-terminal framework, which facilitates the transmission of system data. However, due to the limited communication bandwidth, transmissions of large amounts of data cause congestion on the communication network, which subsequently increases the communication delay [9]. Therefore, it is of great engineering significance to investigate how to reduce the communication delay while ensuring the control performance of the VPP.

To solve the VPP communication delay problem under the cloud-edge-terminal framework, the event-triggered mechanism (ETM) is an effective method [10]. Unlike the time-triggered method, the ETM transmits data only when a pre-defined triggering condition is satisfied [11]. This means that ETM reduces unnecessary packet releases and decreases the likelihood of network congestion, thereby reducing communication delay. Recently, some scholars have proposed dynamic event-triggered mechanisms (DETMs) [12–14], which can flexibly manage the release of packets and further save network resources compared to static event-triggered mechanisms [15–17]. However, the existing DETMs still have room for improvement. For example, for the DETM in [13], the coefficients are fixed and cannot dynamically adjust the packet releases. These provide opportunities for designing improved DETMs to reduce the delay of VPPs further.

Model predictive control (MPC) has the advantages of handling hard constraints and yielding good dynamic performance [18–20]. Some scholars have studied the model predictive LFC with VPP participation and there have been some results [21,22]. The aforementioned studies use a centralized MPC (CMPC) approach, but under the cloud-edge-terminal framework, VPPs with the CMPC strategy for LFC will aggregate all data to an edge-centered processor for processing. Data centralization often leads to network congestion and increases the data processing time, which triggers the latency problem of the VPP. On the contrary, the distributed MPC (DMPC) avoids centralized transmission of data by decentralizing the data to be processed on multiple edge nodes. This shortens the data processing time and reduces the network burden, which helps to reduce communication delay. However, the DMPC problem for VPPs has not been discussed, let alone an improved DETM-based DMPC for VPPs. This presents an opportunity to explore distributed model predictive LFC for VPP under the new DETM.

In this paper, in order to reduce the communication delay of a VPP, a new DETM is proposed for VPPs, and a dynamic event-based DMPC algorithm is developed under the cloud-edge-terminal framework. The main contributions are three-fold: (1) a low-delay communication technique based on novel DETM is proposed, and the proposed DETM is more generalized than some existing ETMs, which further reduces the network burden and thus reduces the communication delay of the VPP; (2) under the cloud-edge-terminal framework, the distributed model predictive controllers were designed for the edge servers of each VPP, which further reduces the communication delay by shortening the data processing time compared to the centralized control strategy while guaranteeing the control performance; and (3) a dynamic event-based distributed model predictive LFC algorithm is developed to reduce the consumption of communication resources and shorten the data processing time, which reduces the communication delay of the VPP.

2. Problem Formulation

2.1. General Structure of VPP

VPPs serve as a way to realize the aggregation and coordinated optimization of distributed energy resources such as distributed power sources, energy storage systems, controllable loads, electric vehicles, and other distributed energy resources through advanced information and communication technologies and software systems. The general structure of a VPP is shown in Figure 1, and this paper focuses on a typical VPP that integrates a combination of wind energy, energy storage, and electric vehicles to participate in the frequency regulation of the power system.

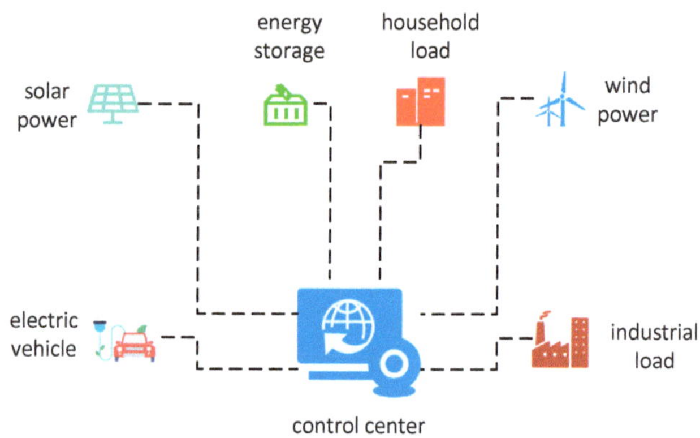

Figure 1. General structure of VPP.

2.2. Wind Storage System Frequency Modulation Model

The expression for the general wind turbine aerodynamic system is shown in [23]. Under the condition of constant wind speed and maintaining the optimal tip speed ratio, the adjustment of the output power of the wind turbine is realized by changing the size of the pitch angle opening. Now, consider the following wind turbine model:

$$\Delta P_{wind} = \Delta\beta G_{wind} = \Delta\beta \frac{k_\beta}{1 + T_{wind}s}, \quad (1)$$

where $\Delta\beta$ is the angular change in paddle pitch angle, G_{wind} is the wind turbine power transfer function, T_{wind} is the wind turbine response time, and k_β is the pitch angle control slope.

The long response time of the pitch angle of the wind turbine leads to a reduction in the frequency modulation capability, and it is necessary to combine the energy storage system with fast response capability to suppress the power fluctuation. The energy storage model of the VPP can be represented as follows [24]:

$$\Delta P_{bess} = \Delta P_s G_{bess} = \Delta P_s \frac{1}{1 + T_{bess}s}, \quad (2)$$

where ΔP_s is the storage instantaneous power, G_{bess} is the storage system transfer function, and T_{bess} is the storage system response time.

2.3. Electric Vehicle Frequency Modulation Model

As a flexible control resource, the charging and discharging processes of electric vehicles can be regarded as load and power supply, respectively, for the power grid. With the number of cars fixed, consider the following frequency modulation model for electric vehicles [25]:

$$\Delta P_{ev} = \Delta U_{ev} G_{ev} = \Delta U_{ev} \frac{1}{1 + T_{ev} s}, \tag{3}$$

where ΔU_{ev} represents the depth of charge and discharge of the electric vehicle battery, and T_{ev} is the electric vehicle system response time.

2.4. Distributed Predictive LFC Model with VPP

In Figure 2, we give the system structure and consider the following dynamic model of LFC with VPP for the ith control area:

$$\begin{cases} \dot{x}_i(t) = A_{ii}x_i(t) + B_{ii}u_i(t) + F_{ii}w_i(t) + \sum\limits_{\substack{j \neq i \\ j=1}}^{N} (A_{ij}x_j(t) + B_{ij}u_j(t)), \\ y_i(t) = C_{ii}x_i(t), \end{cases} \tag{4}$$

where

$$x_i(t) = [\Delta f_i \quad \Delta P_{tie-i} \quad \Delta P_{mi} \quad \Delta P_{vi} \quad \int ACE_i \quad \Delta P_{windi} \quad \Delta P_{bessi} \quad \Delta P_{evi}]^{\mathrm{T}},$$

$$y_i(t) = ACE_i = \beta_i \Delta f_i + \Delta P_{tie-i}, \; w_i(t) = \Delta P_{di},$$

$$A_{ii} = \begin{bmatrix} A_{11} & A_{12} \\ 0_{3\times5} & A_{22} \end{bmatrix}, A_{12} = \begin{bmatrix} \frac{1}{Mi} & \frac{1}{Mi} & \frac{1}{Mi} \\ 0 & 0 & 0 \\ 0 & 0 & 0 \\ 0 & 0 & 0 \\ 0 & 0 & 0 \end{bmatrix}, A_{22} = \begin{bmatrix} -\frac{1}{T_{windi}} & 0 & 0 \\ 0 & -\frac{1}{T_{bessi}} & 0 \\ 0 & 0 & -\frac{1}{T_{evi}} \end{bmatrix},$$

$$A_{11} = \begin{bmatrix} -\frac{D_i}{M_i} & -\frac{1}{M_i} & \frac{1}{M_i} & 0 & 0 \\ 2\pi \sum\limits_{j=1, j\neq i}^{N} T_{ij} & 0 & 0 & 0 & 0 \\ 0 & 0 & -\frac{1}{T_{chi}} & \frac{1}{T_{chi}} & 0 \\ -\frac{1}{RT_{gi}} & 0 & 0 & -\frac{1}{T_{gi}} & 0 \\ \beta_i & 1 & 0 & 0 & 0 \end{bmatrix},$$

$$B_{ii} = \begin{bmatrix} 0 & 0 & 0 & \frac{1}{T_{gi}} & 0 & 0 & 0 & 0 \\ 0 & 0 & 0 & 0 & 0 & \frac{k_{\beta i}}{T_{windi}} & 0 & 0 \\ 0 & 0 & 0 & 0 & 0 & 0 & \frac{1}{T_{bessi}} & 0 \\ 0 & 0 & 0 & 0 & 0 & 0 & 0 & \frac{1}{T_{evi}} \end{bmatrix}^{\mathrm{T}},$$

$$F_{ii} = [-\frac{1}{M_i} \quad 0 \quad 0 \quad 0 \quad 0 \quad 0 \quad 0 \quad 0]^{\mathrm{T}}, C_{ii} = [\beta_i \quad 1 \quad 0 \quad 0 \quad 0 \quad 0 \quad 0 \quad 0],$$

$$B_{ij} = \begin{bmatrix} 0 & 0 & 0 & 0 & 0 & 0 & 0 & 0 \\ 0 & 0 & 0 & 0 & 0 & 0 & 0 & 0 \\ 0 & 0 & 0 & 0 & 0 & 0 & 0 & 0 \\ 0 & 0 & 0 & 0 & 0 & 0 & 0 & 0 \end{bmatrix}^{\mathrm{T}}, A_{ij} = \begin{bmatrix} 0 & 0 & 0 & 0 & 0 & 0 & 0 & 0 \\ -2\pi T_{ij} & 0 & 0 & 0 & 0 & 0 & 0 & 0 \\ 0 & 0 & 0 & 0 & 0 & 0 & 0 & 0 \\ 0 & 0 & 0 & 0 & 0 & 0 & 0 & 0 \\ 0 & 0 & 0 & 0 & 0 & 0 & 0 & 0 \\ 0 & 0 & 0 & 0 & 0 & 0 & 0 & 0 \\ 0 & 0 & 0 & 0 & 0 & 0 & 0 & 0 \\ 0 & 0 & 0 & 0 & 0 & 0 & 0 & 0 \end{bmatrix},$$

$$T_{ij} = T_{ji}.$$

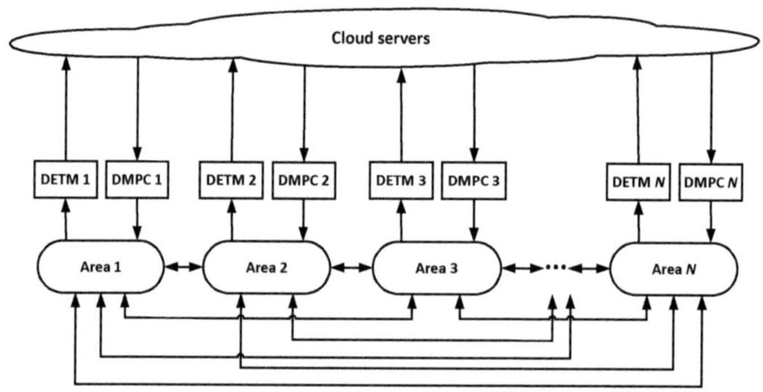

Figure 2. Structure of multiple VPPs using DETM-based DMPC under loud–edge–terminal framework.

For explanations regarding the parameters in Equation (4), please refer to [3] for details.

The discrete-time model of LFC for a VPP for the ith control area is given as follows:

$$\begin{cases} x_i(k+1) = \overline{A}_{ii}x_i(k) + \overline{B}_{ii}u_i(k) + \overline{F}_{ii}w_i(k) + \sum_{\substack{j\neq i \\ j=1}}^{N} (\overline{A}_{ij}x_j(k) + \overline{B}_{ij}u_j(k)), \\ y_i(k) = \overline{C}_{ii}x_i(k), \end{cases} \tag{5}$$

where $\overline{A}_{ii}$, $\overline{B}_{ii}$, $\overline{F}_{ii}$, $\overline{A}_{ij}$, $\overline{B}_{ij}$, and $\overline{C}_{ii}$ are the discretized coefficient matrices.

Remark 1. *In Figure 2, each control area contains distributed energy from wind energy, energy storage, and vehicle clusters in the VPP. The distributed model predictive controllers are designed for the edge servers of each VPP, which reduces the processing time of the data and the communication delay. In addition, each control area is a terminal, and the cloud server is used for uploading and downloading data. Under this cloud–edge–terminal framework, DETMs are utilized to manage the release of packets, thereby reducing network congestion and achieving low-delay communication.*

Suppose that the load disturbance satisfies

$$\|w_i(k)\|^2 \leq p_i, \tag{6}$$

where $p_i \geq 0$ is constant.

Consider the following hard constraints:

$$|[x_i(k)]_\iota| \leq [\bar{x}_i]_\iota, \ \iota \in \mathbb{L} \triangleq \{1, 2, \ldots, n_{x_i}\}, \tag{7}$$

where the vector $\bar{x}_i$ is given with its elements being nonnegative.

To reduce communication delay and increase the packet transmission rate, the following DETM is presented:

$$\begin{cases} s_{mi+1} = \min_{k \in \mathbb{N}^+} \left\{ k > s_{mi} \mid \eta_{1i}\zeta_i(k) + \theta_{1i}\varphi_i(k)x_i^{\mathrm{T}}(k)\Omega_i(k)x_i(k) \right. \\ \qquad \left. -\theta_{2i}e_i^{\mathrm{T}}(k)\Omega_i(k)e_i(k) \le 0 \right\}, \\ \zeta_i(k+1) = \eta_{2i}\zeta_i(k) + \theta_{3i}\varphi_i(k)x_i^{\mathrm{T}}(k)\Omega_i(k)x_i(k) - \theta_{4i}e_i^{\mathrm{T}}(k)\Omega_i(k)e_i(k), \\ \varphi_i(k) = \underline{\varphi}_i + \varepsilon_i \varrho_i^{e_i^{\mathrm{T}}(k)e_i(k)}, \end{cases} \tag{8}$$

where s_{mi} stands for the mth instant for data release, $\zeta_i(k)$ is a dynamic variable (DV) with $\zeta_i(0) \ge 0$, $\varphi_i(k)$ is an adaptive adjustment variable (AAV), $\Omega_i(k)$ is a weighting matrix, $\eta_{1i} \ge 0$, $0 \le \eta_{2i} \le 1$, $\theta_{1i} \ge 0$, $\theta_{2i} > 0$, $\theta_{3i} \ge 0$, $\theta_{4i} > 0$, $0 < \varrho_i \le 1$, $\underline{\varphi}_i \ge 0$, and $\varepsilon_i \ge 0$ are given constants, and $e_i(k) = x_i(k) - x(s_{mi})$ where $s_{0i} = 0$.

Note that in DETM (8), there exist a DV $\zeta_i(k)$ and an AAV $\varphi_i(k)$, which are incorporated to enhance the capability of adjusting releases of data packets. Additionally, we have

$$\underline{\varphi}_i < \varphi_i(k) \le \underline{\varphi}_i + \varepsilon_i \triangleq \overline{\varphi}_i. \tag{9}$$

Let

$$\tilde{x}_i(k) = x(s_{mi}), \ \forall k \in [s_{mi}, s_{mi+1}). \tag{10}$$

Then, we obtain

$$e_i(k) = x_i(k) - \tilde{x}_i(k), \tag{11}$$

which holds for any $k \in [0, \infty)$.

Remark 2. *In (8), a new DETM is designed that contains an AVV with adjustable upper and lower bounds to flexibly regulate packet releases. In addition, DETM (8) can also be degraded to existing DETMs by setting suitable parameters. For example, by setting $\eta_{1i} = 1$, $\theta_{1i} = \theta_{2i}\theta_{3i}$, $\theta_{4i} = 1$, $\underline{\varphi}_i = 0$, $\varepsilon_i = 1$, and $\varrho_i = 1$, DETM (8) becomes the DETM in [13]. As a result, DETM (8) is more general and flexible than existing DETMs, which can further decrease the possibility of network congestion and thus reduce communication delay.*

To further reduce the communication delay, a distributed control strategy is used to shorten the data processing time. A distributed state feedback controller for the ith control area can be designed:

$$u_i(k) = H_i(k)\tilde{x}_i(k). \tag{12}$$

The state feedback controller of the neighboring subsystem $j(j \ne i)$ is expressed as

$$u_j(k) = H_j(k)\tilde{x}_j(k). \tag{13}$$

Applying (11)–(13) to (5), we can obtain the following closed-loop control system:

$$\begin{aligned} x_i(k+1) &= (\overline{A}_{ii} + \overline{B}_{ii}H_i(k))x_i(k) - \overline{B}_{ii}H_i(k)e_i(k) + \overline{F}_{ii}w_i(k) \\ &\quad + \sum_{\substack{j \ne i \\ j=1}}^{N} ((\overline{A}_{ij} + \overline{B}_{ij}H_j(k))x_j(k) - \overline{B}_{ij}H_j(k)e_j(k)) \\ &= (A_i + B_iH(k))x(k) - B_iH(k)e(k) + \overline{F}_{ii}w_i(k), \end{aligned} \tag{14}$$

where

$$x(k) = \left[x_1(k)^T \cdots x_i(k)^T \cdots x_N(k)^T\right]^T, e(k) = \left[e_1(k)^T \cdots e_i(k)^T \cdots e_N(k)^T\right]^T,$$
$$A_i = \left[\overline{A}_{i1} \ldots \overline{A}_{ii} \ldots \overline{A}_{iN}\right], B_i = \left[\overline{B}_{i1} \ldots \overline{B}_{ii} \ldots \overline{B}_{iN}\right],$$
$$H(k) = \text{diag}\{H_1(k) \ldots H_i(k) \ldots H_N(k)\}.$$

For ease of presentation, we define

$$\cdot(k+l \mid k) \triangleq \cdot(l,k), \cdot(k \mid k) \triangleq \cdot(k).$$

The following prediction model is given:

$$\begin{cases} x_i(l+1,k) = (A_i + B_i H(k))x(l,k) - B_i H(k)e(l,k) + \overline{F}_{ii}w_i(l,k), \\ \zeta_i(l+1,k) = \eta_{2i}\zeta_i(l,k) + \theta_{3i}\varphi_i(l,k)x_i^{\mathrm{T}}(l,k)\Omega_i(k)x_i(l,k) - \theta_{4i}e_i^{\mathrm{T}}(l,k)\Omega_i(k)e_i(l,k). \end{cases} \tag{15}$$

Consider the following objective function:

$$\begin{aligned} J_i^\infty(k) = &\sum_{l=0}^{\infty}\left[\|x_i(l,k)\|_{Q_i}^2 + \|u_i(l,k)\|_{R_i}^2 - \rho_i w_i^{\mathrm{T}}(l,k)w_i(l,k)\right] \\ &+ \sum_{\substack{j\neq i\\ j=1}}^{N}\sum_{l=0}^{\infty}\left[\|x_j(l,k)\|_{Q_j}^2 + \|u_j(l,k)\|_{R_j}^2\right], \end{aligned} \tag{16}$$

where $Q_i > 0$, $R_i > 0$, $Q_j > 0$, and $R_j > 0$ are given weighting matrices, and ρ_i is a constant.

For DMPC in each area i, and at each time instant k, we construct the following "min=-max" optimization problem (OP) to design the DMPC controllers for System (14):

$$\mathbf{OP}_1 \begin{cases} \min\limits_{u_i(k)} \max\limits_{w_i(k)} J_i^\infty(k), & \\ |[x_i(l,k)]_\iota| \le [\bar{x}_i]_\iota,\ \iota \in \mathbb{L} & \text{(17a)} \\ \left[x_i^{\mathrm{T}}(l,k), \zeta_i^{\frac{1}{2}}(l,k)\right]^{\mathrm{T}} \in \Lambda_i, & \text{(17b)} \end{cases}$$

where

$$\Lambda_i \triangleq \{[x_i^{\mathrm{T}}(l,k), \zeta_i^{\frac{1}{2}}(l,k)]^{\mathrm{T}} | x_i^{\mathrm{T}}(l,k)P_i(k)x_i(l,k) + \zeta_i(l,k) \le \gamma_i\}, \tag{18}$$

where $P_i(k) > 0$ and $\gamma_i > 0$ give the weighting matrix to be designed and the optimisation index, respectively.

Definition 1. *For System (5), the set Λ_i is a positively invariant set (PIS) if $[x_i^{\mathrm{T}}(l,k), \zeta_i^{\frac{1}{2}}(l,k)]^{\mathrm{T}} \in \Lambda_i$, $\forall l \in \mathbb{N}$.*

Lemma 1. *In DETM (8), where $\zeta_i(0) \ge 0$, if the constants $\eta_{1i} \ge 0$, $0 \le \eta_{2i} \le 1$, $\theta_{1i} \ge 0$, $\theta_{2i} > 0$, $\theta_{3i} \ge 0$, and $\theta_{4i} > 0$ satisfy $\eta_{1i}\theta_{4i} \le \eta_{2i}\theta_{2i}$ and $\theta_{1i}\theta_{4i} \le \theta_{2i}\theta_{3i}$, then $\zeta_i(k) \ge 0$, $\forall k \in \mathbb{N}^+$.*

Proof. It is not difficult to complete the proof by following the method in [14]; the details are thus omitted. □

Since it is difficult to solve $\mathbf{OP}_1$ directly, we need to transform $\mathbf{OP}_1$. Therefore, we next propose an LMI-based auxiliary OP for $\mathbf{OP}_1$ and then develop a DETM-based DMPC algorithm to design the controller gain matrix $H_i(k)$ and the DETM weighting matrix $\Omega_i(k)$.

3. Main Results

3.1. LMI-Based Auxiliary OP of OP_1

In this section, we will design a DMPC algorithm based on the DETM (8) for System (14). To achieve this goal, we design an LMI-based auxiliary OP to solve $\mathbf{OP}_1$.

Consider the following Lyapunov-like function:

$$V_i(\zeta_i(k), x_i(k)) = x_i^T(k)P_i(k)x_i(k) + \zeta_i(k). \tag{19}$$

Rewrite (16) as follows:

$$J_i^\infty(k) = \sum_{l=0}^{\infty}\left[\|x(l,k)\|_Q^2 + \|u(l,k)\|_R^2 - \rho_i w_i^{\mathrm{T}}(l,k)w_i(l,k)\right], \tag{20}$$

where

$$Q = \mathrm{diag}\{Q_1 \ldots Q_i \ldots Q_N\}, R = \mathrm{diag}\{R_1 \ldots R_i \ldots R_N\}.$$

At each time instant k, we suppose the following inequality:

$$\begin{aligned} &V_i(\zeta_i(l+1,k), x_i(l+1,k)) - V_i(\zeta_i(l,k), x_i(l,k)) \\ &\le -\left(\|x(l,k)\|_Q^2 + \|u(l,k)\|_R^2\right) + \rho_i w_i^{\mathrm{T}}(l,k)w_i(l,k). \end{aligned} \tag{21}$$

Adding both sides of (21) from $l = 0$ to $l = \infty$, we obtain

$$-V_i(\zeta_i(k), x_i(k)) \le -J_i^\infty(k) - V_i(\zeta_i(\infty,k), x_i(\infty,k)) \le -J_i^\infty(k). \tag{22}$$

According (18) to (22), we obtain

$$J_i^\infty(k) \le V_i(\zeta_i(k), x_i(k)) \le \gamma_i. \tag{23}$$

Based on (23), we construct the following auxiliary OP:

$$\begin{aligned} &\mathbf{OP}_2 : \min_{H_i(k)} \gamma_i, \\ &\text{s.t.}(14), (17a), (17b), (21). \end{aligned} \tag{24}$$

Theorem 1. *Let the scalars $\eta_{1i} \ge 0$, $0 \le \eta_{2i} \le 1$, $\theta_{1i} \ge 0$, $\theta_{2i} > 0$, $\theta_{3i} \ge 0$, $\theta_{4i} > 0$, $0 < \varrho_i \le 1$, $\underline{\varphi}_i \ge 0$, $\varepsilon_i \ge 0$, and $\rho_i > 0$, and the matrices $Q > 0$ and $R > 0$ be given. If a scalar $\gamma_i > 0$ and matrices $S_i(k) > 0$, $X_i > 0$, $T_i(k) > 0$, $Y_i(k)$, and $G_i(k)$ exist such that, $\iota \in \mathbb{L}$, the following LMIs*

$$\begin{bmatrix} \mathscr{D}_1^{(1,1)} & * \\ \mathscr{D}_1^{(2,1)} & \mathscr{D}_1^{(2,2)} \end{bmatrix} \le 0, \tag{25}$$

$$\begin{bmatrix} X_i & G_i(k) \\ * & \overline{\mathscr{D}}_1^{(2,2)} \end{bmatrix} \ge 0, \tag{26}$$

$$[X_i]_{\iota\iota} \le [\bar{x}_i]_\iota^2, \tag{27}$$

hold, where

$$\mathscr{D}_1^{(1,1)} = \text{diag}\left\{ \widetilde{\mathscr{D}}_1^{(1,1)}, \widetilde{\mathscr{D}}_1^{(2,2)}, -\rho_i\gamma_i L_i^{\mathrm{T}} L_i \right\}, \widetilde{\mathscr{D}}_1^{(1,1)} = E_i^{\mathrm{T}}\left(S_i(k) - G_i^{\mathrm{T}}(k) - G_i(k)\right)E_i,$$

$$\widetilde{\mathscr{D}}_1^{(2,2)} = \theta_{4i}E_i^{\mathrm{T}}\left(T_i(k) - G_i^{\mathrm{T}}(k) - G_i(k)\right)E_i, \mathscr{D}_1^{(2,2)} = \text{diag}\{-S_i(k), -\gamma_i I, -\gamma_i I, -T_i(k)\},$$

$$\mathscr{D}_1^{(2,1)} = \begin{bmatrix} A_iG(k) + B_iY(k) & -B_iY(k) & \gamma_i\overline{F}_{ii}L_i \\ R^{\frac{1}{2}}Y(k) & -R^{\frac{1}{2}}Y(k) & 0 \\ Q^{\frac{1}{2}}G(k) & 0 & 0 \\ \sqrt{\theta_{3i}\overline{\varphi}_i}E_iG(k) & 0 & 0 \end{bmatrix}, \overline{\mathscr{D}}_1^{(2,2)} = G_i^{\mathrm{T}}(k) + G_i(k) - S_i(k),$$

$$G(k) = \text{diag}\{G_1(k)\ldots G_i(k)\ldots G_N(k)\}, Y(k) = \text{diag}\{Y_1(k)\ldots Y_i(k)\ldots Y_N(k)\},$$

$$E_i = \begin{bmatrix} Z_{1i}^- & I_{1i} & Z_{1i}^+ \end{bmatrix} \in R^{n_{x_i}\times(n_{x_1}+\cdots+n_{x_N})},$$

$$L_i = \begin{bmatrix} Z_{2i}^- & I_{2i} & Z_{2i}^+ \end{bmatrix} \in R^{n_{w_i}\times(n_{w_1}+\cdots+n_{w_N})},$$

$I_{1i} \in R^{n_{x_i}\times n_{x_i}}$, $I_{2i} \in R^{n_{w_i}\times n_{w_i}}$, *and* $Z_{1i}^- \in R^{n_{x_i}\times(n_{x_1}+\cdots+n_{x_{i-1}})}$, $Z_{1i}^+ \in R^{n_{x_i}\times(n_{x_{i+1}}+\cdots+n_{x_N})}$, $Z_{2i}^- \in R^{n_{w_i}\times(n_{w_1}+\cdots+n_{w_{i-1}})}$, $Z_{2i}^+ \in R^{n_{w_i}\times(n_{w_{i+1}}+\cdots+n_{w_N})}$ *are zero matrices,*

then conditions (17a) and (21) are satisfied, and the feedback gain matrix $H_i(k)$ and the weighting matrix $\Omega_i(k)$ are designed by

$$H_i(k) = Y_i(k)G_i^{-1}(k), \Omega_i(k) = \gamma_i T_i^{-1}(k). \tag{28}$$

Proof. It follows from (15) that

$$\begin{aligned} \zeta_i(l+1,k) - \zeta_i(l,k) &= (\eta_{2i} - 1)\zeta_i(l,k) + \theta_{3i}\varphi_i(l,k)x_i^{\mathrm{T}}(l,k)\Omega_i(k)x_i(l,k) \\ &\quad - \theta_{4i}e_i^{\mathrm{T}}(l,k)\Omega_i(k)e_i(l,k) \\ &\leq \theta_{3i}\varphi_i(l,k)x_i^{\mathrm{T}}(l,k)\Omega_i(k)x_i(l,k) - \theta_{4i}e_i^{\mathrm{T}}(l,k)\Omega_i(k)e_i(l,k). \end{aligned} \tag{29}$$

Substituting (19) into (21) yields

$$\begin{aligned} &x_i^{\mathrm{T}}(l+1,k)P_i(k)x_i(l+1,k) + \zeta_i(l+1,k) - x_i^{\mathrm{T}}(l,k)P_i(k)x_i(l,k) - \zeta_i(l,k) \\ &+ \|x(l,k)\|_Q^2 + \|u(l,k)\|_R^2 - \rho_i w_i^{\mathrm{T}}(l,k)w_i(l,k) \leq 0. \end{aligned} \tag{30}$$

According to (9), (14), and (29), Inequality (30) holds if

$$\begin{aligned} &[(A_i + B_iH(k))x(l,k) - B_iH(k)e(l,k) + \overline{F}_{ii}w_i(l,k)]^{\mathrm{T}}P_i(k)[(A_i + B_iH(k))x(l,k) \\ &- B_iH(k)e(l,k) + \overline{F}_{ii}w_i(l,k)] - x_i^{\mathrm{T}}(l,k)P_i(k)x_i(l,k) + \theta_{3i}\overline{\varphi}_i x_i^{\mathrm{T}}(l,k)\Omega_i(k)x_i(l,k) \\ &- \theta_{4i}e_i^{\mathrm{T}}(l,k)\Omega_i(k)e_i(l,k) + \|x(l,k)\|_Q^2 + \|u(l,k)\|_R^2 - \rho_i w_i^{\mathrm{T}}(l,k)w_i(l,k) \leq 0. \end{aligned} \tag{31}$$

Further, we obtain

$$\begin{aligned} &[(A_i + B_iH(k))x(l,k) - B_iH(k)e(l,k) + \overline{F}_{ii}L_iw(l,k)]^{\mathrm{T}}P_i(k)[(A_i + B_iH(k))x(l,k) \\ &- B_iH(k)e(l,k) + \overline{F}_{ii}L_iw(l,k)] - x^{\mathrm{T}}(l,k)E_i^{\mathrm{T}}P_i(k)E_ix(l,k) + \theta_{3i}\overline{\varphi}_i x^{\mathrm{T}}(l,k)E_i^{\mathrm{T}}\Omega_i(k)E_ix(l,k) \\ &- \theta_{4i}e^{\mathrm{T}}(l,k)E_i^{\mathrm{T}}\Omega_i(k)E_ie(l,k) + \|x(l,k)\|_Q^2 + \|u(l,k)\|_R^2 - \rho_i w^{\mathrm{T}}(l,k)L_i^{\mathrm{T}}L_iw(l,k) \leq 0, \end{aligned} \tag{32}$$

where

$$w^{\mathrm{T}}(l,k) = \left[w_1(l,k)^T \cdots w_i(l,k)^T \cdots w_N(l,k)^T\right]^T.$$

We write (32) as the following equivalent form:

$$\xi^{\mathrm{T}}(k)\left(\mathscr{C}_1 + \mathscr{C}_2P_i(k)\mathscr{C}_2^{\mathrm{T}} + \mathscr{C}_3R\mathscr{C}_3^{\mathrm{T}}\right)\xi(k) \leq 0, \tag{33}$$

where

$$\xi(k) = \begin{bmatrix} x^{\mathrm{T}}(l,k) & e^{\mathrm{T}}(l,k) & w^{\mathrm{T}}(l,k) \end{bmatrix}^{\mathrm{T}},$$
$$\mathscr{C}_1 = \mathrm{diag}\{Q - E_i^{\mathrm{T}} P_i(k) E_i + \theta_{3i}\overline{\varphi}_i E_i^{\mathrm{T}} \Omega_i(k) E_i, -\theta_{4i} E_i^{\mathrm{T}} \Omega_i(k) E_i, -\rho_i L_i^{\mathrm{T}} L_i\},$$
$$\mathscr{C}_2 = \begin{bmatrix} A_i + B_i H(k) & -B_i H(k) & \overline{F}_{ii} L_i \end{bmatrix}^{\mathrm{T}}, \mathscr{C}_3 = \begin{bmatrix} H(k) & -H(k) & 0 \end{bmatrix}^{\mathrm{T}}.$$

By the Schur complement, (33) is guaranteed if

$$\begin{bmatrix} \mathscr{D}_2^{(1,1)} & * \\ \mathscr{D}_2^{(2,1)} & \mathscr{D}_2^{(2,2)} \end{bmatrix} \leq 0, \tag{34}$$

where

$$\mathscr{D}_2^{(1,1)} = \mathrm{diag}\{-E_i^{\mathrm{T}} P_i(k) E_i, -\theta_{4i} E_i^{\mathrm{T}} \Omega_i(k) E_i, -\rho_i L_i^{\mathrm{T}} L_i\},$$
$$\mathscr{D}_2^{(2,1)} = \begin{bmatrix} A_i + B_i H(k) & -B_i H(k) & \overline{F}_{ii} L_i \\ R^{\frac{1}{2}} H(k) & -R^{\frac{1}{2}} H(k) & 0 \\ Q^{\frac{1}{2}} & 0 & 0 \\ \sqrt{\theta_{3i}\overline{\varphi}_i} E_i & 0 & 0 \end{bmatrix}, \mathscr{D}_2^{(2,2)} = \mathrm{diag}\{-P_i^{-1}(k), -I, -I, -\Omega_i^{-1}(k)\}.$$

Pre- and post-multiply the left-hand side matrix of Inequality (34) by diag $\{\gamma_i^{-\frac{1}{2}} G^{\mathrm{T}}(k), \gamma_i^{-\frac{1}{2}} G^{\mathrm{T}}(k), \gamma_i^{\frac{1}{2}} I, \gamma_i^{\frac{1}{2}} I, \gamma_i^{\frac{1}{2}} I, \gamma_i^{\frac{1}{2}} I, \gamma_i^{\frac{1}{2}} I\}$ and its transpose, respectively. Then, by denoting $\gamma_i P_i^{-1}(k) = S_i(k)$ and (28), we have

$$\begin{bmatrix} \mathscr{D}_3^{(1,1)} & * \\ \mathscr{D}_1^{(2,1)} & \mathscr{D}_1^{(2,2)} \end{bmatrix} \leq 0, \tag{35}$$

where

$$\mathscr{D}_3^{(1,1)} = \mathrm{diag}\left\{\widetilde{\mathscr{D}}_3^{(1,1)}, \widetilde{\mathscr{D}}_3^{(2,2)}, -\rho_i L_i^{\mathrm{T}} L_i\right\},$$
$$\widetilde{\mathscr{D}}_3^{(1,1)} = -E_i^{\mathrm{T}} G_i^{\mathrm{T}}(k) S_i^{-1}(k) G_i(k) E_i, \widetilde{\mathscr{D}}_3^{(2,2)} = -\theta_{4i} E_i^{\mathrm{T}} G_i^{\mathrm{T}}(k) T_i^{-1}(k) G_i(k) E_i.$$

Since $-G_i^{\mathrm{T}}(k) S_i^{-1}(k) G_i(k) \leq S_i(k) - G_i^{\mathrm{T}}(k) - G_i(k)$ and $-G_i^{\mathrm{T}}(k) T_i^{-1}(k) G_i(k) \leq T_i(k) - G_i^{\mathrm{T}}(k) - G_i(k)$, then (35) holds if (25) is satisfied.

Now, we prove that conditions (26) and (27) guarantee (17a), and noting that $\gamma_i P_i^{-1}(k) = S_i(k)$, we obtain

$$\begin{aligned} \max_{n\in\mathbb{N}} |[x_i(l,k)]_\iota|^2 &= \max_{n\in\mathbb{N}} \left| \left[S_i^{\frac{1}{2}}(k) S_i^{-\frac{1}{2}}(k) x_i(l,k) \right]_\iota \right|^2 \leq \max_{n\in\mathbb{N}} \left\| \left[S_i^{\frac{1}{2}}(k) \right]_\iota \right\|^2 \left\| S_i^{-\frac{1}{2}}(k) x_i(l,k) \right\|^2 \\ &\leq \max_{n\in\mathbb{N}} \left\| \left[S_i^{\frac{1}{2}}(k) \right]_\iota \right\|^2 \left(1 - \frac{\zeta_i(l,k)}{\gamma_i} \right) \leq \left\| \left[S_i^{\frac{1}{2}}(k) \right]_\iota \right\|^2. \end{aligned} \tag{36}$$

Then, (17a) holds if there exists a matrix $X_i > 0$ such that

$$X_i - S_i(k) \geq 0 \tag{37}$$

and (27) are met.

By the Schur complement, we obtain

$$\begin{bmatrix} X_i & * \\ I & S_i^{-1}(k) \end{bmatrix} \geq 0. \tag{38}$$

Pre- and post-multiplying the left hand-side matrix of (38) by $\mathrm{diag}\{I, G_i^{\mathrm{T}}(k)\}$ and its transpose, and noting that $G_i^{\mathrm{T}}(k)S_i^{-1}(k)G_i(k) \geq G_i^{\mathrm{T}}(k) + G_i(k) - S_i(k)$, we can obtain (26). Then, (17a) holds if (26) and (27) are satisfied, which completes the proof. □

3.2. Sufficient Conditions for PIS

To ensure that Λ_i is a PIS, we give sufficient conditions and give the corresponding LMIs.

Theorem 2. *Define the scalar α_i satisfying $0 \leq \alpha_i \leq 1 - \eta_{2i}$, and the scalars associated with DETM (8) are the same as in Theorem 1. If there exist a scalar $\gamma_i > 0$ and matrices $S_i(k) > 0$, $T_i(k) > 0$, $Y_i(k)$, and $G_i(k)$ such that the LMIs*

$$\begin{bmatrix} I & * & * \\ x_i(k) & S_i(k) & * \\ \sqrt{\zeta_i(k)} & 0 & \gamma_i \end{bmatrix} \geq 0, \tag{39}$$

$$\begin{bmatrix} \mathscr{E}_1^{(1,1)} & * \\ \mathscr{E}_1^{(2,1)} & \mathscr{E}_1^{(2,2)} \end{bmatrix} \leq 0, \tag{40}$$

hold, where

$$\mathscr{E}_1^{(1,1)} = \mathrm{diag}\Big\{\widetilde{\mathscr{E}}_1^{(1,1)}, \widetilde{\mathscr{E}}_1^{(2,2)}, -\frac{\alpha_i}{p_i}L_i^{\mathrm{T}}L_i\Big\},$$
$$\widetilde{\mathscr{E}}_1^{(1,1)} = (1-\alpha_i)E_i^{\mathrm{T}}\big(S_i(k) - G_i^{\mathrm{T}}(k) - G_i(k)\big)E_i, \widetilde{\mathscr{E}}_1^{(2,2)} = \theta_{4i}E_i^{\mathrm{T}}\big(T_i(k) - G_i^{\mathrm{T}}(k) - G_i(k)\big)E_i,$$
$$\mathscr{E}_1^{(2,1)} = \begin{bmatrix} A_iG(k) + B_iY(k) & -B_iY(k) & \overline{F}_{ii}L_i \\ \sqrt{\theta_{3i}\overline{\varphi}_i}E_iG(k) & 0 & 0 \end{bmatrix}, \mathscr{E}_1^{(2,2)} = \mathrm{diag}\{-S_i(k), -T_i(k)\}$$

(refer to Theorem 1 for the detailed form of matrices L_i, E_i, $G(k)$, and $Y(k)$), then (18) is a PIS.

Proof. Noting that $\gamma_i P_i^{-1}(k) = S_i(k)$, by Schur complement, we know that (39) guarantees $[x_i^{\mathrm{T}}(k), \zeta_i^{\frac{1}{2}}(k)]^{\mathrm{T}} \in \Lambda_i$; further, we obtain $[x_i^{\mathrm{T}}(l,k), \zeta_i^{\frac{1}{2}}(l,k)]^{\mathrm{T}} \in \Lambda_i$. Due to the presence of disturbance, we need to prove that $[x_i^{\mathrm{T}}(l+1,k), \zeta_i^{\frac{1}{2}}(l+1,k)]^{\mathrm{T}} \in \Lambda_i$.

The invariant set (18) holds at $k+1$ if

$$\begin{aligned} &\frac{1}{\gamma_i}\parallel x_i(l+1,k)\|_{P_i(k)}^2 + \frac{1}{\gamma_i}\zeta_i(l+1,k) - \frac{1-\alpha_i}{\gamma_i}\|x_i(l,k)\|_{P_i(k)}^2 \\ &- \frac{1-\alpha_i}{\gamma_i}\zeta_i(l,k) - \frac{\alpha_i}{p_i}w_i^{\mathrm{T}}(l,k)w_i(l,k) \leq 0. \end{aligned} \tag{41}$$

Substituting (6) into (41), and noting that $\parallel x_i(l,k)\|_{P_i(k)}^2 + \zeta_i(l,k) \leq \gamma_i$, we have that $\frac{1}{\gamma_i}\| x(l+1,k)\|_{P_i(k)}^2 + \frac{1}{\gamma_i}\zeta_i(l+1,k) \leq (1-\alpha_i) + \alpha_i = 1$, and thus $\| x_i(l+1,k)\|_{P_i(k)}^2 + \zeta_i(l+1,k) \leq \gamma_i$. This proves that $\left[x_i^{\mathrm{T}}(l+1,k), \zeta_i^{\frac{1}{2}}(l+1,k)\right]^{\mathrm{T}} \in \Lambda_i$. Furthermore, we obtain that (41) is ensured by (40) [the detailed proof is similar to the proof of Theorem 1 and is omitted here]. Finally, the condition (17b) is guaranteed by (39) and (40). This completes the proof. □

From the sufficient conditions given above, we transform $\mathbf{OP}_2$ into the following auxiliary **OP**:

$$\begin{aligned} \mathbf{OP}_3 : &\min_{H_i(k)} \gamma_i, \\ &\text{s.t.}(25), (26), (27), (39), (40). \end{aligned} \tag{42}$$

Based on $\mathbf{OP}_3$, we give Algorithm 1.

Algorithm 1 DETM-based DMPC for System (14)

Step 1. At $k = 0$, set the initial state $x_i(0)$, the matrices Q_i and R_i, and set the scalars η_{1i}, η_{2i}, θ_{1i}, θ_{2i}, θ_{3i}, θ_{4i}, ε_i, ϱ_i, $\underline{\varphi}_i$, α_i, p_i, and $\zeta_i(0)$.

Step 2. If the error $e_i(k)$ between $x_i(k)$ and $x(s_{mi})$ meets the event-triggered condition in DETM (8), then update $x(s_{mi})$ by $x_i(k)$; otherwise, implement $x(s_{mi})$ on the controller.

Step 3. Obtain the feedback gain matrix $H_i(k)$, and the matrix $\Omega_i(k)$ by solving $\mathbf{OP}_3$, and then calculate $u_i(k)$ and $u_j(k)$ by (12) and (13).

Step 4. Execute $u_i(k)$ and $u_j(k)$ on System (5). Set $k = k + 1$ and go to *Step 2*.

4. Case Study

To illustrate the superiority of our proposed approach, we construct the model of an 11-node LFC system in two areas, as shown in Figure 3. Areas 1 and 2 include 600 MW thermal power units, G1 and G2, respectively, and each area contains a VPP aggregating wind power, energy storage, and electric vehicles. In each area, all generation units are simplified as an equivalent generation unit, and the system parameters (see Tables 1 and 2) are borrowed from [26].

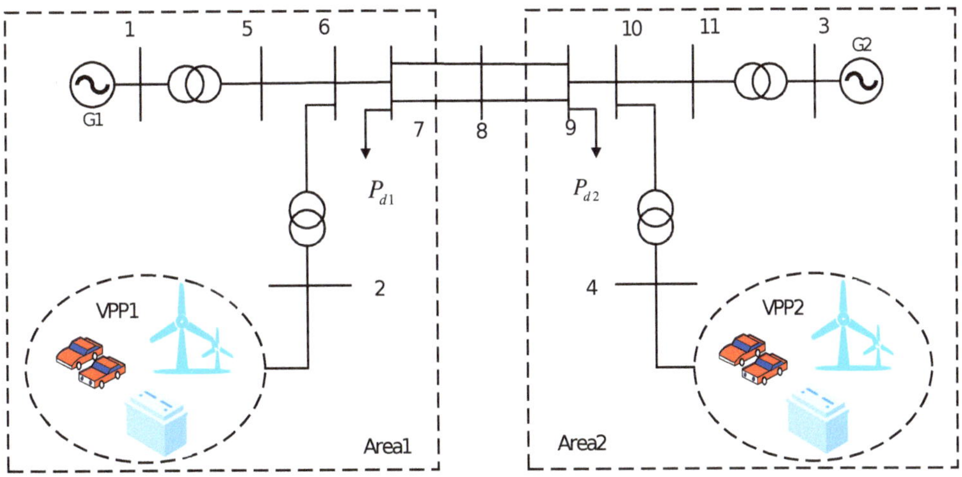

Figure 3. System simulation.

Table 1. Parameters of two-area power system.

Parameters	T_{ch}	T_g	R	D	β	M
Area 1	0.3	0.1	0.05	1.0	21.0	10
Area 2	0.4	0.17	0.05	1.5	21.5	12
			$T_{12} = T_{21} = 0.1986$			

Table 2. Parameters of two virtual power plants.

Parameters	T_{wind}	T_{bess}	T_{ev}
Area 1	1	0.1	0.05
Area 2	1.2	0.15	0.06

Let $\bar{x}_1 = \bar{x}_2 = [0.5, 0.1, 0.5, 0.5, 0.2, 0.5, 0.5, 0.2]^{\mathrm{T}}$, and the scalars $\eta_{11} = \eta_{12} = 1$, $\eta_{21} = \eta_{22} = 0.4$, $\theta_{11} = \theta_{12} = 0.1$, $\theta_{21} = \theta_{22} = 10$, $\theta_{31} = \theta_{32} = 0.11$, $\theta_{41} = \theta_{42} = 1.2$, $\underline{\varphi}_1 = \underline{\varphi}_2 = 0.2$, $\varepsilon_1 = \varepsilon_2 = 1.3$, $\varrho_1 = \varrho_2 = 0.5$, $\alpha_1 = \alpha_2 = 0.6$, $\zeta_1(0) = \zeta_2(0) = 5$, $p_1 = p_2 = 0.01$, and $\rho_1 = \rho_2 = 4$, and the matrices $Q = 15I$ and $R = I$. Set $x_1(0) = [0.07, 0, 0, 0, 0, 0, 0, 0]^{\mathrm{T}}$ and $x_2(0) = [0.05, 0, 0, 0, 0, 0, 0, 0]^{\mathrm{T}}$. Select the discretization period $T_p = 3$ s.

Scenario 1. Suppose that in the first 60 s, there exist step load disturbances with $\Delta P_{d1} = 0.03$ pu and $\Delta P_{d2} = 0.02$ pu.

Figure 4 shows the DETM triggering instants and release intervals for Area 1 (Area 2). Figure 5 shows the trajectories of Δf and ACE. The above simulation results indicate that the designed dynamic event-based DMPC algorithm is able to reduce packet releases while achieving control performance, thus reducing the communication delay of the VPP.

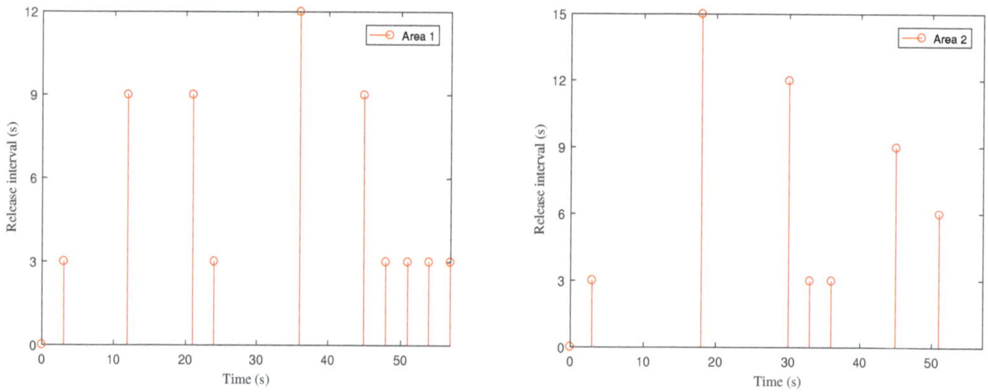

(**a**) Area 1 (**b**) Area 2

Figure 4. DETM evolutions of two areas under step load disturbances.

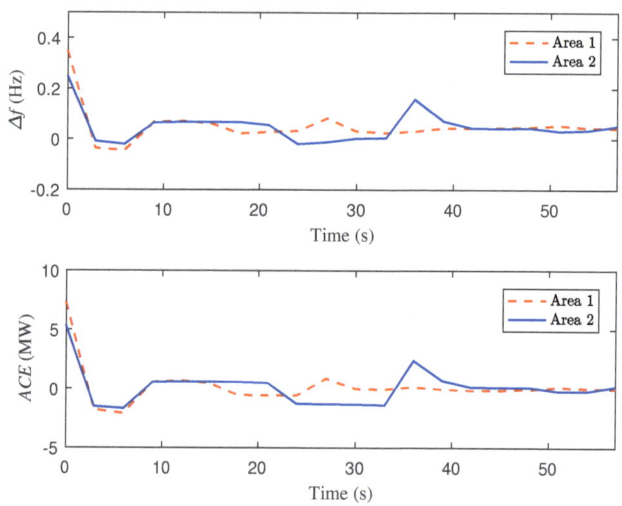

Figure 5. Trajectories of Δf and ACE subjected to step load disturbances.

To further illustrate the superiority of DETM (8), we compare it with DETM in [13]. By setting $\eta_{11} = \eta_{12} = 1$, $\theta_{11} = \theta_{12} = 1.1$, $\theta_{41} = \theta_{42} = 1$, $\underline{\varphi}_1 = \underline{\varphi}_2 = 0$, $\varepsilon_1 = \varepsilon_2 = 1$, and $\varrho_1 = \varrho_2 = 1$ and keeping the other parameters of DETM (8), we can obtain DETM in [13].

Table 3 gives the average triggering rates (ATRs) of the above two DETMs for two areas, and it can be noticed that the ATR of DETM (8) is lower than that of DETM in [13]. This means that the DMPC algorithm under DETM (8) is able to significantly reduce the ATR, thus reducing network congestion and decreasing the communication delay of the VPP at a slight sacrifice of control performance (see Table 4 for the detailed performance comparison of different algorithms, where summation of the absolute value of the error (SAE), summation of the square value of the error (SSE), and summation of the time multiplied by square value of the error (STSE) are defined as $\text{SAE} = \sum_{k=0}^{T_s/T_p} T_p(|ACE_1| + |ACE_2| + |ACE_3|)$, $\text{SSE} = \sum_{k=0}^{T_s/T_p} T_p(|ACE_1|^2 + |ACE_2|^2 + |ACE_3|^2)$, and $\text{STSE} = \sum_{k=0}^{T_s/T_p} kT_p^2(|ACE_1|^2 + |ACE_2|^2 + |ACE_3|^2)$ with $T_s = 60$ s and $T_p = 3$ s).

Table 3. ATRs under DETM (8) and DETM in [13] for the case of step load disturbances.

	DETM (8)	DETM in [13]
ATRs	47.5%	57.5%

Table 4. Performance comparison under step load disturbances between different algorithms.

Performance Criteria	SAE	SSE	STSE
DMPC under DETM in [13]	1.3671	0.1162	0.0887
DMPC under DETM (8)	1.5217	0.1184	0.1269
CMPC under DETM (8)	1.4762	0.1195	0.1295

In Table 5, we list the data processing time under DMPC and CMPC algorithms on the MATLAB R2016b platform with Intel(R) Core(TM) i5-8300H CPU @2.3 GHz. We obtain from Tables 4 and 5 that, under the cloud-edge-terminal framework, compared with the CMPC algorithm under DETM (8), the DMPC algorithm under DETM (8) is able to shorten the data processing time while guaranteeing the control performance. The above results demonstrate that the DMPC algorithm under DETM (8) reduces the network burden and shortens the data processing time, which helps to reduce the communication delay of the VPP.

Table 5. Data processing time under DMPC and CMPC algorithms.

Method	DMPC	CMPC
Times (s)	1.3087	2.2116

Scenario 2. Suppose that in the first 60 s, there exist random load disturbances, which are shown in Figure 6.

Figure 7 shows the DETM triggering instants and release intervals for Area 1 (Area 2). Figure 8 gives the trajectories of Δf and ACE. The above results show that the dynamic event-based DMPC algorithm reduces the packet releases and the communication delay of the VPP while achieving the desired control performance.

To further illustrate that our proposed algorithm is also superior under random load disturbances, we compare the DMPC algorithms under DETM (8) and DETM in [13]. The corresponding ATRs for the above two DETMs are shown in Table 6, from which we can observe that the ATR of DETM (8) is lower than that of DETM in [13]. In other words, the DMPC algorithm under DETM (7) can significantly reduce the ATR at a slight sacrifice of control performance (see Table 6 for the detailed performance comparison of different algorithms). Therefore, the DMPC algorithm under DETM (8) is able to reduce the release of packets and thus decrease the communication delay of the VPP.

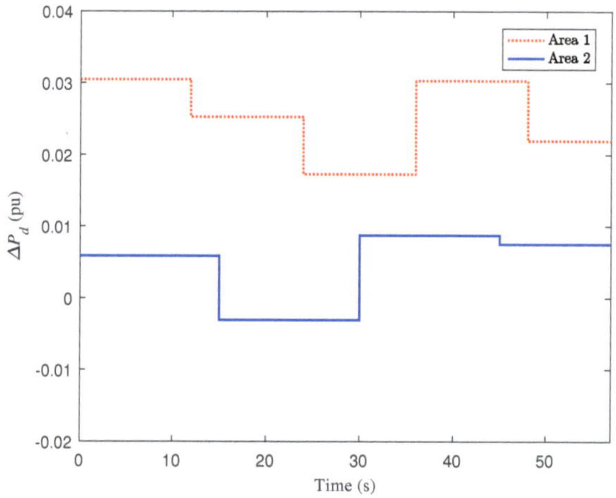

Figure 6. Random load disturbances.

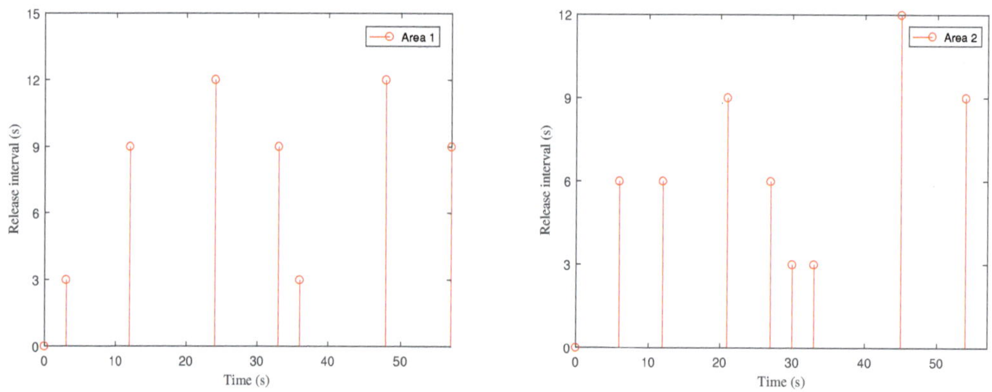

Figure 7. DETM evolutions of two areas under random load disturbances.

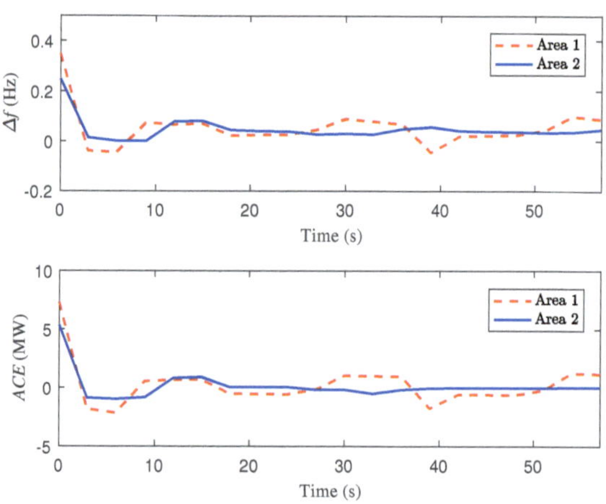

Figure 8. Trajectories of Δf and ACE subjected to random load disturbances.

In addition, under the cloud–edge–terminal framework, we give the performance criteria of the DMPC algorithm and CMPC algorithm under random load disturbances in Table 7, and the result shows that the DMPC algorithm under DETM (8) can guarantee the control performance. The data processing time under the two MPC algorithms is almost identical to the corresponding results in Table 5 and is thus omitted here. The above experimental results show that the DMPC algorithm under DETM (8) can shorten the data processing time while guaranteeing the control performance, which contributes to a reduction in the communication delay of the VPP.

Table 6. ATRs under DETM (8) and DETM in [13] for the case of random load disturbances.

	DETM (8)	DETM in [13]
ATRs	42.5%	50.0%

Table 7. Performance comparison under random load disturbances between different algorithms.

Performance Criteria	SAE	SSE	STSE
DMPC under DETM in [13]	1.3849	0.1143	0.1015
DMPC under DETM (8)	1.4498	0.1156	0.1177
CMPC under DETM (8)	1.4068	0.1165	0.1193

5. Conclusions

The DETM-based distributed model predictive LFC problem is investigated for VPPs under the cloud-edge-terminal framework. We design a new DETM that manages packet release more flexibly than several existing ETMs and reduces network congestion, which further reduces the communication delay of the VPP. Based on the above DETM, we design DMPC controllers for each edge server of the VPP. Specifically, to facilitate the solution of the "min-max" OP of the DMPC, we design an auxiliary OP which utilizes the Lyapunov-like function containing a DV, and then transform it into another auxiliary OP with LMI constraints which provides the design method of the DMPC controller. A case study of two-area VPPs is given to illustrate the effectiveness and superiority of the DETM-based DMPC approach under the cloud-edge-terminal framework. The designed DETM-based DMPC algorithm is able to save network resources and shorten data processing time while maintaining the control performance, which helps to reduce the communication delay of the VPP.

Author Contributions: Conceptualization, K.K. and N.S.; methodology, K.K., N.S., and S.C.; validation, L.Z., X.S., and H.C.; formal analysis, K.K., N.S., and M.F.; investigation, K.K., S.C., S.Z., and X.W.; writing—original draft preparation, L.Z., X.S., and H.C.; writing—review and editing, M.F., S.Z., X.W., and K.K. All authors have read and agreed to the published version of the manuscript.

Funding: This research was supported by the project "Research and Applications of Key Technology for Virtual Power Plants in New Power Systems Subject to Cloud-Edge-Terminal Collaboration and Source-Grid-Load-Storage Integration" from PowerChina Hubei Electric Engineering Co., Ltd (Project Number: DJ-ZDXM-2023-14).

Data Availability Statement: The data is contained within the article.

Conflicts of Interest: Author Kai Kang is employed by the company PowerChina Hubei Engineering Co., Ltd., Authors Nian Shi, Si Cai, Liang Zhang, Xinan Shao, Haohao Cao, Mingjin Fei are employed by the company PowerChina Hubei Electric Engineering Co., Ltd. The remaining authors declare that the research was conducted in the absence of any commercial or financial relationships that could be construed as a potential conflict of interest.

Abbreviations

The following abbreviations are used in this manuscript:

VPP	virtual power plant
LFC	load frequency control
DMPC	distributed model predictive control
CMPC	centralized model predictive control
DETM	dynamic event-triggered mechanism
OP	optimization problem
DV	dynamic variable
AAV	adaptive adjustment variable

References

1. Zhang, Y.; Shi, X.; Zhang, H.; Cao, Y.; Terzija, V. Review on deep learning applications in frequency analysis and control of modern power system. *Int. J. Electr. Power Energy Syst.* **2022**, *136*, 107744. [CrossRef]
2. Lan, Y.; Illindala, M.S. Robust distributed load frequency control for multi-area power systems with photovoltaic and battery energy storage system. *Energies* **2024**, *17*, 5536. [CrossRef]
3. Shangguan, X.C.; Zhang, C.K.; He, Y.; Jin, L.; Jiang, L.; Spencer, J.W.; Wu, M. Robust load frequency control for power system considering transmission delay and sampling period. *IEEE Trans. Ind. Inform.* **2022**, *17*, 5292–5303. [CrossRef]
4. Chen, X.; Qu, G.; Tang, Y.; Low, S.; Li, N. Reinforcement learning for selective key applications in power systems: Recent advances and future challenges. *IEEE Trans. Smart Grid* **2022**, *13*, 2935–2958. [CrossRef]
5. Chen, J.; Liu, M.; Milano, F. Aggregated model of virtual power plants for transient frequency and voltage stability analysis. *IEEE Trans. Power Syst.* **2021**, *36*, 4366–4375. [CrossRef]
6. Wang, Z.; Wang, Y.; Xie, L.; Pang, D.; Shi, H.; Zheng, H. Load frequency control of multiarea power systems with virtual power plants. *Energies* **2024**, *17*, 3687. [CrossRef]
7. Li, X.; Li, C.; Liu, X.; Chen, G.; Dong, Z.Y. Two-stage community energy trading under end-edge-cloud orchestration. *IEEE Internet Things J.* **2023**, *10*, 1961–1972. [CrossRef]
8. Lin, W.T.; Chen, G.; Zhou, X. Distributed carbon-aware energy trading of virtual power plant under denial of service attacks: A passivity-based neurodynamic approach. *Energy* **2022**, *257*, 124751. [CrossRef]
9. Hu, G.; Zhu, Y.; Zhao, D.; Zhao, M.; Hao, J. Event-triggered communication network with limited-bandwidth constraint for multi-agent reinforcement learning. *IEEE Trans. Neural Netw. Learn. Syst.* **2023**, *34*, 3966–3978. [CrossRef]
10. Bu, X.; Yu, W.; Cui, L.; Hou, Z.; Chen, Z. Event-triggered data-driven load frequency control for multiarea power systems. *IEEE Trans. Ind. Inform.* **2022**, *18*, 5982–5991. [CrossRef]
11. Wang, M.; Cheng, P.; Zhang, Z.; Wang, M.; Chen, J. Periodic event-eriggered MPC for continuous-time nonlinear systems with bounded disturbances. *IEEE Trans. Autom. Control* **2023**, *68*, 8036–8043. [CrossRef]
12. Wan, X.; Wei, F.; Zhang, C.K.; Wu, M. Networked output-feedback MPC: A bounded dynamic variable and time-varying threshold-dependent event-based approach. *IEEE Trans. Cybern.* **2024**, *54*, 2308–2319. [CrossRef]
13. Shi, T.; Shi, P.; Chambers, J. Dynamic event-triggered model predictive control under channel fading and denial-of-service attacks. *IEEE Trans. Autom. Sci. Eng.* **2024**, *21*, 6448–6459. [CrossRef]
14. Wan, X.; Han, T.; An, J.; Wu, M. Fault diagnosis for networked switched systems: An improved dynamic event-based scheme. *IEEE Trans. Cybern.* **2022**, *52*, 8376–8387. [CrossRef]
15. Zhang, M.; Dong, S.; Shi, P.; Chen, G.; Guan, X. Distributed observer-based event-triggered load frequency control of multiarea power systems under cyber attacks. *IEEE Trans. Autom. Sci. Eng.* **2023**, *20*, 2435–2444. [CrossRef]
16. Fang, X.; Xu, M.; Fan, Y. SOC-SOH estimation and balance control based on event-triggered distributed optimal kalman consensus filter. *Energies* **2024**, *17*, 639. [CrossRef]
17. Shen, H.; Wang, D.; Park, J.H.; Sreeram, V.; Wang, J. Switching-like event-triggered sliding mode load frequency control for networked power systems under energy-limited DoS attacks. *IEEE Trans. Syst. Man Cybern. Syst.* **2024**, *54*, 1589–1598. [CrossRef]
18. Yang, Q.; Chen, G.; Guo, M.; Chen, T.; Luo, L.; Sun, L. Model predictive hybrid PID control and energy-saving performance analysis of supercritical unit. *Energies* **2024**, *17*, 6356. [CrossRef]
19. Zhang, B.; Song, Y. Model-predictive control for Markovian jump systems under asynchronous scenario: An optimizing prediction dynamics approach. *IEEE Trans. Autom. Control* **2022**, *67*, 4900–4907. [CrossRef]
20. Tang, X.; Li, Y.; Yang, M.; Wu, Y.; Wen, Y. Adaptive event-triggered model predictive load frequency control for power systems. *IEEE Trans. Power Syst.* **2023**, *38*, 4003–4014. [CrossRef]

21. Bolzoni, A.; Parisio, A.; Todd, R.; Forsyth, A.J. Optimal virtual power plant management for multiple grid support services. *IEEE Trans. Energy Convers.* **2021**, *36*, 1479–1490. [CrossRef]
22. Bao, P.; Zhang, W.; Zhang, Y. Secondary frequency control considering optimized power support from virtual power plant containing aluminum smelter loads through VSC-HVDC link. *J. Mod. Power Syst. Clean Energy* **2023**, *11*, 355–367. [CrossRef]
23. Wang, C.; Liu, X.; Lee, K.Y. Two-layer robust distributed predictive control for load frequency control of a power system under wind power fluctuation. *Energies* **2023**, *16*, 4714. [CrossRef]
24. Ma, Y.; Hu, Z.; Song, Y. A Reinforcement learning based coordinated but differentiated load frequency control method with heterogeneous frequency regulation resources. *IEEE Trans. Power Syst.* **2024**, *39*, 2239–2250. [CrossRef]
25. Qiao, S.; Liu, X.; Wang, Y.; Xiao, G.; Wang, P. H_∞ load frequency control of power system integrated with EVs under DoS attacks: Non-fragile output sliding mode control approach. *IEEE Trans. Intell. Transp. Syst.* **2024**, *25*, 4565–4577. [CrossRef]
26. Peng, C.; Zhang, J.; Yan, H. Adaptive event-triggering H_∞ load frequency control for network-based power systems. *IEEE Trans. Ind. Electron.* **2018**, *65*, 1685–1694. [CrossRef]

MDPI AG
Grosspeteranlage 5
4052 Basel
Switzerland
Tel.: +41 61 683 77 34

Energies Editorial Office
E-mail: energies@mdpi.com
www.mdpi.com/journal/energies

www.ingramcontent.com/pod-product-compliance
Lightning Source LLC
Chambersburg PA
CBHW041356210926
51850CB00013B/3101
9783725838837